Engaging the Avatar

New Frontiers in Immersive Education

A Volume in
Research in Management Education and Development

Series Editor:
Charles Wankel, *St. John's University, New York*

Research in Management Education and Development

Charles Wankel, Series Editor

Rethinking Management Education for the 21st Century (2002)
edited by Charles Wankel and Robert DeFillippi

Educating Managers with Tomorrow's Technologies (2003)
edited by Charles Wankel and Robert DeFillippi

Educating Managers Through Real World Projects (2005)
edited by Charles Wankel and Robert DeFillippi

The Cutting Edge of International Management Education (2006)
edited by Charles Wankel and Robert DeFillippi

New Visions of Graduate Management Education (2006)
edited by Charles Wankel and Robert DeFillippi

University and Corporate Innovations in Lifelong Learning (2008)
edited by Charles Wankel and Robert DeFillippi

Management Education for Global Sustainability (2009)
edited by Charles Wankel and James A. F. Stoner

Being and Becoming a Management Education Scholar (2010)
edited by Charles Wankel and Robert DeFillippi

Cutting-Edge Social Media Approaches to Business Education: Teaching with LinkedIn, Facebook, Twitter, Second Life, and Blogs (2010)
edited by Charles Wankel

Effectively Integrating Ethical Dimensions Into Business Education (2011)
edited by Charles Wankel and Agata Stachowicz-Stanusch

Engaging the Avatar: New Frontiers in Immersive Education (2012)
edited by Randy Hinrichs and Charles Wankel

Engaging the Avatar

New Frontiers in Immersive Education

edited by

Randy Hinrichs
2b3d

and

Charles Wankel
St. John's University

Information Age Publishing, Inc.
Charlotte, North Carolina • www.infoagepub.com

Library of Congress Cataloging-in-Publication Data

Engaging the avatar : new frontiers in immersive education / edited by Randy Hinrichs and Charles Wankel.
p. cm. -- (Research in management education and development)
Includes bibliographical references.
ISBN 978-1-61735-751-0 (pbk) -- ISBN 978-1-61735-752-7 (hc) --
ISBN 978-1-61735-753-4 (ebk)
1. Virtual reality in education. 2. Shared virtual environments. I. Hinrichs, Randy J. II. Wankel, Charles.
LB1044.87.E55 2012
371.33'468--dc23

2011052474

Printed in the United States of America

CONTENTS

PART IV: IMMERSING THE AVATAR

PART V: 360 DEGREE FEEDBACK

ACKNOWLEDGMENTS

We gratefully acknowledge the world-class expertise of the 2b3d professional staff, whose powerful guidance in grounding this volume's contributions in cutting-edge immersive interface development and deployment knowhow made it immensely improved. This team includes Janice, Colin, Anna, Claire, Evan, and Kathryn. The professional editorial acumen of Neil Washington and Kyle Miller is evident in the stringent conformance of bibliographic citations, style, and format to the highest standards. We also wish to acknowledge the wonderful experience of working with Information Age Publishing's many supportive professionals, who are incomparable in bringing book projects to strikingly superior conclusions agilely.

Engaging the Avatar: New Frontiers in Immersive Education, pp. vii–vii

PART I

INTRODUCTION

CHAPTER 1

EMERGING VISTA OF IMMERSIVE EDUCATION

Introduction to *Engaging the Avatar*

Randy Hinrichs and Charles Wankel

INTRODUCTION

Whether you use 3D virtual worlds for learning, business/management, entertainment, or social networking, the objective for building an experiential virtual world environment is to understand what the avatar wants, needs, and would delight in. Of course, there is a physical person behind the avatar. Success in such a context depends on understanding the point of view of an avatar. The question emerges: How do you engage the avatar? This book takes a scholarly approach to looking at current theories and practices associated with engaging avatars.

We exhort you to join the fray and create and manage virtual environments in which avatars are immersed in for long periods of time. Engage them from the perspective of being on the inside. As an exercise, we ask you to present several kinds of problems avatars have to solve. You should create activities for avatars, requiring presentations, critical thinking, collaboration and cohabitation. In this exercise, create 360 degree feedback

Engaging the Avatar: New Frontiers in Immersive Education, pp. 3–11

so the avatar can receive and provide reflections on their experience. Develop measureable communities of practice. We will look specifically at some relevant examples in virtual communities that perpetuate living among peers and experts in the field. We shall examine cases studies in genetics, nursing, civics, stuttering pathologies, virtual economies, and workforce education. We examine considerations for design, problem solving, and how the eye works and what it expects to see. As a developer of virtual worlds, how do you pull the avatar in, motivate it to react, stimulate its curiosity, answer its questions, embrace it with experience, and teach it how to see again?

A BOLD AGENDA

The avatar is the first neutral, geopolitical representation of who we want to be when interacting on the 3D Internet. We have the opportunity to represent ourselves in any way we choose—we can pick our ideal, our antithesis and experiment in various habitats. We can venture into places unfamiliar, sometimes even scary or dangerous, and not endure the consequences as intensely as we might in the physical world. There is a certain liberation in engaging the world via the Internet as a 3D character who wanders about seeking any kind of experience or information available. The fundamental element of a virtual world is that there is a physical person behind the cursor. Each person, be they a student, teacher, businessman, librarian, salesman, politician, entertainer or any profession, enters into the virtual world prepared to participate. The persona may or may not represent themselves as their true self, but they engage in the virtual world looking for some action. If they do not find any, they leave.

This volume has a bold agenda, in which academics create immersive worlds where the avatar is the center of the universe. As the virtual world grows, avatars move away from quasi-human interactions within virtual domiciles, gardens, and businesses to being blood cells in the blood stream, or to be firing neurons in the human brain, or creatures competing on the ocean floor or the surface of Mars or just about anything that can be imagined using the magic of photographic and artistic images, programming, narrative and avatars. What are the frameworks and strategies for building these environments? What are the things the avatar adapts and learns from in its environment? This book will examine such frameworks, strategies, examples and feedback systems to explore what it takes to create a global education environment for learning.

FRAMEWORK FOR ENGAGING THE AVATAR

Andreas Schmeil, Martin Eppler, and Sara de Freitas, in their chapter "A Framework for the Design of Avatar-Based Collaboration," provide a model for avatar engagement grounded in research. Though 3D virtual worlds are being increasingly used as interactive environments for collaborative learning, there is still no structured approach for the combined design of 3D virtual environments and collaborative activities within them. Integrating structural, including collaborative elements, of virtual worlds with teaching approaches can foster more fruitful learning and work experiences. Towards this, they present an Avatar-Based Collaboration Framework (ABC Framework). Based on semiotics theory (Hamel, 2011), the framework puts the collaborating groups into the center of the design and emphasizes the use of distinctive features of 3D virtual worlds in collaborative learning environments and activities. The authors have drawn from best practices in instructional and game design, research in human computer interaction (HCI), and specific findings from their own empirical research investigating collaboration patterns in virtual worlds. Along with the framework, they present a case study of its first application for a global collaborative learning project. This chapter particularly addressed to virtual world designers and other practitioners, thoroughly describing the process of creating rich collaboration and collaborative learning experiences in virtual worlds with the ABC Framework.

In "Architectural Evolution of E-Learning Virtual Worlds: Proposed Design Measures to Enhance the E-Learning Experience Within 3D Learning Spaces," Noha Saleeb and Georgios Dafoulas examine architecture as a formal environment for interaction. As in the physical world, the play between the physical and the meaningful provides information on how an avatar may perceive the world. Numerous approaches are currently being adopted to efficiently utilize emergent 3D Virtual Worlds (VWs) for delivering e-learning, concentrating mainly on teaching and learning applications. Equally significant, this research demonstrates the importance of customizing architectural design characteristics of 3D virtual learning spaces within virtual worlds as another complementary and essential user-centered approach to enhancing students' e-learning experiences. This is achieved by displaying higher education students' propositions and requirements from the design elements of their 3D educational spaces best suited for promoting their skills, satisfaction and participation during their e-learning sessions. These recommendations can be used by educators and designers to enhance the design features of existing and future 3D virtual educational facilities within 3D VWs to augment students' e-learning processes.

STRATEGIES FOR ENGAGING THE AVATAR

A strategy provides a plan of action and an understanding of how the engagements in the action plan are linked. Central to designing for the avatar is a fundamental understanding of how the avatar's physical world persona sees that world and translates it into the virtual world. The first chapter in this part of the book takes us into the eye of the avatar and reflects on how we see and how it relates to the virtual world. The next chapter opens our eyes to thinking about how users apply problem-solving strategies for acquiring virtual information and discusses how users seek, evaluate and use information in the virtual world. It examines new ways to develop trust in virtual worlds. The final chapter in this part examines the strategy of teacher training across multiple locations in a Mexican school system. This chapter is from the perspective of a practitioner who has applied similar strategies. The strategies are applicable at all educational levels.

In Joseph Trachtman's chapter, we look at both the cognitive and non-cognitive elements in the processing visual information. Understanding these physiological fundamentals helps us think about how avatars are engaging in the virtual world. Once you spend some time in this chapter, you wonder if the virtual the same as the physical world, in which we are sensitive to color, brightness, contrast, depth, and perspective. Trachtman, a Doctor of Optometry, and a PhD in experimental psychology, explains how sensory information might work in a virtual world. His insights suggest that what we see in a virtual world have an effect on our emotions, our immune system responses, and how we focus our attention. Think of the effects of the visual components placed in a virtual world just to have a mouth watering effect, so consumers think about buying food for example, or imagine how the visual image of a meal might be stored in our memory affecting the way we think of dieting or nutrition. Imagine placing objects in a virtual world and trying to use depth perception to attract and maintain attention in simulation based environments. If you want a user to move from one location to another, how is their vision affecting their next decision? Trachtman asks, "How can we take advantage of the perceptual factors involved in visual accommodation in a 3D world when the images are on a 2D surface? Trachtman's description into how our vision and brain work together is fundamental for understanding the science of virtual world design, and how to engage an avatar in that world. He concludes, there should be a balance among the various perceptual factors that make a virtual world community. Colors, size, shape, and so forth, must all be carefully considered. It's more than just the aesthetics; it's about managing the avatar's experience.

Mike Eisenberg, John Marino, Natascha Karlova, and Peyina Lin, in their chapter "Strategies for Exploring Information Behavior in Second

Life," examine virtual information behavior environments. They start with an overview of their investigation into how users seek, evaluate, and use information in a variety of contexts within Second Life. Based upon the work of the Virtual Information Behavior Environments (VIBE) project at the University of Washington Information School, they focus on findings, insights, and recommendations that may be useful in guiding learning and working practices within virtual worlds. Although their interest lies in the information behavior of SL residents, they believe that their experience will benefit those seeking to use SL for a variety of purposes. Therefore, they have divided their chapter into four parts, each with strategies designed to be of practical use and grounded in their experience with conducting information behavior research in SL.

Caroline Avey and Lyn Maize in their "Virtual Worlds and Workforce Education: Real Strategies for Engagement and Learning" offer perspectives and practical examples of how multi-dimensional, virtual worlds can be used for adult learning in corporate and other work related environments. They share their experiences in building and facilitating 3D virtual worlds and learning programs, and why these immersive learning environments are particularly powerful in impacting performance and proficiency. They also facilitate the capture of the "informal" or collaborative knowledge exchanges that occur every day in their organizations and provide insight into how this can be done for everyone. They begin by looking at 3D virtual worlds as part of a strategy for a high-performing organization.

IMMERSING THE AVATAR

Engaging the Avatar requires immersion of the student in a learning space, surrounding him or her with activities, people, feedback, and content. In this learning culture, students ideally adapt to this new environment by learning content in a community context. Heidi Trotta and Marian Glenn explore a problem based learning scenario for teaching environmental science. Grant Meredith, Charlynn Miller, and Greg Simmons demonstrate how immersion in health related fields can help those who stutter and those who provide nursing services. Thomas Bryer and Michelle Gardner show how learners can be trained in a civic virtual world to serve as poll workers. And Mary Anne Clark explains her genetics instruction on Genome Island, creating a living lab for genetic experiments. In a virtual world, you have access to global laboratories and programmed environments to simulate just about anything. The avatar has never had a more widely distributed global classroom ever.

Heidi Trotta and Marian Glenn in "Salt Marsh Dynamics ~ a Problem-Based Learning Scenario" discuss how multi-user, persistent virtual environments foster community and situated learning by supporting immersive, collaborative, extended experiences through simulations similar to physical world contexts. Their project developed a problem-based learning environment using the online virtual world of Second Life, for undergraduate environmental science courses. Both process and content promote collaborative learning, problem solving and applying research and theory. Students gather, interpret and analyze data to solve an ill-defined physical-world problem. Specific scaffolding, process modeling, and frequent reflection helped students advance from reliance on miniscripts, acquired through past content-driven courses, to search for new information and to observe characteristics of the scenario beyond what they expected, based on superficial acquaintance with real world sites.

Grant Meredith, Charlynn Miller, and Greg Simmons in "Stuttering Support and Nursing Education: Two Case Studies in Second Life" immerse learners who stutter. This work makes a novel and valuable addition to the body of research in online learning as it focuses specifically on the use of 3D Multi-User Virtual Environments (MUVEs) for simulation-based teaching and learning. Two specific case studies are described, one outlining how the technology can enhance the lives of people who stutter, the other describing how 3D MUVEs can be used to improve the training of nurses. The chapter draws on a broad range of extant research conducted over the past three decades and synthesizes it with newer developments and examples that have come about since the advent and proliferation of the "3D Web." The chapter adopts a research-informed approach to surveying and examining current initiatives and detailing features of development, backed by relevant literature in the areas of online learning, constructivism, and simulation learning. The chapter concludes with a discussion of future initiatives from the point of view of best practice in MUVE-based simulation.

Thomas Bryer and Michelle Gardner take us into "Training for Technological Democracy: A Civic Engagement Class Example." Implementing a civic engagement course in one of the lowest youth-engaged states was a challenge. Adding a layer of technology use to such a program presented the necessary hook to get the students involved. Overall, the two civic engagement courses taught online at the University of Central Florida (UCF) in spring and summer 2010 were successful at introducing students to the potential for utilizing Internet based collaborative technologies for civic engagement. In this chapter they describe a special course offered at UCF in conjunction with the Orange County Supervisor of Elections Office. With an emphasis on civic engagement through technology. Each student was trained in a virtual environment to serve Orange County, Florida as a poll worker for the November 2010 general election.

Mary Anne Clark takes is into "Genome Island: A Virtual Laboratory Environment in Second Life." Genome Island was created as a virtual laboratory environment in the Second Life metaverse. The island contains over 60 interactive experiments and exhibits, from Mendelian genetics to bioinformatics. Although it was first developed to provide a laboratory experience for online students, it is also used as a supplemental learning space for face-to-face students. It is used most intensively in a nonmajors class that meets exclusively in a virtual world. The in-world course includes eight content modules, including a course orientation. Students get activity guidelines and receive and return assignments via a Blackboard interface, and meet weekly on Genome Island for questions and trouble shooting. Data collected from the fall 2009 class includes visitor records for the island, object interaction records for the class, performance on learning objectives, and student comments.

360 DEGREE FEEDBACK

There are many ways to get feedback through a virtual world interface. You can watch yourself and examine your own strategies for solving problems. You can interact and watch others interact with you, read their words, and hear their emotions in voice over IP. You can experience the social consequences simply by engaging and watching how others respond and manage their attention when engaged. You can perform a task and receive feedback from an expert observer. You can relish the immediate and constant feedback of programmed objects that tells you if what you are doing works or not, which is different from being told whether you are right or wrong, or being provided with a numerical analysis of performance. Virtual worlds provide multi-rater systems in which feedback seems to come to the avatar from everywhere and everyone. In this part, we look at the impact of such feedback systems in practical applications such as nursing and virtual economies.

Michelle Aebersold and Dana Tschannen in their chapter, "Using Virtual Simulations in Second Life for Teaching and Learning in Nursing Education," discuss how educating nursing students is a challenge in the complex healthcare environment in which they practice. To be effective, it is critical to assist nurses in training to achieve a high level of skill development in a safe environment. Virtual worlds provide a training environment in which virtual simulations can be conducted to allow students to practice integrating their knowledge and skills in the delivery of patient care. Virtual simulations are effective when developed using a comprehensive approach that incorporates elements of online teaching pedagogies and the five step process outlined in this chapter. Issues and challenges in using virtual environments should also be anticipated and addressed to ensure student learning is achieved.

Sami Ahma-aho and Teemu Surakka in their chapter, "Using Second Life to Teach and Research Virtual Economy," inform us that currently the greatest benefits of 3D virtual environments seem to reside in teaching and training applications. To tap into these unique benefits, an innovative approach combining teaching and research was developed in a virtual world. The goals were to enhance both learning objectives in the collaborating business courses and the quality of research data on digital business. Second Life was selected as a platform because of its diverse business activity. In this chapter, the approach and the utilized technologies are presented, along with examples of the exercise's outcomes and problems encountered. Students largely found this Second Life exercise to be unduly challenging in many ways. This derailed the associated research study to a large extent. The authors discuss the implications of their experience for future teaching and research involving such technologies.

Antonio Santos's chapter "The SimEscuela: An Innovative Virtual Environment for Teacher Training" describes the dynamic relationship among educational theories, the instructional design process, and actual construction of a virtual environment called The SimEscuela in Second Life. This was a simulation of a K-12 Mexican school. The participants, who were Mexican teachers from various parts of that country, from diverse institutions and levels, framed this exercise as an online community of practice. Their common interest was in sharing each other's professional experiences to foster the development of improved teaching skills and to engage collaboratively in problem solving.

Conclusion: You University. Imagine a world in which you are the center of your learning agenda. You create your own corner of the virtual world and gather your artifacts, your libraries, your objets d'art, your dinosaurs, and solar systems, and they are all tagged with how you relate to them. Another person enters into your environment, and instantly they link to the objects where the two of you have common interests and you're both teleported to a location in which to engage. Imagine bringing experts to you—2, 10, 50 each interacting with you on your learning stage. Imagine your own special university the You U, growing up around a single location that is easily accessed by you, and augmented by every experience and performance you engage in. In this case one might create virtual biodomes in every subject area, utilizing the collaborative frameworks and strategic engagements suggested here. The characters you meet along the way are the experts who have pioneered useful, deep dives into their content areas to provide you with a head start into changing the way you learn.

This is a bold agenda that starts with engaging your avatar and is completed in a transformation in how you interact with the internet. Whether using the visual internet to learn or to interact with a customer about a product or service, this immersive interface can be a world that knows you and forms around your unique needs and interests.

REFERENCE

Hamel, S. C. (2011). *Semiotics: Theory and applications.* Hauppauge, NY: Nova Science.

PART II

FRAMEWORK FOR ENGAGING THE AVATAR

CHAPTER 2

A FRAMEWORK FOR THE DESIGN OF AVATAR-BASED COLLABORATION

Andreas Schmeil, Martin J. Eppler, and Sara de Freitas

While 3D virtual worlds are being more and more used as interactive environments for collaborative learning, there is still no structured approach developed specifically for the combined design of 3D virtual environments and collaborative activities within them. We argue that formalizing both the structural elements of virtual worlds and collaboration or didactic aspects of collaborative learning can help to develop fruitful collaborative work and learning experiences. To this aim we present an Avatar-Based Collaboration Framework (ABC Framework). Based on semiotics theory, the framework puts the collaborating groups into the center of the design and emphasizes the use of distinct features of 3D virtual worlds for utilization in collaborative learning environments and activities. In the development of the framework we have drawn from best practices in instructional design and game design, research in HCI, and specific findings from our own empirical research investigating collaboration patterns in virtual worlds. Along with the framework we present a case study of its first application for a global collaborative learning project. This chapter particularly addresses virtual world designers and other practitioners by thoroughly describing the process of creating rich collaboration and collaborative learning experiences for virtual worlds with the ABC Framework.

Engaging the Avatar: New Frontiers in Immersive Education, pp. 15–48

INTRODUCTION

An ideal online, three-dimensional virtual environment would provide a space in which users can move freely, interact intuitively with all kinds of objects, recognize familiar people, and communicate in a natural manner with them—all in the most realistic look-and-feel setting, evoking a feeling of really being part of the virtual world. In addition to that, it would allow displaying complex content or data in innovative and useful ways, neglecting the limitations imposed by physical reality. Such an environment holds the promise of moving remote collaboration and learning to another level of quality. But even if such platforms were available today (and they soon will be): without the right kind of choreography, script or setup, users would not know how to best benefit from such an infrastructure.

We believe that today's available virtual worlds are already capable of adding significant value to collaborative work and collaborative learning. However, an exploration of collaborative activities in the virtual world Second Life has shown that companies, institutions as well as educators for the most part do not make use of the distinct features of a three-dimensional environment (being embodied as customizable avatars, moving around in a configurable responsive space, sharing that 3D space with data and information)—although, there are promising approaches (Schmeil & Eppler, 2009).

Moreover, many of the virtual environments that are currently (early 2011) being advertised as offering great productivity boosts for collaborative work emphasize the collaborative editing of text documents, spreadsheets and presentation slides that are mounted on big virtual walls (e.g., Teleplace, Open Wonderland)—a method of working together that may work similarly well without gathering in a three-dimensional virtual environment.

Our premise, consequently, is that the fact of being embodied in an immersive environment that is configurable at will allows for new, innovative, and valuable forms of working and learning together. Being embodied as the natural form of human existence lets collaborators interact and communicate naturally using their bodies—which, in the case of virtual environments, are virtual bodies or avatars. Our research aims at improving collaboration in these virtual environments or virtual worlds by first systemizing and formally describing the necessary elements for online, visual and spatial collaboration; secondly our research aims at developing and identifying novel and existing collaboration patterns that can be described in that formalism; and thirdly we want to evaluate their effectiveness experimentally and compare them (in terms of added value) to other online collaboration arrangements.

In this paper, we focus on steps one and two and present a framework for embodied collaboration in online 3D virtual environments, based on semiotics theory. Our framework represents a blueprint of how collaborative group interaction patterns in virtual environments can be described or generated. The framework is distinctive with regard to its two ways of use: collaboration patterns can be generated or documented either starting with a goal, or starting with the infrastructure of objects and actions available. We also present a study of the first application case of the framework, showcasing 13 collaborative learning patterns. We believe this framework to form a first important step in the process of formalizing collaboration in virtual environments—a task that is crucial in order to put forward the application of 3D virtual environments for serious and productive uses.

The remainder of this chapter is structured as follows: In section 2, we define online virtual environments and present their advantages for collaboration. In section 3 we then present our framework that helps to formalize the design elements and necessary infrastructure of collaboration patterns in such environments. Section 4 covers the case study. We conclude the article with a review of our contribution and highlight directions of future research.

MULTI-USER VIRTUAL ENVIRONMENTS AND COLLABORATION

Virtual environments in general attempt to provide an environment where the user or spectator feels fully immersed and present. This presence is a psychological phenomenon that has been defined as the sense of "being there" in an environment, based on the technologically—if not physically—founded base of immersion (Slater, Usoh, & Steed, 1994).

Especially when speaking of online virtual environments, the term virtual presence has also been established, putting extra emphasis on the absence of physical presence. Virtual presence in its original, broad sense denotes the feeling of being present via intermediate technologies, such as television, radio, telephone, or the Internet. In 3D virtual world applications, the concept of virtual presence refers to the avatar, the objects, the programs, the textures, the land, and the multimedia that all combine to make up data stored on online servers rather than data stored in a local database. Immersion augments the pervasive presence of the data, by presenting a 3D avatar based user interface that aims to lead the user into feeling present among the data, seeing it, touching it, and reacting to it almost physically through the gestures of the avatar. This experience is best known as the sense of presence which can be achieved to varying degrees, stimulating a variable number of human senses. However, the expression of feeling immersed is often also used for online, desktop-based, 3D virtual environments that are controlled only by keyboard and mouse and address only two sensory channels: the visual and auditory one. In the ongoing scientific discourse in the research community, a virtual world has recently been defined as "a synchronous, persistent network of people, represented as avatars, facilitated by networked computers" (Bell, 2008, p. 2).

To date, there is an abundance of virtual worlds available, for all age groups and for many different areas of interest. The virtual worlds' consultancy K Zero keeps informative graphs up-to-date on their company website. Most available virtual worlds are held online by the operators on proprietary servers. They immersive environments can be accessed on one's own servers directly off the Internet; some refer to this as cloud computing. Usually a client software download is required to access many of the features such as voice over IP, or IM chatting, or navigating through predesigned content. Similarly, software can be installed on your own server intranet, and the virtual world can be accessed behind the firewall in a similar manner.

Some systems like Second Life, OpenSim and Activeworlds enable users to access their software and either experience others virtual world data and interfaces, or allow individual creation in these flexible worlds to create static and interactive content to engage users in innovative ways. Other virtual worlds, such as Open Wonderland and Teleplace, focus on productivity in conventional tasks like the editing of text documents, spreadsheets and presentation slides. Only the up- and download of documents and the repositioning of furniture is possible in these latter worlds. Still others like SAIC's OLIVE (On-Line Interactive Virtual Environment) focus on providing training scenarios in which users access preprogrammed features that allow them to practice simulations and role plays. New virtual worlds are launched almost monthly, and it seems like each new one tries to fill another niche. Nevertheless, for most application domains, it is still unclear what value virtual worlds might add to the existing modes of communication and collaboration, just as it remains unclear which features and enhancements are needed to maximize the benefit of using virtual worlds (Bainbridge, 2007). In a previous paper, we have discussed the advantages (and potential risks) that collaborative virtual worlds bring for knowledge work and education (Schmeil & Eppler, 2009). In this chapter, we present a specific approach to how these advantages can be harnessed in order to engage the avatar in ways that bring value.

THE AVATAR-BASED COLLABORATION FRAMEWORK

The work around using models and frameworks for learning interaction design draws from the two main strands of learning theory and human computer interaction. One framework that tries to bring together these two elements is the *four dimensional framework* (de Freitas & Oliver, 2006). In this earlier work, de Freitas and Oliver aim to bring together four dimensions of learning to help structure the goals of the framework:

1. *the learner*—(how are learners profiled and modelled in virtual worlds differently than in the physical world)
2. *the learning theory adopted*—(what type of theory—associative, cognitive and situative—are being used to frame the learning)
3. *the representation of the environment*—(how high does the fidelity need to be, how interactive does the world need to be, how are the various levels of immersion constructed, and
4. *the context* (which materials are available, especially including those adding discipline to the context and those used as additional resources to support the context).

Together these dimensions can be used to select virtual worlds, choose appropriate games and applications for learning, and leverage e-learning materials designed and implemented specifically for engaging the avatar. Utilizing or integrating gaming concepts and gaming design in virtual worlds can be very useful for engaging avatars. Some of these features include a focus on narrative flows, employing quests and structured interactions within the 3D environment, and creating reward systems that provide feedback to the user. Work extending from the four dimensional framework has developed notions around a game-based learning development approach, which brings design processes, game principles and tools and techniques together (de Freitas & Jarvis, 2009).

In de Freitas and Jarvis work on serious games, design has led to a multidimensional approach that encompasses different human computer interfaces (HCI), game design and learning design elements. In this engagement model, the notion of developing learning rests more with choreographing learning experiences based upon exploratory learning rather than more creating knowledge centered approaches (de Freitas & Neumann, 2009). This strategy should lead to more effective, accelerated learning, and increase learner motivation and engagement.

However, the emergence of the less structured "social learning" in virtual and game-based environments as a central benefit of virtual worlds has led to the need for more focus to be placed upon choreographing or designing well-crafted experiences to ensure competency based outcomes, rather than mere social interactions. This level of effort can be difficult for facilitators, managers or tutors and even more so to learners as it deviates far from normal classroom based interactions and approaches which learning often relies on from habituation. Since learning in virtual worlds requires collaboration in both verbal and nonverbal formats, the use of language and narrative are incredibly important to create the interactive environment to enhance being present and interacting with real people in real situations. The ABC framework serves as a blueprint in which diverse collaborative learning tasks can be designed, formulated and executed.

The ABC framework is based on theoretical key distinctions of semiotics (i.e., the syntactic, semantic, and pragmatic levels of any language). This well-known semiotic triad has already been applied to various contexts within computer science (see, e.g., Price & Shanks, 2005) and relates to concepts from HCI (e.g., interaction typology). We discuss the use of syntax, semantics and pragmatics in more detail when presenting in the subsections below. Semiotics is a theory about how people communicate with each other, and how they create and derive meaning out of that communication, or in this case collaboration and communication. It also encompasses how people use signs and symbols to convey meaning. Virtual worlds highly leverage the visual aspect of the physical environment. The interactions between users communicating in virtual world is relied on heavily in designing engagement, so semiotics provides an excellent backdrop to the discussion of the framework.

In the following subsections, we describe the various steps that we have taken in the development of the framework.

Describing Virtual Embodied Collaboration Through Patterns

While identifying group interaction patterns of collaborative work and learning in the virtual world Second Life (Schmeil & Eppler, 2009) we have noticed the need for a solid formal framework that is capable of describing collaboration in virtual worlds in all its aspects. Patterns are a useful and concise approach to classify and describe different forms of online collaboration. Manninen states that the utilization of real-world social patterns as basis for virtual environment interactions might result in more usable and acceptable solutions, after having successfully applied a social theory framework for categorizing interactions in virtual environments (Manninen, 2000).

An alternative approach to using patterns is to describe collaborative situations as *scenarios*. A scenario is an "informal narrative description" (Carroll, 2000). However, comparing this with definitions of patterns (e.g., "a description of a solution to a specific type of problem" Gottesdiener, 2001), reveals that the pattern concept has been conceived with more focus to solve a problem or to reach a goal. In addition to that, a look at the work of Smith and Willans (2006), who implement the concept of scenarios for a requirements analysis of virtual objects, makes it clear that the scenario-based approach would be too fine-grained and at a too low, functional level to describe whole collaborative tasks in flexible multi-user settings.

Hence, we have decided to use the pattern approach. We adapt the collaboration pattern definition from (Gottesdiener, 2001) by adding the notions of tools and a shared meeting location, to give us the following definition: A collaboration pattern is *a set of tools, techniques, behaviors, and activities for people who meet at a place to work on a common goal, together in a group or community*. How exactly this definition fits with the resulting framework will be explained by means of an illustration in subsection 3.5.

The Semiotic Triad as an Organizing Schema

From a theoretical point of view, one can conceive of collaboration activities as interpretive actions and of collaboration spaces as sign systems in need of joint interpretation. Visual, spatialized on-screen events have to be interpreted by users of a virtual environment as relevant, meaningful, context-dependent signs that contribute towards joint sense making and purposeful coordination. As in any sign interpretation system or (visual) language, semiotic theory informs us that three different levels can be fruitfully distinguished, namely the syntactic, semantic, and pragmatic ones (Eco, 1978; Morris, 1938). This threefold distinction has already been applied effectively to various forms of information systems or social online media (see, e.g., Shanks, 1999, or Schmid & Lindemann, 1998). These three distinct interpretive layers can be applied as follows to immersive virtual worlds:

The *syntactic* dimension contains the main visible components of a collaboration pattern that can be understood as the fundamental building material of which patterns are constructed. The syntactic dimension ensures visibility and readability of a collaboration pattern by providing these basic elements (digital artifacts and actions) organized in a clear structure.

The *semantic* dimension refers to the acquired meaning of elements and to the conventions used in a collaboration pattern. It outlines which operations or artifacts assume which kind of meaning. While the syntactic dimension reveals what elements a collaboration pattern consists of, the semantic dimension puts these basic elements into meaningful combinations that make sense in respect to the goals. It thus aligns the available visual vocabulary to the desired objectives or contexts. In this sense the semantic level is a liaison layer between the virtual world and the participants' objectives.

The *pragmatic* dimension reflects the social context of the participants, and their practices, goals and expectations. It is these notions that need to be supported through the choreography or dramaturgy (semantic dimension) and the infrastructure (syntactic dimension). This dimension clarifies the situation(s) in and for which the dramaturgy and infrastructure in the particular setup make sense.

ACTION AND INTERACTION IN 3D VIRTUAL ENVIRONMENTS

In our understanding, the support of action and interaction forms a major part of a virtual environment's infrastructure. It determines how users can act and affects their behavior in both lonely jaunts and group settings. Moreover, the way users can control their avatars and perform actions heavily influences the level of satisfaction of the user and thus in the end determines whether or not collaborative work or other planned tasks in the virtual environment succeed or fail, continue or are abandoned. In order to manage action and interaction, we believe that a formalization of the concept in various forms in virtual environments on a high abstraction level is required.

Manninen (2000) applied a social theory framework in order to create a taxonomy of interaction, resulting in a classification consisting of eight categories: *Language-Based Communication*, *Control & Coordination*, *Object-based Interactions*, *World Modifications*, *Autonomous Interactions*, *Gestures*, *Avatar Appearance*, and *Physical Contacts*. While this is a useful and accurate description of the expressive repertoire available in virtual worlds, this classification is only based on studies in multi-player online action and role-playing games, where different requirements regarding interaction must be assumed than for serious collaborative tasks. Also, as the author also concedes himself, the social theory framework might have put the study too much in a language-centered perspective and might have neglected some of the genuinely visual aspects of virtual worlds. Furthermore, the end result (i.e., the classification) lacks a formal rationale and does not entirely meet the criteria of a taxonomy (i.e., empirically derived, disjunctive and exhaustive groups, single classification principle per level, etc.).

In the field of Human Computer Interaction there is a generally accepted distinction among *navigation* and *manipulation* techniques (Bowman, Kruijff, Poupyrev, & LaViola, 2005). Navigation techniques consist of techniques for moving one's position and for changing one's view. Manipulation techniques designate all interaction methods that select and transform or modify objects in a virtual space. In some cases, the side category *System Control* is used, consisting of all actions that serve to change modes and modify parameters, as well as other functions that alter the virtual experience itself. 3D user interface expert Doug Bowman and colleagues refine this classification by adding a fourth category *Symbolic Input*, describing the communication of symbolic information (text, numbers, and other symbols or marks) to the system (Bowman et al., 2005).

For our purpose of formalizing (inter)actions for collaboration, we build on this clear classification and make adjustments to align it with the requirements of the area of Online 3D Multi-User Virtual Environments:

The importance of communicating text, numbers, symbols, and also speech to the system (and thus to other avatars or users, interactive objects, or the environment itself) has increased significantly. We call this first category *Communicative Actions*. A subdivision differentiates between verbal (i.e., text and voice chat) and nonverbal communication (i.e., nodding, gesturing).

Having combined navigation techniques and methods for changing the view in one shared category results from the fact that HCI and VR systems do not necessarily assume the existence of an avatar as a personalization device in the virtual environment; without this embodiment, navigating and changing the viewpoint can be considered as one and the same action. In our classification, changing one's view falls into the communicative actions category, as a nonverbal form of letting others know where the user's current focus of attention is, or to communicate a point or object of interest to others in the virtual environment. As a result, our second category, *Navigation*, merely comprises walking, flying and swimming, and teleporting.

We rename the manipulation techniques category (selecting and modifying objects) to *Object-Related Actions*. Actions referring to the creation or insertion of virtual objects also belong to this category, along with selection and modification techniques. By insertion we mean the result of uploading or purchasing virtual objects, for instance.

All interactions concerning system control are much less important in contemporary virtual worlds than they are in classic virtual reality systems. Due to the often customized or prototype forms of virtual reality applications, system control is in many cases developed and tailored to a single application. In virtual worlds, by contrast, the viewer software (i.e., the client application to enter the virtual environment) is usually standardized and provides a predefined set of system control options. Hence, we do not use the system control category for the description of collaboration patterns.

If one were to put these actions on a continuous spectrum, they could also be distinguished in terms of their virtual world *effects* or their level of invasiveness or (space) intrusion. Chatting or changing one's avatar appearance, view, or position is far less intruding than moving an object, triggering a rocket, or blocking a door.

It has to be noted that these distinctions and the resulting classification do not include virtual objects. These digital objects require a separate classification that takes their manifold types and functions into account. In the following subsection, we discuss this important element of virtual environments.

CLASSIFYING VIRTUAL OBJECTS

In his successful book *The Design of Everyday Things*, Donald Norman (1988) postulates that people's actions and human behavior in general profits from everyday objects being designed as to provide straight forward affordances, that is, they should communicate how they should be used. He argues that less knowledge in the head is required (to perform well) when there is, what he calls "knowledge in the world." This insight can be fruitfully applied to virtual worlds by building on latent knowledge that users have and by providing cues that reuse appropriate representations (Smith & Harrison, 2001). This not only gives motivation for practitioners to utilize virtual environments for collaborative tasks, but implies that objects in virtual environments and their design are of great importance. Hence, we understand virtual objects as to form another major part of a virtual environment's infrastructure that goes along with the previously discussed part of action and interaction. Affordances can (and should) be used to signal users how to interact with a particular object, or how objects with built-in behaviors may act without any direct influence from the user.

One problem that arises is that those mental representations—or mental models, as Norman calls them—may trigger more expectations as to how objects and the environment may behave than current virtual worlds may be able to provide. In the opposite case users do not anticipate any functionality when acting in a virtual environment, and may get easily confused or disoriented when things happen without a direct command. As a result, for the time being, two extreme types of users of virtual environments are possible: underestimating and overestimating ones. This might in fact be part of an explanation to why it takes (or took) so long for virtual worlds to become accepted and be viewed as being more than just games, although they have been debated in research and practice for so many years.

But that left aside, it seems like for the longest time virtual environments researchers and developers have focused largely on graphical representation and rendering issues. With the launch (and most of all with the hype) of Second Life, a new era of accessible online virtual environments has begun. Following the trend of enabling users to create content, which is also an essential element of the definition of the Web 2.0, Second Life users could for the first time create and edit virtual objects, and also customize the appearance of their avatars, directly in a persistent virtual world (if we ignore Activeworlds, which provided content creation features long before, but unfortunately was never widely used).

With the possibility of scripting objects, that is, programming them in order to make them responsive to user actions, execute animations or follow behaviors, or simply update their own states continuously, virtual objects have become a powerful instrument in designing memorable user experiences in virtual worlds. In fact, interactive virtual objects represent technology in virtual environments; without active and interactive objects, any virtual environment would be nothing more than a virtual version of a world without technology. This comparison should illustrate the need for a formalization regarding virtual objects.

In spite of their crucial functional importance, little research has been conducted on classifying virtual objects so far. More work has been done on the technical side; for instance, the approach of including detailed solutions for all possible interactions with an object into its definition has been proposed (Kallmann & Thalmann, 1998). These so-called *smart objects* integrate descriptions for subobjects on how to behave, on positions for avatars or agents to interact with, and provide gestures up to the precision of finger splay. Another later presented framework takes up on this idea and adds interobject interaction definitions (Jorissen & Lamotte, 2004). Currently—to the authors' knowledge—at least the two virtual world platforms Second Life and OpenSim support defining avatar positions for interaction within an object definition, as well as interobject communication.

A first informal classification of virtual objects was proposed by Smith and Willans (2006) while investigating the requirements of virtual objects in relation to interaction needs: the authors state that the task requirements of the user define the behavioral requirements of any object. Consequently, they distinguish between *background objects*, which are not critical to the scenario, *contextual objects*, being part of the scenario but not in the focus, and *task objects*, which are central to the scenario and the actions of the user (Smith & Willans, 2006). While this distinction may be useful for determining the level of importance of virtual objects, that is, in requirements analysis phase, it does not distinguish objects based on their actual functioning. Hence, we present a classification of virtual objects into the following three categories:

Static Objects solely exist; they do not follow any kind of behavior and do not respond to any of the user's actions. We distinguish between fixed objects that are not meant to be moved (such as statues, furniture, architecture) and portable objects, meant to be picked up or carried around, held, or worn (such as handheld objects, items that hover over the avatar, and hats and distinctive clothes in general). This quality does not have to be persistent; the categorization should work in order to describe the collaboration pattern.

Automated Objects either execute animations repeatedly or by being triggered. Alternatively they follow a behavior (ranging from simple behaving schemes such as, e.g., following an avatar, to highly complex autonomous, intelligent behaviors). We further separate the most rudimentary of all object behavior forms into an extra subcategory—the behavior of merely continuously updating its state or contents.

Interactive Objects represent generally the notion of a tool or *instrument*; either they produce an output as a response to a given input, or they execute actions on direct user commands (like, e.g., a remote control), or they act as vehicles, meaning that the user directly controls their movement (with or without the user's avatar on it), using the primary navigation controls.

The border between automated and interactive objects may seem fuzzy at first, but it is clearly delineated by the differentiation *whether* a user triggers an object to act deliberately or not.

Considering alternative classification properties, for example the distinction of size, whether virtual objects can be entered or *not*, or follow physical laws, for example, move in the wind, is in our belief of secondary importance—especially for the use cases we try to support with our contribution (collaboration tasks and collaborative learning).

A BLUEPRINT FOR AVATAR-BASED COLLABORATION

The ABC framework supports and fosters the development of innovative collaboration patterns for virtual worlds by providing formalization that *connects* the distinct features of the medium (the infrastructure) with specific collaboration and learning goals in given contexts. To establish this connection it provides a dramaturgy layer that adds a semantic to the syntactic elements of the medium (i.e., available actions and objects), defining macroactions, settings, roles, steps, timing, and so on.

Figure 2.1 illustrates the framework for Avatar-Based Collaboration (ABC Framework) based on the distinctions described in the previous

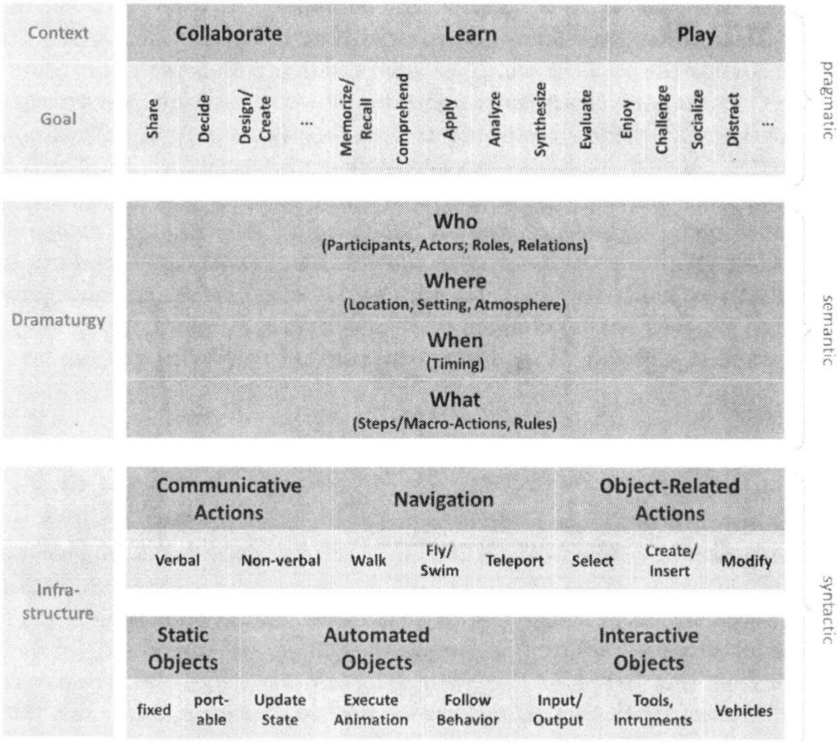

Figure 2.1. The Avatar-Based Collaboration Framework.

sections. It is intended as a blueprint for embodied collaboration in virtual environments. As such, it can be used as a basis to develop or describe collaboration patterns in virtual worlds such as Second Life or OpenSim. Its three-tier architecture reflects the syntactic, semantic, and pragmatic level of a collaboration medium, as discussed in subsection 3.2. In the following, we explain the parts of the framework, following a top down order.

CONTEXT AND GOAL

The context describes the application domain of a collaboration pattern, while the goal defines more specifically what kind of activity a pattern aims to support. A first category comprises patterns that aim for collaborative work in the traditional sense, that is, having main goals such as to *share* information or knowledge, collaboratively *design or create* a draft, a product, or a plan, *assess or evaluate* data or options, or make *decisions*, and so on. Since these goals do not necessarily have to be associated with work in the narrow sense of the word, we label the first context category *Collaborate*. The category *Learn* frames the domain of education. We assigned six goals to it, selected according to Bloom's Taxonomy of Educational Objectives (Bloom, 1956). Bloom distinguishes between different levels of learning goals starting with simple *memorizing or recalling* information, to the more complex tasks of *comprehending*, being able to *apply*, *analyzing*, and being able to *synthesize* or even *evaluate* new knowledge regarding its limitations or risks. For the category *Play* we allude to such usual game oriented goals as feeling challenged by *competition*, *distracting* oneself (losing oneself in a game), or *socializing* with others in a playful manner.

Dramaturgy

The term dramaturgy in this context designates the way in which the infrastructure in virtual world is used to reach a specific collaboration goal or in other words support a group task. It consists of the necessary *participants* and their roles and relations (the "who"), their interaction spaces and repertoire (the "where"), as well as the timing and sequencing of their interactions (the "when"). The dramaturgy also specifies the actions (the "what") taken by the participants and the social norms and rules they should follow within a given collaboration pattern. While the goals and contexts specify the why of a collaboration pattern, and the infrastructure the how, this level consequently addresses the who, where, when and what of a purpose-driven online interactions. The dramaturgy defines in which ways the infrastructure of a virtual world can be used by the participants to achieve a common goal.

Infrastructure

The final, most basic level of the blueprint contains the previously discussed elements *Actions* and *Objects*. As explained above, we think it is useful (for the design of patterns) to distinguish among communicative, navigational, and object-related actions and among static, automated, and interactive virtual objects.

We refined a definition of a collaboration pattern in subsection 3.1, as being a set of tools, techniques, behaviors, and activities for people who meet at a place to work on a common goal, together in a group. Using the wording of the framework, this translates to *a set of objects, actions, rules, and steps for participants with roles who meet at a location to collaborate on a com-*

mon goal in a given context. A specific collaboration pattern is then an instance of the framework and can be defined using the parameters positioned within the framework.

There are two distinct ways in which the above blueprint can be used for the creation of collaboration patterns: it can be used in a *top-down* manner from goal to infrastructure in order to specify how a given goal can be achieved using an online 3D virtual environment. Here, the given context(s) and the goal(s) of the pattern are specified first, then the dramaturgy is developed, before finally the required actions and objects are filled in. Alternatively, the blueprint can be used *bottom-up* in order to explore how an existing virtual world infrastructure can enable innovative dramaturgies that help achieve a certain collaboration (or learning) goal. Using this approach, existing objects and available actions are filled in first, a dramaturgy is developed on top of it, before finally the goal(s) and the context(s) is/are specified. The bottom-up approach seems more feasible for situations in which a virtual world including objects and tools is already available (or virtual world developers and 3D modelers to create infrastructure are unavailable), while the top-down approach may be the better choice to go about creating collaboration patterns when specific goals are to be achieved (or developers and modelers are available to create new content and functionality). The top-down and bottom-up approaches can also be mixed or used in combination or succession, as an iterative development process.

We do not strive for mutually exclusive and collectively exhaustive categories in the framework—the categories are meant to serve as a structure to describe and develop collaboration patterns rather than to provide an entirely unambiguous way of formalizing collaboration patterns.

The case study in the next section will illustrate the application of the ABC Framework in more detail, also pointing out advantages and difficulties in using it to develop collaboration tasks, in the case more precisely collaborative learning exercises.

CASE STUDY: DESIGNING COLLABORATIVE LEARNING EXPERIENCES FOR THE SHANGHAI LECTURES

As a first use case the ABC Framework was deployed in 2009 and 2010 for the development of a dozen exercises for a 3D virtual world to accompany a global series of lectures on natural, artificial, and embodied intelligence, the ShanghAI Lectures. This section describes the entire process from the ideation of collaborative learning patterns in the given context to the output of specific requirements to 3D designers and modelers on the one hand and virtual world developers and scripters on the other.

THE SHANGHAI LECTURES

The ShanghAI Lectures (http://shanghailectures.org) is a cross-reality global teaching and international student collaboration project. Its core components are a lecture series on natural, artificial, and embodied intelligence presented by Professor Rolf Pfeifer (director of the Artificial Intelligence Lab at the University of Zurich) and exercise assignments for multicultural student teams from all over the globe. The lectures are broadcasted via videoconference. Both in the first and the second year of the project, approximately 300 students collaborate in self-managed global virtual teams on weekly group assignments, view videorecorded lectures and expert talks together, and meet online with their peers and tutors, all embodied as avatars in a virtual world. As an environment for this purpose of work and socializing we created UNIworld, building on the virtual world platform Open Wonderland (open-source; formerly developed by Sun Microsystems, now independent), which enables the customized design of the virtual environment, collaboration features such as shared applications, the extension of communication tools, and the implementation of tailored extensions, such as authentication schemes and social software features. We further chose to create ShanghAI Island, a second virtual environment, using OpenSim (open-source; came out of the Second Life platform). The idea behind this was for the students to "travel" to a virtual seminar location twice in the semester, in order to work on additional exercises, for extra credits. For the project it was an option to test a different virtual world platform, for a possible move away from Open Wonderland (OpenSim offers more features for a straightforward creation of interactive objects, using scripting, and has proven to be more stable as a platform in general).

MAIN DEVELOPMENT TASK AND KEY CHALLENGES

The fact that students from all around the globe work together in small teams for the exercises of the ShanghAI Lectures, and the fact that a virtual world was chosen as the single place to facilitate and foster this intercultural collaborative learning, make it a compelling case to be looked at in detail here.

The main development task was to create an appropriate number of exercises on the contents of the lecture series for teams of students to work on, together in a virtual world. Since these exercises constitute the key activity in UNIworld, the architectural design was heavily influenced by it. The final design of UNIworld was created around team "arenas": team rooms that facilitate collaborative work in a private setting. In total, we developed 13 virtual world exercises to accompany seven lectures over the whole semester. With many of the students being first-time virtual world users, the first exercise needed to serve as a virtual world training task, to make the students familiar with navigating, communicating, and interacting with tools in UNIworld. As a second exercise not related to the topic of the lecture, a team-building task was required, as the teams were assigned randomly from all students (from 23 different universities). The eleven remaining exercises were developed on contents of the lecture, utilizing the ABC Framework as a scaffold to create innovative patterns (the creation of collaboration patterns was initiated by filling in the available contexts and goals, then adding available objects and actions, and connecting them with a fitting dramaturgy).

Key challenges in the development of the exercises were to transfer exercises that accompanied an earlier (traditional) version of the same lecture from their paper-based form to a format for use in a virtual world, harnessing the distinct features of the medium. The decision to deploy a virtual world as a platform in which to conduct the exercises had come out of the thought of illustrating embodied intelligence in a more comprehensible way than having exercises solely on paper sheets or flat web pages. The idea was to use virtual embodiment (avatars) to help illustrate embodied intelligence. After this was achieved however, problems arose concerning the virtual world platform that had been chosen: promised functionality was not available, additional developers to solve this could not be hired, and system stability in general was not guaranteed. This led to a fundamental modification of the already developed collaboration patterns for the exercises, toward using much less interactive tools (or none at all)—which meant a significant decrease in quality, and in originality of the use of the virtual world. Eventually, the decision was made to include a second virtual world in order to host additional exercises in a "remote" location. This implied an additional key challenge: to modify/redesign the collaboration patterns for the additional exercises in a way that respects the environment to be merely a place to work on the current exercise, not to socialize or have a team venue. This second virtual world, and the exercises in it, finally had to be designed as sort of a seminar getaway.

APPROACH AND DESIGN PROCESS

As the main source of information and contextual input to the design of the collaboration patterns for the exercises, the lecture slides of an earlier (face-to-face) version of the lecture series were available, the (paper-based) exercise sheets to that lecture, and further contextual information about the domain and the contents of the lecture in form of a textbook that served as a basis for the lecture. As a source of inspiration for the development of innovative dramaturgies for the collaboration patterns, data from previous research was utilized: virtual world collaboration pattern descriptions from an exploratory investigation in Second Life (see Schmeil & Eppler, 2009). The combination of these two sets of sources of input eventually led to novel ideas and innovative patterns for exercises to accompany the lecture series in the ShanghAI Lectures.

The design process was not entirely defined at the beginning of the project, but rather evolved during this first application of the ABC Framework. It unfolds as follows: as a first step for the development of a pattern, context(s) and goal(s) are specified roughly. Then, the designated content of the exercise is drafted into the agenda section of the dramaturgy layer. Having this as a scaffold, existing patterns and other sources are used as inspiration to ideate a dramaturgy for the exercise. While the dramaturgy layer gets filled in, single action and object elements are added in the infrastructure layer whenever their requirement for any of the sections in the dramaturgy gets apparent. As soon as the dramaturgy layer is completed, the agenda and the context/goals sections are updated, if required. As a last step, with the context(s), goal(s) and the entire process in mind, the infrastructure layer is traversed section by section, with the goal to find out whether or not any additional actions or objects can be deployed for the pattern to make it more engaging or exciting. Whenever any amendments or modifications are made, any affected sections are updated.

This fully completed formalization of a collaboration pattern is then used to deduce requirements for 3D modelers (or architects), virtual world designers and developers/scripters. These are outlined as notes at the right edge of the framework sheet or formalized and given out on separate documents.

IMPLEMENTATION

Table 2.1 lists the final 13 exercises that were developed. It has to be noted here that due to the lack of functionality and stability of Open Wonderland as a virtual world platform and the lack of dedicated virtual world programmers and scripters in the ShanghAI Lectures project, a number of exercises had to be modified in order for them to be implemented.

In the following, three of these 13 exercises are described in detail, as examples of innovative collaboration patterns. A focus is put on which features of the medium are utilized by the particular exercises. The appendix of this chapter shows these three collaboration patterns in the structure of the ABC Framework. There it is clearly visible how collaboration patterns are created, namely by filling in readily available (and

Table 2.1. The 13 Collaboration Patterns That Were Developed for the ShanghAI Lectures 2010

Exercise Name	*Description*	*Comments*
WL Tutorial (WL)	This "first contact with Wonderland (WL)" exercise aims to make the students get to know and learn to use basic WL functionality, such as navigation, camera control, communication, and the use of different tools. Tasks and instructions are provided along a path in-world.	Individual exercise; basic tutorial
Design Team Room (WL)	The student teams work together to furnish their team room, in order to make it unique, to feel home, and to convey the spirit of the team in the room's design. Ideally, the team rooms should look different from each other also from the outside. Drag and drop import of 3D objects from the 3D model library Google Warehouse is supported.	Team-building exercise
Embodied Memory (WL)	Students stand in front of a 4-point-path and have to point in given directions. After, they move on the 4-point path and have to point in the same directions again, now being turned by moving on the 4-point path. Students should learn to orient themselves in this exercise.	Orientation in the environment
Anticipate Robot Behavior (WL)	After getting a robot design (as images in WL) the student team has to anticipate how the robot would behave in certain situations. The situations should be provided as images as well, for example a parcours the robot is (imaginatively) put into. The students use a whiteboard and a sticky notes board to put together their solution, alternatively they could be asked to perform a sort of role play for their tutor, as a more active solution.	Exercise simplified for use of whiteboard
Redesign Robots (WL)	Student teams are shown videos (or schematic designs) of robots that are not designed perfectly. For each video they discuss in their group and develop a redesign of that robot that is supposed to overcome the flaws students discover with the original design and resolve shortcomings. Whiteboard and sticky notes are used for this exercise.	Exercise simplified for use of whiteboard
Experience Situatedness (OS)	The student team mounts a robot. The robot starts driving around in a parcours, showing behavior typical for its model/type. The students get a situated experience, a "first-person view." After the performance, the students get off the robot and walk up a hill to watch the robot behave again, now as spectators from outside. They discuss the differences between the two experiences. With all this fresh in mind, they log out of OpenSim and get back to their team room in UNIworld, where they write down their comparison on a sticky note board.	

(Table continues on next page)

Table 2.1. Continued

Exercise Name	*Description*	*Comments*
Tag and Annotate Videos and Images (WL)	In this exercise, student teams are asked to watch videos and images, discuss each of them and write down notes, comments, or comparisons on a sticky note board, or sketch something on a whiteboard. This is rather a scaffold for an exercise—content is fairly open here.	Exercise simplified for use of whiteboard
Which Robot Am I? (OS)	The student team watches a robot behave/move/act in an arena (a parcours) and guesses which kind of robot it is (the performing robot is initially "cloaked": the team cannot tell what robot it is by appearance). After the student team comes to an agreement and makes their guess, the robot reveals its identity by "decloaking" itself, illustrating whether or not the team's guess was correct. This is done for several robots. The student team makes their decision by "voting by feet," that is by moving their avatars to robot prototypes (an agreement is made when all team members have moved to the same robot).	
Anticipate Self-Assembly (WL)	After getting a design of a self-assembly robot (as images in WL) the student team has to anticipate how the robot would self-assemble. The student team uses a whiteboard and a sticky notes board to put together their solution.	Exercise simplified for use of whiteboard
Categorize Robots in Framework (WL)	On a whiteboard with a framework as background image that allows categorizing robots according to different classifications, the student team marks the different robots they encounter during this exercise (robots can wait along a path, drive by, or simply park in a line).	Exercise simplified; not realized
Robot Colorization (WL)	The student team walks along a path at which (single-colored) robots are aligned. For each robot, they have to colorize its parts, according to a given categorization (e.g., sensors—actuators). This can be repeated for several robots, and for different categorizations per robot.	Exercise simplified; not realized
Robot Pantomime/ Role Play (WL /OS)	Each student team designs and practices a choreography to illustrate typical behavior of a particular swarm of robots. One after another, the teams perform their play on stage in front of the other teams. The other teams have to guess what swarm or what swarm situation is illustrated.	Not realized
Develop Own Virtual World Exercise (?)	The student teams ideate and formalize their own virtual world exercise, making use of their experience in the virtual world, and choosing any content of the lecture as a contextual input.	Not realized

unchangeable) information first and then connecting these elements through an engaging dramaturgy, which in turn is constant subject to change when adding or removing elements from the pattern in the framework.

Experience Situatedness

This exercise demonstrates situatedness for the students. Using a 3D virtual world with an orchestrated immersive experience, the exercise aims to illustrate not only the meaning, but also the feel of situatedness. To this extent, students experience both a first-person and a third-person view of the same robot movement, and are asked to directly make a comparison between the two. In order to make the experience a memorable (and hopefully an exciting) one, the virtual world features of immersion and spatiality are harnessed: the sensation of sitting on a "real" robot while it moves around on a parcours would be very difficult to deliver in a face-to-face setting (and close to impossible with hundreds of globally dispersed students); and videos of first-person and third-person views of a robot lack the interactive character of the experience.

Which Robot Am I?

This exercise is one of a couple of exercises on robot behavior and artificial intelligence. It has a game-like character, incorporating interactive elements and a responsive environment. Thus, it harnesses the feature of a virtual world allowing for an extensively configurable and scriptable environment that can support and foster collaboration and increase engagement. Among other sources, we were inspired by a children's TV show (for the voting by feet part) and aimed to develop a fun activity and an overall enjoyable experience for the students. Furthermore, the decloaking effect gives memorable immediate feedback to the student team's decisions, and the fact that the virtual world automatically sends an e-mail containing the team's results for the exercise to their respective tutor demonstrates the interoperability of the medium.

Anticipate Robot Behavior (+ Robot Pantomime/Role Play)

This exercise requires the student teams to synthesize their knowledge in order to develop a performance about robot behavior. It was implemented as a combination with the not realized exercise Robot Pantomime/Role Play, as there was no significant surplus in implementation effort. Thus, the student teams deliver embodied experiences to their respective tutors, themselves profiting by both orchestrating the performance and participating in it by immersing into roles of robots. Moreover, the exercise emphasizes collaboration on various levels, requiring the students to work together as a team for both staging and delivering the performances.

FINDINGS AND EXPERIENCES

First, it should be noted that the author of the case study was also this chapter's first author, and thus not a neutral user of the ABC Framework, as he was heavily involved in its development in the first place. In the application of the framework—the ideation and development of the 13 exercises—up to nine people were involved. Comments of users and otherwise involved people on the framework were generally positive, but also pointed out some shortcomings that should be looked at in future modifications of the framework. On the positive side, the framework was found to help harness the distinct features of virtual worlds in order to develop more engaging ways of collaborating and learning together. The breakdown of the infrastructure of virtual worlds helps think in innovative ways and not overlook any options the medium offers to support and foster interaction. The framework was considered as a checklist, referring to this point. Another point was the fact that the framework serves as a tool for fast-prototyping, especially in combination with existing collaboration patterns that provide inspiration and guidance; after having gotten used to the structure, patterns can be developed in short time. Furthermore, the framework is self-explanatory for the description/ development of patterns and comes ready-to-use, without any training necessary. On the negative side, the most notable point was that there is no description of where to start and how to go through the framework (for first use); a path to guide first-time users through the framework would be a valuable add-on to it. Then, while the developer or designer of a pattern naturally can read and remember their own created patterns, a person not involved in the development might have difficulties understanding them—guidelines for how to fill in the different sections to make patterns comprehensive for others might help resolve this issue. Other comments were that the process of constructing the output for 3D modelers and virtual world developers and scripters is not structured at all and that there is no option to connect the content to the patterns. These latter two points might be taken into consideration for future work on and around the framework.

CONCLUSIONS AND FUTURE RESEARCH INITIATIVES

In this chapter, we have presented the ABC Framework, a systematic framework that organizes the necessary elements for the design and implementation of collaboration patterns in virtual worlds. The framework is based on three levels, namely the pragmatic or contextual level, including the goals of an online interaction, the semantic or dramaturgic level that defines how elements and actions are used (and interpreted) in time to achieve the collaboration goal, and the syntactic or infrastructure level consisting of the actual objects and online actions that are combined to implement a collaboration dramaturgy. We have presented a case study to describe an application of the framework that covered the whole process of developing collaboration patterns and showcased a list of 13 exercises that were developed for a global education project, and a detailed description of three of them. First steps of future work are to integrate the findings and experiences from the first use case into the framework and possible extensions to it, and to further evaluate the framework also for related ideation and documentation purposes.

Having established a systematic map of the elements required to devise and implement innovative and engaging collaboration patterns for virtual worlds, the question nevertheless remains which patterns are more effective than others in terms of their benefit in supporting knowledge-intensive collaboration tasks in groups. To this end, we are preparing controlled experiments in order to empirically investigate which aspects and factors of virtual world collaboration patterns and designs are beneficial for the process and the outcome of collaboration in virtual worlds.

APPENDIX: COLLABORATION PATTERNS DEVELOPED USING THE ABC FRAMEWORK

(Appendix figures begin on next page)

Anticipate Robot Behavior (+ Robot Pantomime / Role Play)

Goal/Context

Collaborate			Learn						Play			
Share	Decide	Design/ Create	Memorize/ Recall	Compre-hend	Apply	Analyze	Synthesize	Evaluate	Enjoy	Challenge	Socialize	Distract
					Sketch, draw, dramatize		Hypotesize					

Dramaturgy

Who	PARTICIPANTS • Team of 4-5 students	ACTIONS	ROLES	RELATIONS
Where	LOCATION Team arena	SETTING Screens with robot designs along a path, with one whiteboard each	ATMOSPHERE	
When	DATE	TIMING		
What	AGENDA I. Discussion of case II. Cartoon development	STEPS / MACRO ACTIONS 1. Student team gets a robot design (or walks to a board displaying it) 2. They discuss and take notes of how the robot would behave in a given situation (eg. on the provided sticky notes board) 3. They stage a short performance for one or more members of the team to deliver to the tutor at the next tutor meeting 4. The tutor assesses the performance and gives feedback 5. [Repeat 1-4] for more robots	RULES	

Infrastructure

Communicative Actions		Navigation			Object-Related Actions		
Verbal	Non-verbal	Walk	Fly / Swim	Teleport	Select	Create / Insert	Transform
• Discuss the robot design		• Simulate robot behavior					• (move the obstacles to demonstrate movement?)

Static Objects		Automated Objects			Interactive Objects		
Fixed	Portable	Update State	Execute Animation	Follow Behavior	Input/Output	Tools, Instruments	Vehicles
• Designs of robots (on screens?) • Obstacles			• (Robot design and whiteboards move away after work on them is finished?)			• Whiteboard with several frames → cartoon • Also noteboard?	

Experience Situatedness

Goal/Context

Collaborate			Learn						Play			
Share	Decide	Design/ Create	Memorize/ Recall	Compre-hend	Apply	Analyze	Synthesize	Evaluate	Enjoy	Challenge	Socialize	Distract
Describe experiences						Compare						

Dramaturgy

	Collaborate	Learn	Play
Who	• Team of 4-5 students	• One subgroup rides on robot, the other watches from outside • Then the roles switch	
Where	Team arena	- Vast arena/stage for the robot to navigate and show some typical and impressing behavior - Spots to watch from outside	Colosseum feeling?
When	2010-11-04	10 mins per robot?	
What	I. Experience robot behavior situated or not II. Share and discuss III. Have other experience IV. Share and discuss again 1. One subgroup mounts the robot, the other one stays outside to watch 2. Robot behaves, with one subgroup in it → subgroups get different experiences 3. The students share their experiences and discuss	4. The subgroups switch (stay out / ride) 5. After the second round the students meet again to discuss and refine their notes 6. [Repeat 1-5] for more robots 7. Write a comparison, back in UNIworld	

Infrastructure

Communicative Actions		Navigation			Object-Related Actions		
Verbal	Non-verbal	Walk	Fly / Swim	Teleport	Select	Create / Insert	Transform
• Share experience, discuss differences	• Express emotions while sharing their experiences • Turn towards focus of attention		• (Fly up to view robot's performance from viewing plaform or from air?)				

Static Objects		Automated Objects			Interactive Objects		
Fixed	Portable	Update State	Execute Animation	Follow Behavior	Input/Output	Tools, Instruments	Vehicles
• Viewing platform • Obstacle course			• Mountable robot (see on right)		• Board to take notes?		• Mountable robot with space for up to 3 students to ride with it

Which Robot am I?

Goal/Context

Collaborate			Learn						Play			
Share	Decide	Design/ Create	Memorize/ Recall	Compre- hend	Apply	Analyze	Synthesize	Evaluate	Enjoy	Challenge	Socialize	Distract
	Agree on an answer			Match, explain		Identify, differentiate						

Dramaturgy

Who	PARTICIPANTS • Team of 5 students	ACTORS	ROLES	RELATIONS
Where	LOCATION	SETTING Robot acting arena: course with obstacles , on which robots perform / show their behavior Robot selection area: a place (circle arc?) where robots stand ‚in line' to be selected by the team		ATMOSPHERE Automated world. Robots feel when you are about to select them and come closer (reacting to proximity) // Confrontation
When	DATE	TIMING		
What	AGENDA	STEPS / MACRO-ACTIONS 1. Cloaked robot performs in acting arena 2. Students discuss and make a choice positioning themselves between the robots in the selection area → robots move towards student team, one of them is selected 3. Behaving cloaked robot from acting arena and the selected one meet decloak → either unite or not 4. [Repeat 1-3] for more robots		RULES →

Infrastructure

Communicative Actions		Navigation			Object-Related Actions		
Verbal	Non-verbal	Walk	Fly / Swim	Teleport	Select	Create / Insert	Transform
• Discuss		• Students get closer to robots to make them feel proximity and come closer					

Static Objects		Automated Objects			Interactive Objects		
Fixed	Portable	Update State	Execute Animation	Follow Behavior	Input/Output	Tools, Instruments	Vehicles
• Robots for the students to make their choice („1, 2, oder 3")			• Robots that execute their typical behavior	• Robots come closer when approached			

Creating Dramaturgies Using the Avatar-Based Collaboration Framework

Objective: Use a form to identify the components of an avatar based collaboration for the design of a virtual world.

Task One: Explain the Current Collaboration Practice in Virtual Worlds

- The design of many virtual world environments and activities today are not only heavily influenced by the physical world and established activities in it, but often copied par for par into the virtual world.
- This approach of designing virtual worlds does not consider the distinct features the medium offers.
- Explain: Magazine design (layout and activities) are fundamentally different from web page design, as it targets an entirely different medium -the reader. Insert any magazine for the students to examine. Ask them to describe a magazine design. Look for answers like the following:

    - "we read exclusively in a magazine—text dominates"
    - "we look at font size changes to call attention to changes"
    - "we flip through pages in a magazine, usually in a linear way"
    - "ads using large ads with people or places are predominant in a magazine"
    - "nothing is moving in a magazine"
    - "you can't do much in the magazine, except crossword puzzles"

- Explain: Website design (layout and activities) are more engaging and interactive than a magazine layout. Insert any website here to compare to the virtual world. Examine a regular website. What do you see about the website that is different from the magazine layout? Look for answers like the following:

    - "we read menus and groups of words first, then we drill down in text"
    - "we look at different pages on the website, it is not linear, but very noncontinuous"
    - "we look at still images but we can also watch animations and videos on the website"

  - "we can write on a web page, i.e, filling out a form"
  - "we can play games on a webpage, i.e., a Flash game"

- Explain: Each medium has its purpose. Magazines are designed around a single content area: (i.e., Food, Architecture, People, etc.). Websites are designed around informing, direct purchasing, and experience. What we do in the physical world: collaborating with other people, experiencing environments, manipulating objects is not treated in magazine formats or websites easily.
- Explain: The virtual world medium needs a design method in its own right—a design method that emphasizes the utilization of its features and specific possibilities.

Task Two: Hand Out the Avatar-Based Collaboration Framework and introduce it

- Explain: The ABC Framework (see Appendix figure at the end of lesson plan) is a blueprint for the design of collaboration patterns that lead to fruitful collaboration and memorable experiences in virtual worlds.
- It is a tool for the ideation and development of innovative patterns that make use of the specific infrastructure virtual worlds provide.
- The framework was designed in a three-tier architecture.
- An instantiation of it, a collaboration pattern, connects selected elements from the infrastructure (the possibilities of the medium) to one or more collaboration and learning goals that are to be worked on.
- This connection is made through the dramaturgy layer.
- The dramaturgy layer gives meaning to all included elements of the infrastructure in a way that the collaboration goals are reached.

Task Three: Filling Out the ABC Framework

- Explain: The framework can be used in different ways, there is no fixed step-by-step method.
- Draw the student's attention to the framework and explain: "however, two main approaches to designing collaboration patterns using the blueprint (see next page) have emerged in our first use cases"
- Explain: The top-down approach starts at the Goal/Context layer where designers fill in the particular goals they want to address with

the collaboration pattern. There are three goal types: collaborate, play and learn. Start with one of the choices at a time, but work to create three separate exercises to see how the information changes with each goal type.

- Choose a goal for "collaborate"
- Choose another one for "play"
- Choose another one for "learn"

- Tell the Student: "Once each goal is described with one or a few words in its respective field, fill out the dramaturgy for the goal." Remember, the dramaturgy are the components of the story: the what, when, where, and who.
- Start from the top here, the participants and their roles and relations in the new collaboration pattern are described first.
- Next, describe the setting and location. Remember as you are doing this to think about what the virtual world environment will look like. What is the "environment," what are some of the "objects" in the environment? What kinds of "sounds" do you hear, what can of animations do you "see."
- Then envision someone coming onto the island to engage you in a conversation or an activity.
- Write down the date and the timing of the encounter, and how the collaboration will be played out between you two. Where are you going to stand? Where are you going to move to? What happens in the environment to create an incident, event, or a conflict that has to be dealt with?

Task Four: Prioritizing What Gets Done First

- Explain: once the framework is completely outlined, the most creative part of the process begins. This is when you take the ideas and begin to design what gets developed first.
- Start with your avatar in front of one of the static objects or in front of another avatar.
- What is this object doing? Is it stationary? Such as a tree, or a house? How is your avatar going to interact with the object? Is it going to walk toward it, turn around and walk away from it? Think of it like a movie script and think of the drama that you're creating just with this first object.

- Is the object portable, can you pick it up, like a ball, or a tennis racket, or a vehicle that you get into?
- Then, think about how the object is going to change. Do you lift it, touch it, and rotate it? How is the object going to change state? If it's a door, do you click on it and it opens? If it's a bot, do you click on it and it turns and moves 5 feet away, beeps, and asks you, "how may I help you"?
- Then, look at how the other objects around you will be automatically engaged. For example if you enter into a car, does the light change in front of you? If you walk into a restaurant, does an avatar bot approach you and ask you, "May I help you." What are the surrounding objects doing when you move around them?
- Then, picture how you're going to make that moveable objects continue to operate. If you're in a car, do you drive it with the steering wheel, turning it right and left, using your mouse? If it's an assembly line, do you click on an interface to make it start, then click on another object to put a car part on the assembly line, then click on a robot arm to assemble the part?
- You have begun creating the infrastructure and the elements that are necessary to realize the designed dramaturgy. All objects and actions that are included in the pattern are described with one or a few words in their respective fields.
- During and after this, the dramaturgy (especially agenda) gets updated, and necessary changes are made to the rest of the pattern. This approach is best for designers who have specific goals in mind and know what they want to do and be done in the virtual world.

This formulaic bottom-up approach is more appropriate in situations in which not all common virtual world infrastructure is available, or when certain features are needed to be used or tested. Designers start by filling in all the available objects, avatars and actions in their particular virtual world environment, then develop an agenda using these building blocks, then continue moving up in the blueprint and think of collaboration outcomes last.

Inspiration for Dramaturgies

Inspiration for innovative collaboration patterns can be found in science-fiction or fantasy movies, video games, cartoons, comic books, dreams, and every other source that transcends physical reality and helps lateral thinking.

End of Exercise

This exercise is the beginning of developing an avatar based collaborative framework. You can see that it is actually more detailed that you might have imagined at first. Some people often jump right into the virtual world and start by creating a single object, then building the next object and not thinking about the interactions or the way people are going to collaborate around the object. This exercise is designed to create an appreciation of how you envision this process in a more advanced model, thinking about how the idea of a virtual unfolds like a story, a narrative, a movie. Once you have this kind of framework, you can begin to engage your avatar in very creative ways.

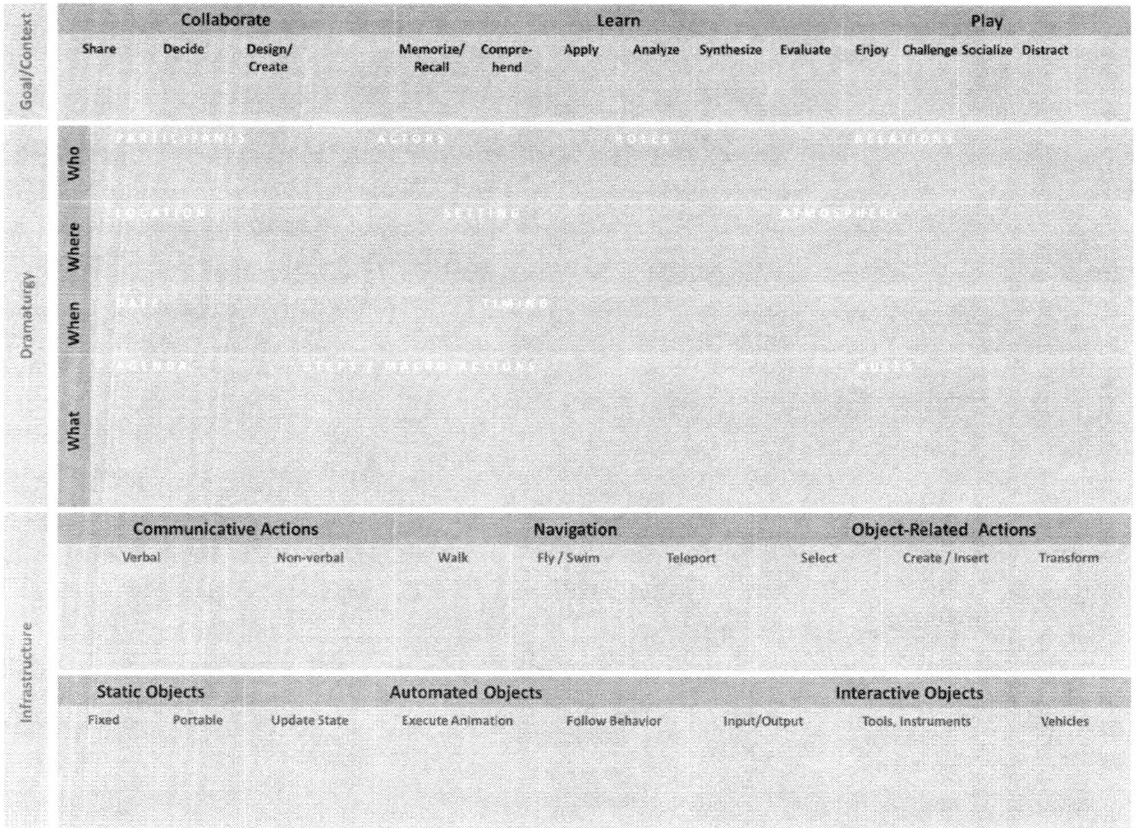
Goal/Context
Collaborate
Share
Decide
Design/ Create
Learn
Memorize/ Recall
Compre-hend
Apply
Analyze
Synthesize
Evaluate
Play
Enjoy
Challenge
Socialize
Distract
Dramaturgy
Who
PARTICIPANTS
ACTORS
ROLES
RELATIONS
Where
LOCATION
SETTING
ATMOSPHERE
When
DATE
TIMING
What
AGENDA
STEPS / MACRO ACTIONS
RULES
Infrastructure
Communicative Actions
Verbal
Non-verbal
Navigation
Walk
Fly / Swim
Teleport
Object-Related Actions
Select
Create / Insert
Transform
Static Objects
Fixed
Portable
Automated Objects
Update State
Execute Animation
Follow Behavior
Interactive Objects
Input/Output
Tools, Instruments
Vehicles

REFERENCES

Bainbridge, W. S. (2007). The scientific research potential of virtual worlds. *Science, 317*(5837), 472- 476.

Bell, M. (2008). Toward a definition of "virtual worlds." *Journal of Virtual Worlds Research, 1*(1). Retrieved from http://www.jvwresearch.org/v1n1_bell.html

Bloom, B. S. (1956). *Taxonomy of educational objectives: The classification of educational goals.* New York, NY: McKay.

Bowman, D., A., & Kruijff, E., & Poupyrev, I., & LaViola, J. J., Jr. (2005). *3D user interfaces: Theory and practice.* New York, NY: Addison-Wesley.

Carroll, J. M. (2000). Introduction to the special issue on "Scenario-Based Systems Development." *Interacting with Computers, 13*(1), 41-42.

de Freitas, S., & Jarvis, S. (2009). Towards a development approach for serious games. In T. M. Connolly, M. Stansfield, & E. Boyle (Eds.), *Games-based learning advancements for multi-sensory human-computer interfaces: Techniques and effective practices* (pp. 215-231). Hershey, PA: IGI Global.

de Freitas, S., & Oliver, M. (2006). How can exploratory learning with games and simulations within the curriculum be most effectively evaluated? *Computers and Education, 46*(3), 249-264.

de Freitas, S., & Neumann, T. (2009). The use of 'exploratory learning' for supporting immersive learning in virtual environments. *Computers and Education, 52*(2), 343-352.

Eco, U. (1978). *A theory of semiotics.* Bloomington, IN: Indiana University Press.

Gottesdiener, E. (2001). Decide how to decide: A collaboration pattern. *Software Development Magazine, 9*(1). Retrieved from http://ebgconsulting.com/Pubs/Articles/DecideHowToDecide-Gottesdiener.pdf

Jorissen, P., & Lamotte, W. (2004). A Framework Supporting General Object Interactions for Dynamic Virtual Worlds. *Smart Graphics,* 154-158.

Kallmann, M., & Thalmann, D. (1998). *Modeling objects for interaction tasks.* In EGCAS'98—9th Eurographics Workshop on Animation and Simulation, pp. 73-86, Lisbon, Portugal.

Manninen, T. (2000). *Interaction in networked virtual environments as communicative action: Social theory and multi-player games.* Proceedings of CRIWG2000 Workshop, IEEE Computer Society Press, Madeira, Portugal.

Morris, C. (1938). *Foundations of the theory of signs.* Chicago, IL: University of Chicago Press.

Norman, D. (1988). *The design of everyday things.* New York, NY: Basic Books.

Price, R., & Shanks, G. (2005). A semiotic information quality framework: Development and comparative analysis. *Journal of Information Technology, 20*(2), 88-102.

Schmeil, A., & Eppler, M. J. (2009). Knowledge sharing and collaborative learning in Second Life: A classification of virtual 3D group interaction scripts. *International Journal of Universal Computer Science, 15*(3), 665-677.

Schmid, B. F., & Lindemann, M. A. (1998). *Elements of a reference model for electronic markets.* Proceedings of the 31st Annual Hawaii International Conference on Systems Science (pp. 193-201), Honolulu, HI.

Shanks, G. (1999). *Semiotic approach to understanding representation in information systems*. Proceedings of the information Systems Foundations Workshop, Ontology, Semiotics and Practice.

Slater, M., Usoh, M., & Steed, A. (1994). Depth of presence in virtual environments. *Presence: Teleoperators and Virtual Environments, 3,* 130–144.

Smith, S. P., & Harrison, M. D. (2001). Editorial: User centered design and implementation of virtual environments. *International Journal of Human-Computer Studies, 55*(2), 109-114.

Smith, S. P., & Willans, J. S. (2006). Virtual object specification for usable virtual environments. *Annual Conference of the Australian Computer-Human Interaction Special Interest Group (ACM OzCHI 2006).*

CHAPTER 3

ARCHITECTURAL EVOLUTION OF E-LEARNING VIRTUAL WORLDS

Proposed Design Measures to Enhance the E-Learning Experience Within 3D Learning Spaces

Noha Saleeb and Georgios Dafoulas

Numerous approaches are currently being adopted to efficiently utilize emergent 3D virtual worlds (VWs) for delivering e-learning, concentrating mainly on teaching and learning applications. Equally significant, this research demonstrates the importance of customizing architectural design characteristics of 3D virtual learning environments (VLE) within these VWs as another complementary and essential user-centered approach to enhancing students' e-learning experiences. This is achieved by displaying higher education students' propositions and requirements from the design elements of their 3D educational spaces best suited for promoting their skills, satisfaction and participation during their e learning sessions. These recommendations can be used by educators and designers to enhance the design features of existing and future 3D virtual educational facilities within 3D VWs to augment the students' e-learning process

Engaging the Avatar. New Frontiers in Immersive Education, pp. 49–80

INTRODUCTION

The fundamental aim of this research is to enhance students' blended learning experiences comprising of both face-to-face and online courses, by investigating factors and procedures that can help achieve maximum assimilation, achievement and enjoyment from e-learning within 3D Virtual Learning environments (VLEs) such as Second Life. Best practices in 3D pedagogy have essentially entailed development of educational methods, 3D applications and activities to support different learning modes and learning styles of students in 3D VLEs (Calongne, 2008; Scopes & Lesley, 2009). However as Philip Johnson, a renowned architect, believed, "All architecture is shelter, all great architecture is the design of space that contains, cuddles, exalts, or stimulates the persons in that space." Since one of the vital factors evidenced to have an impact on students' physical learning in is the design characteristics of the learning space (Evans, 1980), elaborated consequently, this research has invested into finding the effect of 3D virtual architectural design of educational spaces on e-learning of students, by capturing learners' preferences and propositions for design elements to be used in their 3D virtual learning spaces that would augment satisfaction, retention, participation and enjoyment. This would hence allow reaching best user-centered practices in virtual architectural design of 3D educational building spaces, for the benefit of students and supplementing their e-learning experience within virtual worlds. Best practices in delivering practical, effective teaching and learning would therefore be achieved with the help of both (1) innovative virtual educational methods and (2) applying most suitable 3D design features to educational spaces during e-learning sessions.

While this study pertains to multiple sectors of students in higher education (namely, undergraduate, postgraduate, and adult learners participating within Second Life as a representative of 3D VLEs), findings of this research are not specific to Second Life only but rather general to 3D virtual worlds in general since the psychological impact of the design of a 3D virtual learning space on its users is universal in any virtual world.

Along with the pedagogical immigration trend of many educational institutions towards 3D VLEs (Joseph, 2007), arose creative opportunities for erecting buildings that cross the realms of reality and explore the imagination of the designer. This is because of the fundamental discrepancy between the physical and the virtual world where there are no restraints on budgets, no engineering natural forces and material strength limitations, no infrastructure requirements, sound, ventilation regulations or even gravity which can be disregarded to have 3D virtual buildings floating in midair or immersed under water (Bridges & Charitos, 1997). Such novel construction techniques have been used to construct virtual university campuses in 3D VLES to produce a wide variety of designs that range between realistic portrayals or imitations of physically existing campuses, and completely imaginative embodiments (Alarifi, 2008).

However there is no academically conducted research that directly associates between the new e-learning techniques sprouting within 3D VLEs (Saleeb & Dafoulas, 2010a), and the architectural design specifications of the 3D virtual spaces within which this e-learning is taking place (Minocha & Mount, 2009), and thus whether these design specifications have an impact on the effectiveness of e-learning on student users of 3D VLEs (Saleeb & Dafoulas, 2010c). Such design features include color, texture, dimensions of space, lighting, and ventilation among others. Even more, sparse study explores the effect of 3D architecture in virtual worlds in general on any genre of users, not just students in 3D VLEs, and their fulfillment and comfort towards it (Saleeb & Dafoulas, 2010b). The current research thus initiates closure of this gap by capturing students' design suggestions and proposals to achieve satisfaction and contentment from architectural design elements of virtual educational buildings within 3D VLEs, hence giving the opportunity to issue recommendations for future learning space enhancement: the technique used for data collection explained hereafter.

This study offers insights in preliminary stages for defining the effect of environmental factors on a student's e-learning experience within 3D virtual learning environments, and provides opportunities for further developing the 'supporting teaching and learning' strand of activities. More specifically the use of 3D VLEs could address issues relating to (1) learning & technology, (2) open educational resources and (3) inclusion.

BACKGROUND

Previous literature indicates that, in the physical environment, characteristics such as color, dimensions, shape, textures, ventilation, sound, lighting and other factors of the physical learning space affect the degree of achievement and assimilation of students from education within these spaces. For example it has been established that classes smaller than 900 sq. ft. in area do not allow for adequate movement between tables without bumping into students and their belongings; crowded classrooms contribute to disciplinary problems (Eberhard, 2008). The Ohio University Facilities Commission also noted that students participated twice as much in discussions in classrooms with warm colors, soft furniture, and textured floor coverings. Students rated these classrooms higher than traditional classrooms. Another study found soft colored classrooms associated with better attendance and positive attitudes toward class, instructor and classmates, while an "ugly" environment gives feelings of discontent, the desire to escape, and fatigue. Light (especially natural) has been shown to affect blood pressure, pulse, respiration rates, and brain activity. Exposure to full-spectrum lighting is associated with better attendance, more positive moods, great concentration, and better scholastic performance (Fink, 2002). On a separate note, narrow hallways, that are too small for student traffic between classes, have been found to encourage fighting and hinder evacuation in emergencies (Hall, 2001).

However, as stated earlier, the analogous effects of learning space architectural and environmental design characteristics in 3D virtual worlds on students' e-learning are currently underresearched. One of the scarce studies relating to this topic conducted by an individual market research, within Second Life, depicting users' reactions to preferences between realistic buildings and imaginative style buildings, only reveals that users prefer realistic style buildings with a percentage of 60% more than imaginative style 3D buildings (Market Truths, 2007, 2009). Furthermore, there are no devised building codes for designing 3D virtual educational facilities, analogous to those available for building physical educational facilities. There are only some general recommendations or guidelines offered by previous researchers interested in design of virtual environments, based on observation and interviews (not on interaction of the learners with the environment during the e-learning process), to aid design 3D virtual educational spaces.

For example Dickey (2004) suggested using architectural and environmental elements such as landmarks, signs, thresholds (e.g., doorways expressing relationship of the space with the surroundings) and boundaries (fences, walls, etc.) to aid students' way finding, or using large open spaces (but with no detailed specifications provided) (Charitos, 1998). Bridges and Charitos (1997) noted that physical world elements, for example, doors, roofs, columns, structural or ornamental details, should only be used if there is a functional use for them (e.g., no door if the walls are penetrable). Minocha and Reeves (2009) further proposed using "open spaces as much as possible" to accommodate flying, wide corridors, realism in design, and arrangement of spaces to follow activities performed in them. As for the factors affecting the level of engagement and participation experienced by the learners, only pedagogical factors were identified not architectural factors (Minocha & Mount, 2009). A research on user orientation within 3D VLEs conducted by Charitos (1999) shows that the application of any rotation on the 3D build in relation to the path clearly decreases the easiness with which a person orientates in this place, although this is unlikely to occur during an e-learning session. Furthermore, based on other conducted experiments, Bridges and Charitos (2001) affirmed that in general design of virtual environments, avatar movement in a virtual environment is significantly enhanced by the use of dynamic textures and rhythmically repeated elements in paths. Charitos also confirmed that ratio of dimensions of a space can induce avatar movement towards the centre or the boundary of that place (if square) or along its main axis (if horizontal or vertical)—hence a virtual space with one dimension more than 2 or 3 times the other dimensions induces movement towards the elongated direction (Charitos, 2005).

As noted by Drettakis, Roussou, Reche, and Tsingos (2007), while the current state of VLE maturity is developing its techniques to support multiple tasks, it is rare to find complete VLEs that attain both a high-quality realistic, immersive real-time environment and the level of interactivity required to carry out adequately intricate physical-world tasks, for example, teaching and learning. Appleton and Lovett (2005) further elaborate that, contrary to expected, a high level of realistic details renders a space unbelievable or unconvincing for users. Mixing abstract and realistic data is hence more beneficial to prevent "information-overload" for the users, because insignificant details are left out and a clearer view is established. Furthermore, this is anticipated to improve user engagement, because the mix of abstract and realistic data stimulates creative thinking, that is, both right-brained-mode and left-brained-mode are stimulated in the thought process. The addition of view-dependent texturing of objects in VLEs, realistic vegetation, 3D sound, shadows and consistent lighting, all contribute to an increased sensation of realism and immersion enabling better perception of space and physical objects as well as the sense of scale (deBoer, Voorbij, & Breure, 2009). For example, smaller trees conveyed a sense of being "hotter" than that with large trees and larger shadows (Drettakis et al. 2007). Together with student interactivity, all of the prior mentioned elements can augment users' capacity to learn.

COINING A NEW FIELD

As evident from preceding literature, there is no recorded research of impact of specific architectural elements for example, color, texture, shape, dimensions, seating arrangements, lighting, and so forth, on students or users in general, nor students' specific preferences and proposals for these different architectural design features of virtual learning spaces. In addition while Bridges and Charitos (1997) also state that virtual building design should not imitate physical building design to detail, no comparisons are available showing the difference between presence of a certain architectural characteristic or dimension and so forth. in the physical world and its counterpart in the virtual world. Thus as can be seen, 3D virtual educational facilities are currently being created mainly in ad hoc fashion, according to each designer's perceptions or taste, with no specific design guidelines, without taking into consideration how this affects the learning of students in this space. Or at best practices, 3D virtual learning spaces are being designed in accordance with real-life physical architectural conventions for building such spaces, not knowing whether or not these same design conventions are suitable in virtual worlds for the e-learning experience of a student (Saleeb & Dafoulas, 2010d).

The authors have accordingly found it imperative to coin or establish a new field term to address the above research area, namely "Architectural Evolution of E-learning Virtual Worlds."

This evolution requires the cooperation of all stake-holders within these virtual worlds and environments to occur. For example, as affirmed by Swan, Gabbard, Hix, Schulman, and Kim (2003), involving users and designers from the beginning in the process of design and evaluation of VLEs, improves the effectiveness of the VLE: hence the importance of capturing students' design suggestions and propositions strategy as employed by the authors within this study. Furthermore, Sowizral et al. (1995) indicate that to work effectively in a virtual environment (VE), the application content must include the ability to access or change environmental/system/meta parameters, create and manipulate particular objects, perform analyses, and export changes to permanent storage. Also detailed user requirements analysis with architects and urban planners can confirm the suitability of choice and lead to a thorough study of these domains (Roussou et al., 2004). Ultimately a key element would be the establishment of a close collaboration with the end-users.

This was done within this research by subjecting different groups of student participants to different 3D sites in Second Life containing diverse variations of a multitude of architectural design characteristics. The evaluation instruments included direct observation of students, think-aloud protocol and postexperimental questionnaires (Ericsson & Simon, 1985). Subjects were consequently asked to reflect on best and worst perceived characteristics, and asked to suggest design features that they would prefer to be used in their customized learning space as elaborated subsequently.

RESEARCH RATIONALE

A qualitative research approach comprising preliminary surveys, focus groups and interviews was identified as suitable (Johnson & Christensen, 2004) to discover students' preferences regarding presence of different environmental and architectural elements within 16 selectively chosen 3D university virtual campuses within Second Life. Partaking in this study were 84 participants from the school of Engineering in Middlesex University, United Kingdom. These were divided into the following categories which correspond to the different clusters of learners utilizing 3D virtual university campuses to participate in online e-learning sessions.

The participants comprised 31 undergraduate students, 33 postgraduate students, and 20 members of faculty—adult learners—from different age groups (30 to 60 years old). The selected 3D virtual campuses were nominated since they represent a variety of building design specifications with regard to architectural style (e.g., modern, baroque, etc.), dimensions of learning space and their ratios, shape of learning space (e.g., circular, rectangular), seating arrangements (e.g., curved, linear rows, etc.), window styles (e.g., arched, bow, etc.) and percentage of open walls for lighting, wall floor and ceiling finishing (e.g., wood, marble, etc.), and environmental elements surrounding the learning space (e.g., fountains, underwater, etc.). The participants were first "teleported in world" to each of the chosen sixteen 3D virtual university campuses, and asked to talk about their reactions, opinions and feelings while navigating interactively for 5 minutes in each site (using think-aloud protocol). Participants were furthermore asked to use different viewpoints during navigation that is, first person view (where the user sees the world through the eyes of the avatar), and third person view (where the user follows the avatar's movement "in world"). Third person view also includes three identified perspectives as identified by Leigh, Johnson, Vasilakis, and DeFanti (1996): the "mortal" (ground-level) viewpoint, the "deity" (global above-ground while flying) viewpoint, and "balcony" (elevated view—recognized as most popular), any of which users can assume in order to interact collaboratively with a virtual environment. These navigational sessions were observed and recorded by a facilitator for later analysis of factors adding to student satisfaction within these learning spaces (Ericsson & Simon, 1985). A questionnaire consisting of 10 open ended questions was then answered to capture student's preferences for educational space design, from each of the sixteen 3D campuses shown to them. Prior to conducting the surveys, the aim of the project was explained to students from several classes, and only those volunteering to contribute remained in the survey sessions, producing the participant numbers mentioned above. As for faculty (adult learners), each member was invited in person to participate, and volunteers were assigned dates and times at their convenience to conduct the survey. The open ended questions used were:

- What interior design aspects did you like most in this learning space?
- How did they make you feel (optional)?

- What interior design aspects did you dislike in this learning space?
- How did they make you feel (optional)?
- What exterior design aspects did you like most in this learning space?
- How did they make you feel (optional)?
- What exterior design aspects did you dislike in this learning space?
- How did they make you feel (optional)?
- What interior design features would you recommend for this learning space?
- What exterior design features would you recommend for this learning space?

The open ended questions were used to allow students to think freely with no inhibitions on their desires, thus opening up points for discussion that the researchers might have overlooked and not specifically asked about or anticipated. After collection and analysis of the preliminary data from the open ended answers, two focus groups were arranged (Kontio, Lehtola, & Bragge, 2004) with eight members from each of our two undergraduate and postgraduate groups of previous participants, and five individual interviews were arranged with members of contributing staff (Guest, Bunce, & Johnson, 2006). These numbers comprised a quarter of the whole survey sample, allowing us to discuss in more detail, the users' perceptions of appropriate architectural design elements for learning spaces, which they proposed earlier.

The following guidelines were used to conduct focus groups, as recommended by Nielsen (1997):

- Each group contained between 6 to 12 members (smaller groups can be controlled by some of the members, and larger groups can lose concentration)
- Each session lasted around 60 minutes
- Results were recorded by manual note-taking
- Participants were preinformed of goals
- Five to six major open-ended questions were prepared for discussion to allow participants to contribute their opinions freely, with flexibility in the questions according to outcome.
- Individuals were chosen who are highly representative of the total population
- Both authors were present as evaluators: one to ask the questions and the other to record conversation and observations of group behavior.

- Questions were started with an "ice-breaker" for example, introductions.
- Questions were clear, easily understandable and not directive or indicative of a particular answer.
- A summary of major discussed points was given to participants at the end.

The interviews conducted within this research, with members of educational staff, used the "interview guide approach." This was a structured method, with a prepared protocol listing the open-ended questions used above and which were to be discussed in more detail with the interviewees. However the questions were asked in random order and their wording was sometimes changed to adapt to the current situation with the interviewee. This was done to achieve flexibility, but at the same time to eliminate any effect on the participants' answers due to ordering the questions in a particular manner. The objectives of the interview were reached through answering the main ideas behind the required open-ended questions even if their diction was adjusted (Johnson & Christensen, 2006).

The different approaches followed in data collection between student sampling focus groups and staff interviews, allowed the authors a more comprehensive understanding of how the two groups would perceive the environment differently. Although student participants were initially approached in groupings who would experience the 16 environments, at the same time the research team ensured that each individual was engaged in one-to-one brief discussions while answering the open-ended questions. The objective of this technique was to ensure that the purpose of each question was clear and establishing that the interpretation of the participant responses was accurate.

It became evident that a key difference between student and staff approaches to questions was due to their different agendas while engaging with the environments. Students approached the exercise keen to share their ideas of what a learning space should look and feel like. It was obvious that their drive was to share their views for design principles that should be followed during the creation of their own space in the future. On the other hand staff members, while also learners in their own respect, were motivated to reflect on what could affect the delivery of certain learning activities or support academic related and administrative tasks.

Data collected using survey, interview and focus group techniques were then analyzed and categorized into groups of architectural design element recommendations, as detailed consequently, to demonstrate students' perception of the visual qualities of the spaces, and preferences for a better learning environment.

DISCUSSION OF RESULTS

The results obtained from the questionnaires and transcribed from the focus groups and interviews were matched and compared. The resulting propositions offered by the three named categories of students (undergraduates, postgraduates, and adult learners—staff), to enhance the interior and exterior design of learning spaces within 3D VLE university campuses, were defined as over 100 design features that were consequently grouped into 11 major categories, as follows:

- The architectural style (e.g., modern, classic, gothic) and shape of the building (e.g., circular, square, use of columns, etc.)
- Wall design, finishing and colors
- External environment elements of design
- Seating arrangements and shapes
- Window styles, shapes and lighting intensity
- Internal space design factors (e.g., dimensions)
- Roof and ceiling design, finishing and colors
- Floor design, finishing, and colors
- Circulation design specifications (e.g., stairs, corridor width, etc.)
- Internal design elements (e.g., availability of desks, screens, boards, etc.)
- Entrance design (e.g., width, height, shape, doors, ease of accessibility, etc.)

These categories represent the foremost design features of a 3D virtual educational building that are of interest to the student or teacher within a 3D VLE to provide satisfaction and contentment during an e-learning session within that space. It was made clear to all participants that the aim of this research was not to focus on just a specific number of design features of the educational facilities, so as not to limit the scope of the research outcomes. The reason is that after identifying design features that could be grouped according to the previously defined categories, the ultimate objective would be to map out the effects of such features on the learning experience of participants, with emphasis on investigating how such innovative environments would contribute to the transformation of e-learning supportive technologies.

The following figures represent the average findings for all participants within this study.

STUDENT RECOMMENDATIONS FOR ARCHITECTURAL STYLE

Figure 3.1 below illustrates the architectural design style characteristics proposed or requested by students, overall and from each individual group, to be present in their ideal 3D virtual learning space. As evident, these propositions are divided into 20 features with a total of 97 votes, indicating a high significance for this factor on student satisfaction from the virtual learning space. The highest achieving propositions were those related to using modernistic styles with few details and more realistic than imaginative or futuristic designs, that is, similar to "physical life" buildings; while using ornate, classical architectural styles with abundant details were generally less in demand along with unfamiliar industrial or mechanical styles with angular or blocky protrusions. There were also some suggestions to use organic and fluid design lines with no merge between opposing styles, for example, modern and classic. All of the previous suggestions

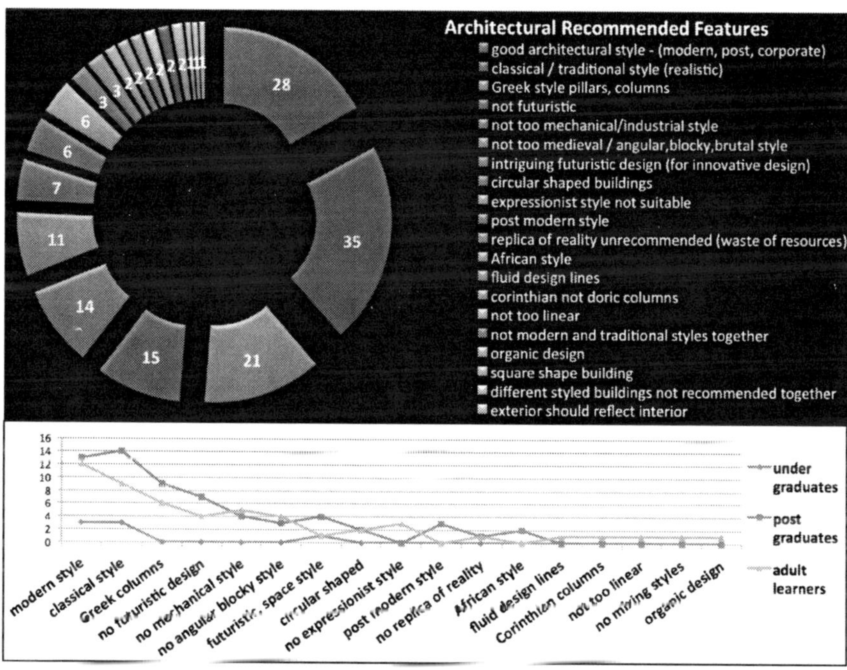

Figure 3.1. Number of votes for Architectural Style features of a 3D virtual learning space proposed by (1) all students (2) different categories of students.

coincide with literature findings mentioned earlier (Appleton & Lovett, 2005) enforcing user preferences in 3D virtual spaces for simple plain designs that depict the physical world. According to discussions during focus groups and interviews, this can be attributed to the fact that this creates a more familiar environment for the students to work in similar to what they are used to in "physical-life," in addition to the elimination of distraction that can be aggravated by using too many details in designing the 3D builds. Further proof of the latter can be seen through requesting Doric style instead of Corinthian style columns which are more elaborate in detail.

Figure 3.1 also demonstrates that while all student categories are consistent in preferring modern, realistic styles, there is a clear discrepancy between the number of votes provided by undergraduates and the other two groups who show more enthusiasm towards suggesting architectural style propositions, which is a fact discussed consequently.

STUDENT RECOMMENDATIONS FOR ENVIRONMENTAL FEATURES

Figure 3.2 illustrates the importance of the surrounding environmental design elements of the 3D virtual learning space on student satisfaction. One of the two highest recommended features was abundance of adjoining greenery, plants, flowers, and trees, and so forth, which according to individual discussions are most preferred to be seen through large windows in the surroundings, not inside the learning space so as not to cause distraction. This is the only feature that all student groups agreed upon in equal high amounts of votes stressing the significance of this feature in student contentment during e-learning sessions. The other overall highly achieving feature, particularly from postgraduates, was presence of the 3D virtual learning space underwater. This was an unexpected result especially since it is alien to the concept of "realism" established as preferred by students in the previous section. However on further discussions with students, and also as shown in Figure 3.2, some students objected to this notion as this increases feelings of discomfort, claustrophobia and even suffocation from the virtual sense of being underwater. Presence of water elements, such as fountains, pools, waterfalls and surrounding sea, were also recommended. Some students also expressed preference for wide open areas and patios between buildings. However, during interviews, some members of staff (adult learners) expressed their apprehension towards using excessive amounts of large trees which can obstruct sight during navigation especially if the user is not proficient using the "in world" camera controls, and which also cause lag due to high rendering demand on the online connection.

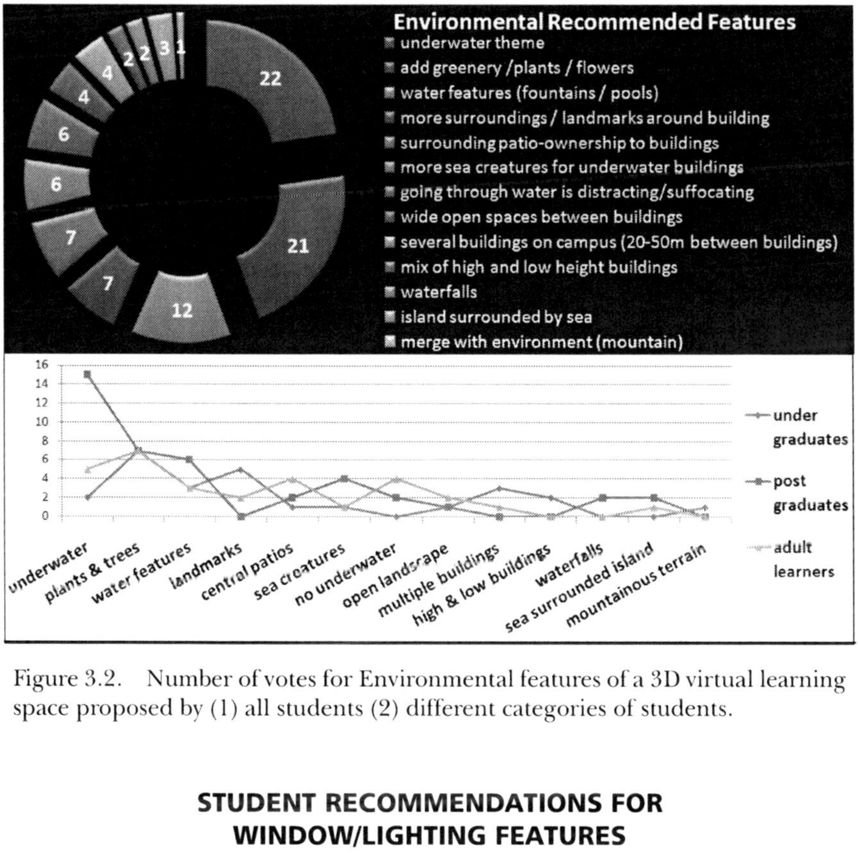

Figure 3.2. Number of votes for Environmental features of a 3D virtual learning space proposed by (1) all students (2) different categories of students.

STUDENT RECOMMENDATIONS FOR WINDOW/LIGHTING FEATURES

The highest voted feature, for lighting and windows design of 3D virtual learning spaces in Figure 3.3 (see next page), was presence of strong internal lighting and extensive use of glass windows. Even though there is no natural lighting in 3D virtual environments, according to discussions, students indicated that utilizing large window areas can induce this feeling of enhanced natural lighting. However students, especially postgraduates and adult learners, were in opposition of removing surrounding boundaries of a space completely or even just the walls and leaving the ceiling intact but supportless. According to individual dialogues, students expressed feelings of insecurity in presence of supportless ceilings, and also refrained from being entirely exposed to the outdoors during e-learning so as not to be distracted by the surroundings. This is consistent with prior findings by the authors (Saleeb & Dafoulas, 2010c) depicting preference of students for a range of 50-60% of open walls and ceiling in the surrounding surface area of the virtual learning space. Other recommendations for window

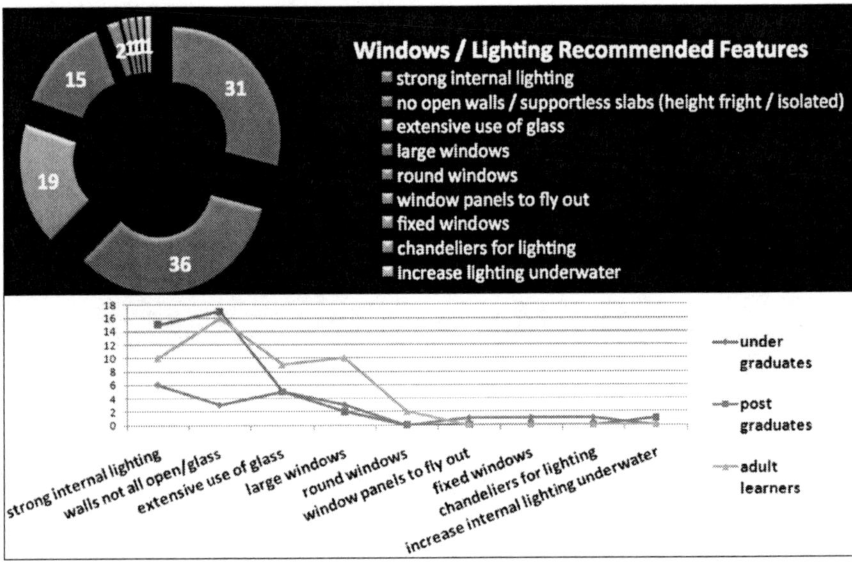

Figure 3.3. Number of votes for Window and Lighting Design features of a 3D virtual learning space proposed by (1) all students (2) different categories of students

types include using round, arched windows and fixed panel windows since using panels has no significance in 3D VWs due to nonpresence of ventilation, and also since panels can be created as "phantom," that is, penetrable for users to fly through which eases navigation to class, as indicated in Figure 3.3.

It is again evident from Figure 3.3 that undergraduates provided the least number of overall propositions which might be due to intimidation and disorientation from being new at using the 3D VLEs technology for their e-learning sessions.

STUDENT RECOMMENDATIONS FOR FLOOR DESIGN

The most prominent recommendations for floor design offered by students, according to Figure 3.4, were usage of wood as a finishing material but not darker shades of it. Carpet flooring was also in demand; however there was a considerable variation between the different student groups in favor of each finishing material except for usage of wood, on which they all approved. There were also some suggestions for usage of tiled floors

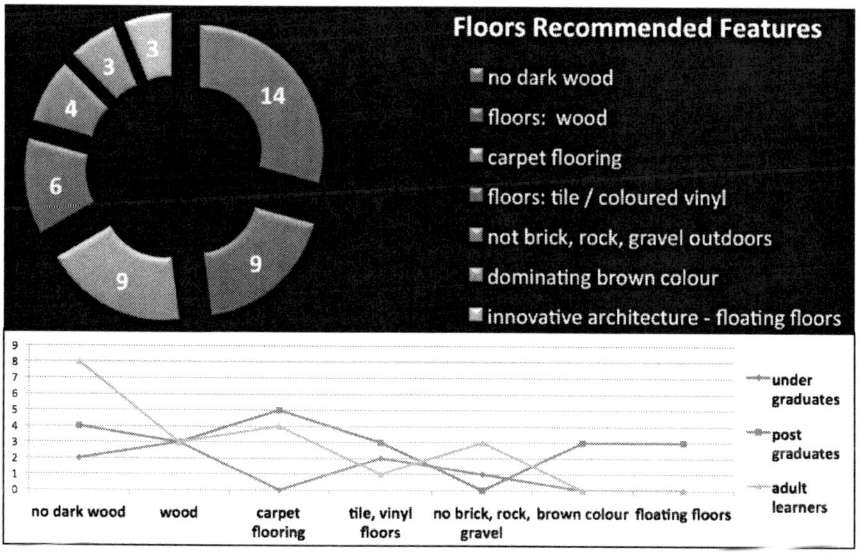

Figure 3.4. Number of votes for Floor Design features of a 3D virtual learning space proposed by (1) all students (2) different categories of students

but not rock and gravel. According to the focus group and interview results, utilization of the latter finishing materials added to the roughness and ruggedness of the space which is not the best design for educational spaces. There was also a conflict around presence of floating floors. Apparently postgraduates were most in favor of them, hence showing flexibility and desire for innovation maybe since they are more adept at using 3D VLEs and thus open to trying more novel imaginative experiences that are not possible in "physical life" educational spaces.

STUDENT RECOMMENDATIONS FOR WALL DESIGN

The number of propositions provided for wall design features, as evident from Figure 3.5 (next page), was almost double those suggested for floor and roof design. This can be attributed to the fact that one of the elements of space most seen directly at student's eye level are the walls, thus having the most impact on them and therefore demanded most; while floors, ceilings and seats are below and above direct eye perspective, hence perceived and required less by students. Results confirm that light, bright and even neutral finishing colors with no dark hues are the mostly

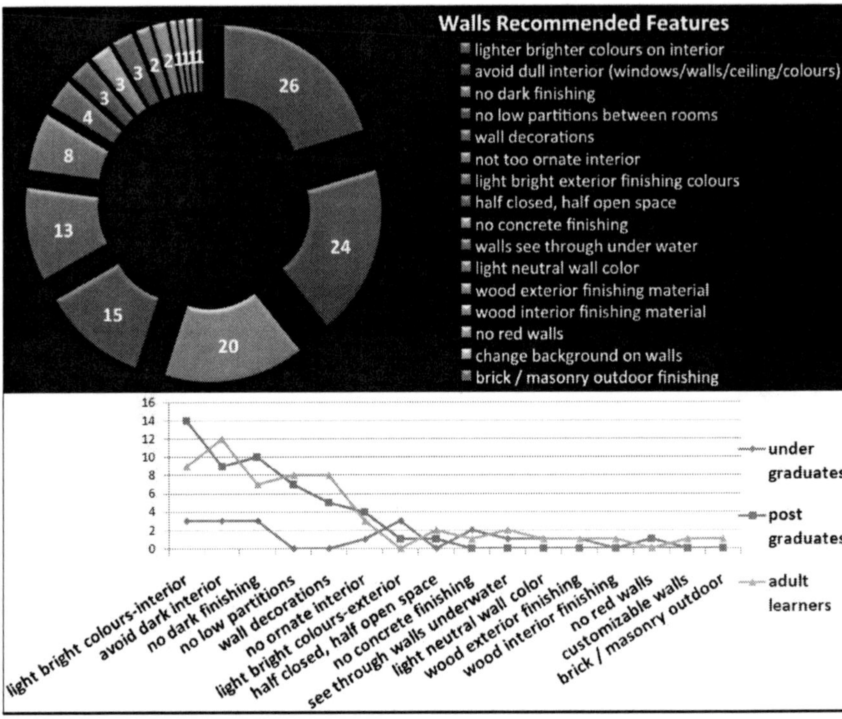

Figure 3.5. Number of votes for Wall Design features of a 3D virtual learning space proposed by (1) all students (2) different categories of students.

recommended features. Also, in accordance with results from previous sections, students expressed their disagreeability with ornate details in walls. Again here in accordance with previous results, there was a requirement for half open, half closed walls not entirely open space. Regarding usage of specific finishing materials, wood, brick, and masonry (outdoors) were indicated as accepted choices unlike concrete. As for colors, red was contraindicated, corresponding with findings from Wilder (2008) reporting that red causes emotional arousal and can have an innerving effect on users of a physical space. Presence of customizable walls was also a recommendation for 3D virtual walls.

Consistent with previous figures, it is noticeable that, while in accordance with results from other student groups, undergraduates provided the least number of propositions for wall design enhancement.

STUDENT RECOMMENDATIONS FOR ROOF/CEILING DESIGN

Figure 3.6 (see below) illustrates student preferences for enhancement of ceiling design for 3D virtual learning spaces. The most required feature was concerned with creating low height buildings. Students were hence divided between their preference for open or retractable roofs, and closed ceilings, even though postgraduates seemed to favor closure. Those in favor of open ceilings commented during discussion that this augments their feeling of brightness and strong lighting in the atmosphere. On the contrary, others remarked that open ceilings made them uncomfortable or insecure due to feeling of exposure to the atmosphere which psychologically relates them to physical spaces under adverse weather conditions, for example, rain or snow. It can be concluded that this fact probably differs from one user to the other according to the "physical life" location in which each individual student lives in the physical world. Large ceiling heights were also commendable. As explained during discussion, this helps avatars fly and eases use of the camera controls "in world." Last, dome and vault shaped roofs were proposed as an agreeable form and shape for the virtual learning space ceiling.

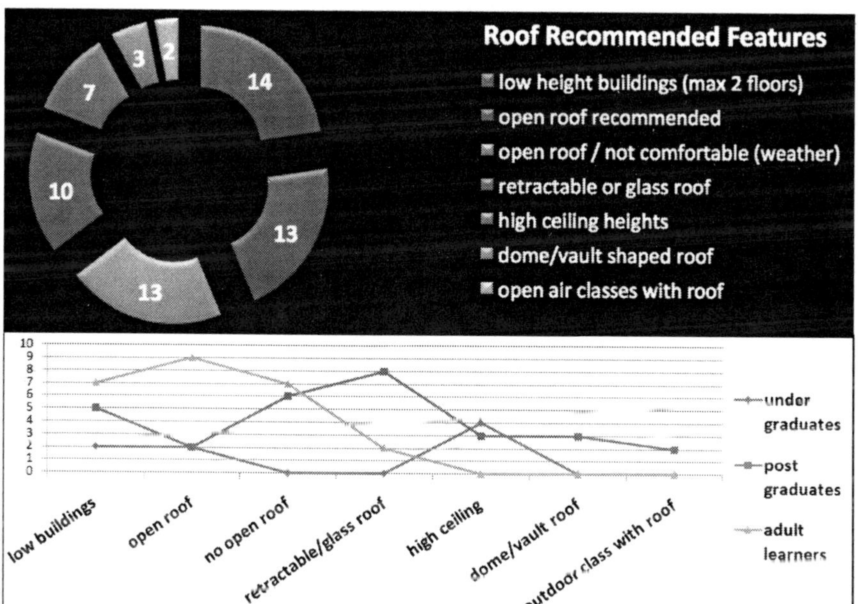

Figure 3.6. Number of votes for roof/ceiling design features of a 3D virtual learning space proposed by (1) all students (2) different categories of students.

There was a considerable discrepancy in votes between the three student groups with regards preferred floor design features, indicating an expected diversity in personal taste between them.

STUDENT RECOMMENDATIONS FOR SEATING ARRANGEMENTS

As evident from Figure 3.7, the most preferred seating arrangements were either using semi circular or circular rows. Students, with the exception of undergraduates, also showed fondness towards using a more comfortable or leisurely arrangement, for example, in the form of pool and random seating. Employing stepped curbs or open floor space for seating, that is, with no designated individual seats, was disfavored. As for the design style, nonwooden seats with backs and sofas were considered as agreeable, with refraining from using floating seats which add to the students' sense of instability or insecurity while floating in air. Position of the instructor was suggested to be either at the centre of the circular rows or in front of the seating rows in case of square shaped spaces.

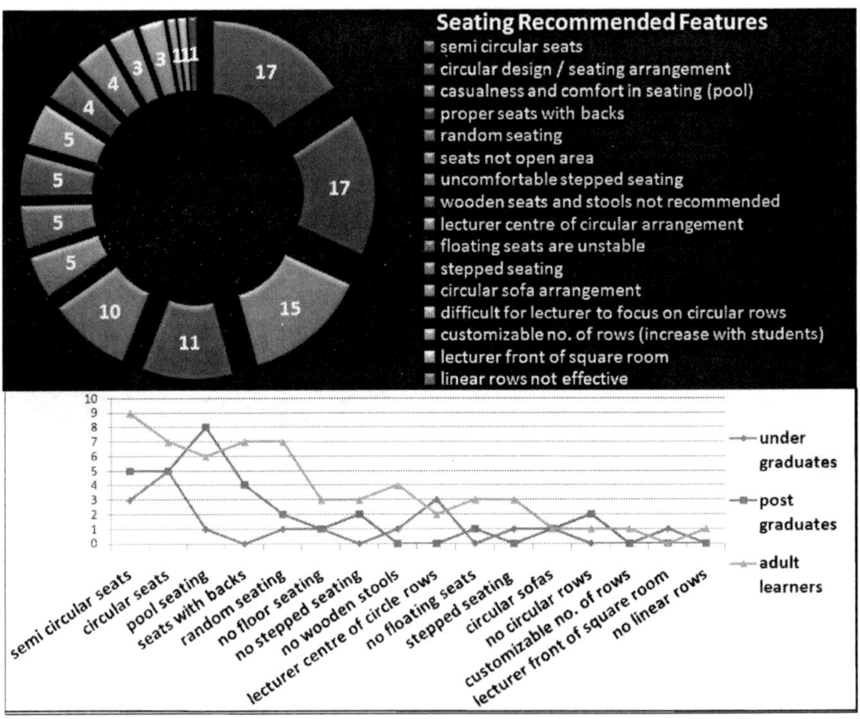

Figure 3.7. Number of votes for Seating Arrangements within a 3D virtual learning space proposed by all students (2) different categories of students.

STUDENT RECOMMENDATIONS FOR SPACE DESIGN

Recommending features for enhancement of the overall space in 3D educational facilities, as shown in Figure 8, indicates a difference between student perception of virtual space and physical space. This is because while many of the Second Life educational sites, visited by participants of this study, were created according to architectural guidelines used in the physical environment as stated earlier, students still commented that they perceived the dimensions of the virtual learning spaces as small, crowded and in need of more enlargement and spaciousness, even though the exact same dimensions in the physical world would be considered as ideal for classrooms and lecture halls. Additional student propositions enforce the need to avoid learning in outdoor virtual spaces to eliminate distraction. Furthermore, while using rectangular shaped spaces were commended, long rectangular proportions between width and length of the space were discouraged so as not to resemble corridor ratio dimensions which, according to Charitos (2005), imposes movement in the elongated direction hence can disturb the e-learning process.

It is worthwhile noting that all student groups were almost in agreement regarding importance of each suggested design feature in Figure 8, with undergraduates providing a slightly less number of votes than post-

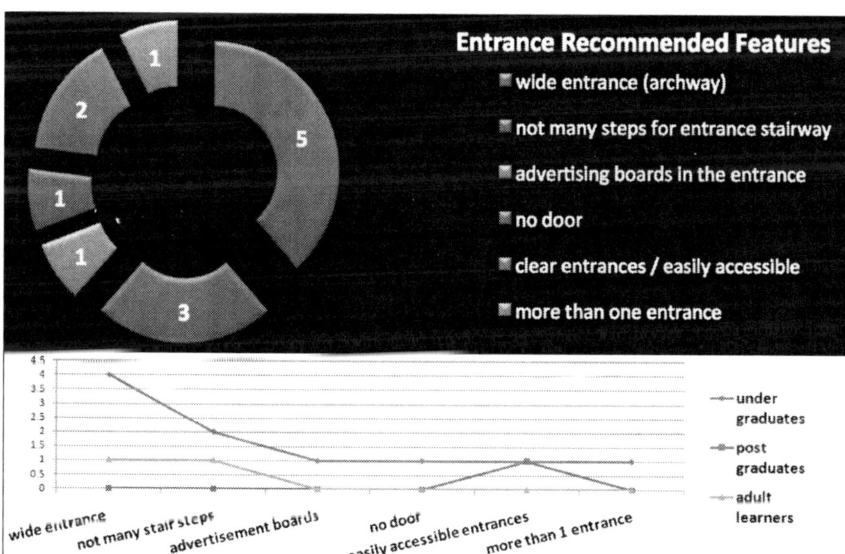

Figure 3.8. Number of votes for Space Design features of a 3D virtual learning space proposed by (1) all students (2) different categories of students.

graduates and adult learners. This emphasizes the significance of space design features for contentment of students and enhancement of the e-learning experience within 3D virtual learning spaces.

STUDENT RECOMMENDATIONS FOR ENTRANCE DESIGN

Figure 3.9 illustrates architectural propositions for improvement of 3D virtual educational facility entrance design. Students highly recommended creating wider entrances than those created using "physical life" design code specifications, preferably with no doors or open door panels (so as not to obstruct navigation of avatars), and with as few entrance steps as possible. Furthermore discussion revealed preference for presence of ramps where possible instead of entrance steps to virtual buildings for better accessibility especially for avatars who prefer using wheelchairs "in world'. More than one entrance to the building was also suggested to be an asset. Moreover students commended having advertisement or bulletin boards beside the building entrance as guide for new students who are not familiar with the virtual space.

Unlike the previous figures, undergraduate students provided most votes for entrance design features compared with postgraduate and adult learners, which implies the importance of this design element for novice

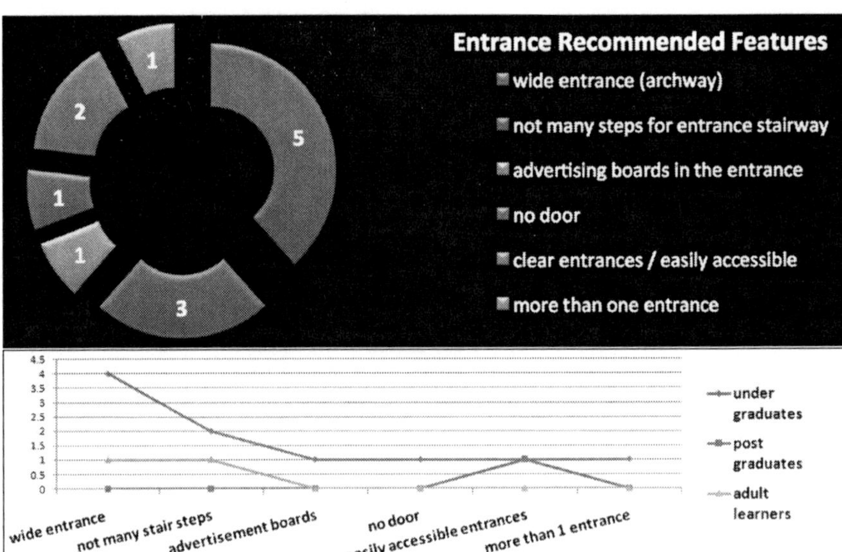

Figure 3.9. Number of votes for Entrance Design features of a 3D virtual learning space proposed by (1) all students (2) different categories of students.

users to 3D virtual learning environments, especially for wayfinding, ease of locating navigational directions and landmarks to familiarize new students with the 3D virtual space.

STUDENT RECOMMENDATIONS FOR SPACE CIRCULATION DESIGN

Features demonstrated in Figure 3.10 complement those in the prior Figure 3.9 by suggesting propositions for enhancement of the circulation to, from and within the virtual learning space. For example in order to ease navigation on winding stairways, not too many steps should be added per flight, and large dimensions for flight width should be used to be more accessible especially for novice users still struggling with navigation skills. Further discussions revealed the need for banisters on stair flights to prevent avatars from falling off during ascent and descent. This is in addition to the nonpopularity of using many winding flights for a staircase. Instead navigation can be relieved using instant "teleporting" stations to transfer avatars instantaneously between floors or rooms, and use of easily accessible and open elevators. Reduction of vertical navigation was also an agreeable notion among students.

While all the previous features were almost voted for equally by students, special fondness was given for flying to class by entering through

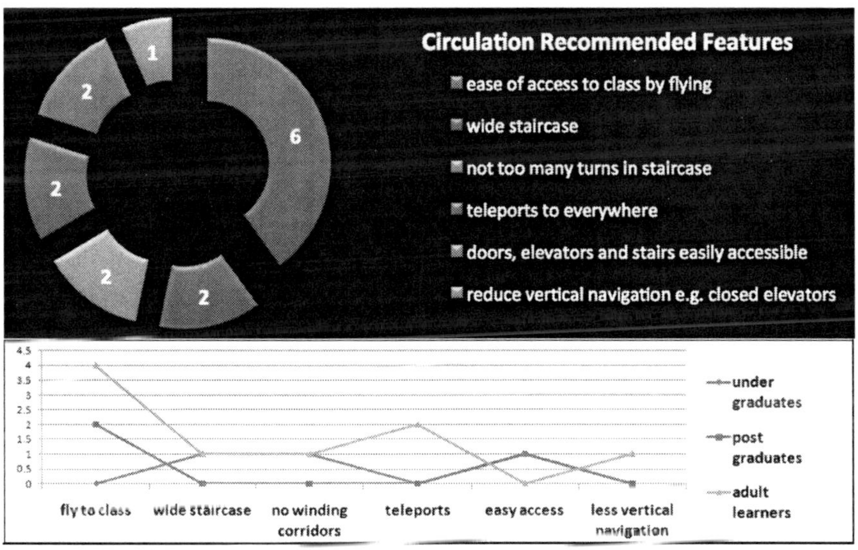

Figure 3.10. Number of votes for Circulation Design features of a 3D virtual learning space proposed by (1) all students (2) different categories of students.

windows. According to student elaborations, this would entail creating large penetrable window areas. However this feature attained no recognition from undergraduate students which again may be attributed to their cautious nature as new users, as opposed to postgraduates and adult learners who are more flexible in trying innovative ideas only possible within virtual worlds not the physical world.

TOTAL NUMBER OF VOTES AND STANDARD DEVIATION PERCENTAGE BETWEEN STUDENT CATEGORIES PER DESIGN ELEMENT

Figure 3.11 compares between overall number of votes provided by each group of students (undergraduates, postgraduates, and adult learners) for each category of architectural recommendations for 3D virtual learning space design. Along the same vein, the standard deviation between the numbers of votes provided by each student group for each design category was also calculated to determine the variation or discrepancy between the results for each design category. The lower the standard deviation, the more the similarity between the numbers of votes provided by each student group, hence implying that the significance of a design category for each student group is similar to the significance for the other groups.

Standard deviation measures spread of data around the mean value and thus how widely dispersed they are from maximum to minimum value. To calculate standard deviation, the mean value of votes per category is first calculated. Next, the deviation of each group votes from the average is calculated by finding the difference between them. Each deviation is squared, and the individual squared deviations are averaged together; this value is the variance. Standard deviation is the square root of the variance (Stanley, 2010).

The bar lines in Figure 3.11 emphasize previous results depicting the low percentage of participation in voting by the undergraduates compared to the postgraduates and adult learners, with the exception of the entrance design features where undergraduates' participation is very high compared to the other groups. This divergence directly affected the values obtained for standard deviation. The least standard deviation was found with the environmental features category indicating a similarity in number of votes and hence significance of this design category for all students. This trend was followed by window design and wall, roof and floor design categories. A larger deviation difference was recorded for architectural style, space and circulation categories, which are evidently of most consequence to the adult learner students followed by the postgraduates.

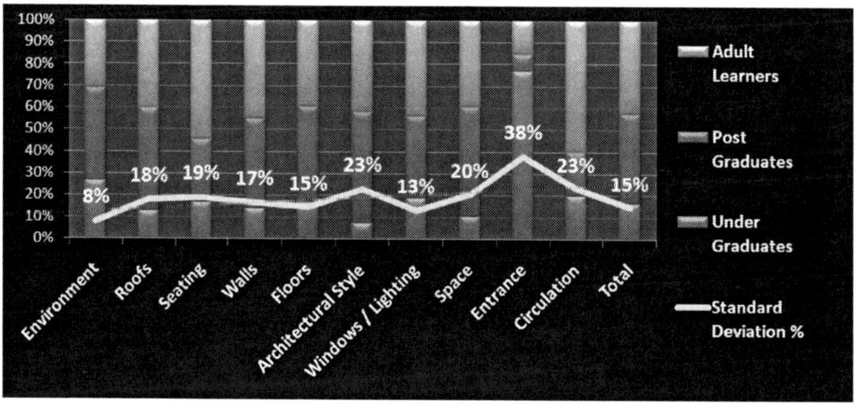

Figure 3.11. Comparison and standard deviation between numbers of votes for design features of a 3D virtual learning space proposed by different categories of students.

This result may be due to the fact that as students become more familiar and comfortable with the environment, their attention becomes increasingly directed towards the surroundings and how they would prefer it to be. The highest standard deviation was found for the entrance design category, which as elaborated earlier was of great interest for undergraduate students possibly due to difficulties they might have encountered with navigation, directions and orientation of the space on first arriving "in world."

FUTURE RESEARCH

Findings within this research are essential in providing preliminary guidelines for enhancement of the architectural design of 3D virtual learning environments to augment students' e-learning experiences. Further research involves conducting experiments to investigate the effects of individual architectural design components on students' retention rates, participation rates and enjoyment during e-learning sessions. Further work is also underway to establish best design practices for 3D VLEs to support students of different disciplinary fields of study, gender and culture in addition to different age groups. An essential strategy for triangulation of the aspired data entails conducting interviews with 3D VLE designers and architects to capture their design attempts to address learning needs in such environments.

CONCLUSION

As evident from prior findings within this study, some design preferences of students in the virtual world were similar to those in the physical world adding to the notion of realism in the design of 3D virtual educational spaces and desire for emulating the "physical life" learning environments for comfort and contentment during e-learning sessions. The most prominent requirements involved usage of plain architectural styles with spacious nonelongated dimensions, light bright finishing, semicircular seating arrangements and plentiful lighting through usage of intensive areas of windows. Additionally, additive features in virtual environments emerged in the data, including the need for external vegetation, wide corridors and entrances, and usage of as little navigational means as possible, for example, stairways. This suggests that architecture as well as the environment collaborate together to provide an aesthetic that students regard as important in feeling comfortable in virtual world learning environments.

Data analysis in this study involved comparing results attained from the identified three groups of learners, subsequent to examining the impact on student learning experience. The total number of votes provided by all students for all propositions, for enhancement of 3D virtual learning spaces, was 776 comprising of 102 features. Hence the average number of suggestions given by each student equalled ten suggestions implying a high interest by the participants in the quality of design of their learning space to reflect on their overall satisfaction during e-learning sessions. Moreover, by investigating the effect of specific design elements of an educational building, this research helps to initiate the development of a framework or recommendations for building codes for educational facilities within 3D virtual learning environments, to complement existing codes for erecting such facilities in the physical world.

The results of this study disseminate findings from a series of pilot studies in the use of 3D VLEs to specifically support practical user-centered teaching and learning in a virtual campus fostering student skills. Furthermore these are recommendations for possible applications of the technology through the use of a variety of educational scenarios in 3D virtual learning environments. The key contribution is to be useful to others as examples of executable approaches and designs to improve teaching and learning content in a 3D virtual world. Additionally, initiated discussions, triggered debates and offered brainstorming opportunities for students during sessions, with respect to the use of such assistive technologies in Higher Education, were highly interactive and required the students to engage in a series of activities, including hands-on experience.

APPENDIX: LESSON PLAN

Guideline Steps to Building an Educational Space in a Virtual Learning Environment (VLE)

1. Does your teaching / learning activity involve:

(a) Field trips, exploration of different sites, constant movement, teleporting or flying in a VLE?

- YES: you may not need to create an educational building to teach in a VLE. Go to question (5)
- NO: go to question (1b)

(b) Lecturing, lab work, creating objects, seminars, debates, role play, in-class training in a VLE?

- YES: you can create an educational building to teach in a VLE. Go to question (2)
- NO: go to question (1c)

(c) Hosting for a large number of students simultaneously not possible to gather in one space in physical world campus?

- YES: you can create an educational building to teach in a VLE. Go to question (2)
- NO: go to question (1d)

(d) Teaching distant learning students, or holding sessions or conferences with distant speakers

- YES: you can create an educational building to teach in a VLE. Go to question (2)
- NO: you may not need to create an educational building to teach in a VLE. Go to question (5)

2. Is the majority of your audience:

(a) Undergraduate students?

- NO: go to question (2b)

- YES: recommended architectural design elements for a 3D VLE include:
    - Architectural style—modern style
    - Environmental elements—plants & trees, patio with fountains
    - Wall design—light bright colors, paneling above windows
    - Floor design—mixed color wood panels, dark smooth carpeting
    - Roof design—high ceilings
    - Window design—bow windows
    - Seating arrangement—semicircular and circular rows
    - Interior lighting—50% open walls/roof, strong internal lighting
    - Dimensions—circular space, width:height = 4:1
    - Entrance—wide entrance
    - Circulation—wide staircases, no winding corridors
- Go to question (3)

(b) Postgraduate students?

- NO: go to question (2c)
- YES: recommended architectural design elements for a 3D VLE include:
    - Architectural style—classical, modern style
    - Environmental elements—underwater theme, space surrounded by water
    - Wall design—light bright colors, paneling above windows
    - Floor design—smooth carpet flooring, rough outdoor tiles
    - Roof design—retractable glass roofs
    - Window design—arched windows
    - Seating arrangement—semicircular seating, leisurely (pool) seating
    - Interior lighting—60% open walls/roof
    - Dimensions—large rectangular space, width:length:height = 1:2:1
    - Entrance—easily accessible entrances (direct wayfinding)
    - Circulation—fly to class

- Go to question (3)

(c) Adult learners/researchers?

- NO: the age of your students is outside the scope of this research. Go to question (3)
- YES: recommended architectural design elements for a 3D VLE include:

  - o Architectural style—modern style
  - o Environmental elements—plants and trees
  - o Wall design—avoid dark interior, use light bright colors
  - o Floor design—light wood
  - o Roof design—open roof
  - o Seating arrangement—
  - o Interior lighting—equal proportions of closed and open wall areas
  - o Dimensions— semicircular rows
  - o Entrance—wide entrances, few stair steps per flight
  - o Circulation—fly to class

- Go to question (3)

3. Is the majority of your audience:

(a) Male students?

- NO: go to question (3b)
- YES: recommended architectural design elements for a 3D VLE include:

  - o Color—light green, light blue
  - o Shape—circular shape, spaciousness
  - o Wall design—stained glass
  - o Floor design—light wood
  - o Roof design—glass

- Go to question (4)

(b) Female students?

- YES: recommended architectural design elements for a 3D virtual learning space include:
    - o Color—white, light blue
    - o Shape—circular shape, spaciousness
    - o Wall design—stained glass
    - o Floor design—light wood
    - o Roof design—decorative wooden panels
- Go to question (4)

4. Are you teaching students with disabilities:

- NO: go to question (5)
- YES: go to question (4a)

(a) Do your students have visual disabilities?

- NO: go to question (4b)
- YES: recommended architectural design elements for a 3D VLE include:
    - o Architectural style—plain modern style
    - o Space dimensions—large space width
    - o Shape—rectangle, square
    - o Navigation—use ramps, one floor single space
    - o Colors—Light colors
    - o Textures—Plain textures
    - o Environmental features—exterior water elements
    - o Lighting—no walls and roof (open space)
    - o Seat arrangements—semicircular, circular rows
- Go to question (5)

(b) Do your students have physical disabilities?

- NO: go to question (4c)

- YES: recommended architectural design elements for a 3D VLE include:

    - Architectural style—plain modern style
    - Space dimensions—large space height
    - Shape—square
    - Navigation—use ramps, one floor single space
    - Colors—Light colors
    - Textures—Plain textures
    - Environmental features—interior, exterior water elements, underwater theme
    - Lighting—extensively large windows
    - Seat arrangements—semicircular, circular, and linear rows

- Go to question (5)

(c) Do your students have cognitive disabilities?

- NO: your students' type of disability is outside the scope of this research. Go to question (5)
- YES: recommended architectural design elements for a 3D VLE include:

    - Architectural style—plain modern style
    - Space dimensions—large space height
    - Shape—circle
    - Navigation—one floor single space
    - Colors—Light colors
    - Textures—Plain textures
    - Environmental features—exterior water elements
    - Lighting—extensively large windows
    - Seat arrangements—semicircular rows

- Go to question (5)

5. Do you need to have the students follow you from one place to another?

- YES: recommended VLE tools to use include:
    - o Teleporting—to transfer students from their location to that of the instructor
    - o In-world bus or coach—to seat students collectively and have them follow instructor from location to another
    - o Script—provide students with scripted code to allow instructor to locate them inside the virtual
- NO: end of this tutorial

REFERENCES

Alarifi, S. A. (2008). *An exploratory study of higher education virtual campuses in Second Life.* Nottingham: University of Nottingham.

Appleton, K., & Lovett, A. (2005). GIS-based visualisation of development proposals—Reactions from planning and related professionals. *Environment and Urban Systems, 29,* 330-332.

Bridges, A. H., & Charitos, D. (1997). On architectural design of virtual environments. *Design Studies, 18*(2), 143-154.

Bridges, A. H., & Charitos, D. (2001). The impact of form on movement within virtual environments. *Automation in Construction, 10*(5), 589-596.

Calongne, C. M. (2008). Educational Frontiers: Learning in a Virtual World. *EDUCAUSE Review, 43*(5), 37-48.

Charitos, D. (1998). *The architectural aspect of designing space in virtual environments* (PhD thesis submitted to the Dept. of Architecure and Building Science, University of Strathclyde. Glasgow).

Charitos, D. (1999). Human Spatial orientation in virtual worlds. In R. Ascott, (Ed.), *Reframing consciousness* (pp. 223-228), Bristol, England: Intellect Books.

Charitos, D. (2005). Communicating environmental meaning through designing space in virtual environments. In M. Y. Yla-Kotola, S. Inkinen, & H. Isomaki (Eds.), *The integrated media machine: Aspects of future interfaces and cross-media culture.* Rovaniemi, Finland: University of Lapland.

De Boer, A., Voorbij, H., & Breure, L. (2009). *Towards a 3D Visualization Interface For Cultural Landscapes And Heritage Information.* Williamsburg, VA: Computer Applications to Archaeology.

Dickey, M. D. (2004). An architectural perspective for the design of educational virtual environments. *Journal of Visual Literacy, 24*(1), 49-66.

Drettakis, G., Roussou, M., Reche, A., & Tsingos, N. (2007). Design and evaluation of a real-world virtual environment for architecture and urban planning. *Presence, 16*(3), 318-332.

Eberhard, J. P. (2008). A place to learn: How architecture affects hearing and learning. *The ASHA Leader, 13*(14).

Ericsson, K. A., & Simon, H. A. (1985). *Protocol analysis: Verbal reports as data*. Cambridge, MA.: MIT Press.

Evans, G. W. (1980). Environmental cognition. *Psychological Bulletin, 88*, 259–287.

Fink, I. (2002). Classroom use and utilization. *Facilities Manager—APPA, 18*(3), 13-24.

Guest, G., Bunce, A., & Johnson, L. (2006). How many interviews are enough? An experiment with data saturation and variability. *Field Methods, 18*(1), 59-82.

Hall, L. D. (2001). *Effect of school architecture on its students* (Environmental psychology). Retrieved from Design Community Architecture Discussion: http://www.designcommunity.com/discussion/26903.html

Johnson, B., & Christensen, L. (2004). *Educational research: Quantitative, qualitative, and mixed approaches* (2nd ed.), Needham Heights, MA: Allyn & Bacon.

Johnson, B., & Christensen, L. (2006). *Validity of research results. Educational research, qualitative, quantitative and mixed*. Retrieved from http://www.southalabama.edu/coe/bset/johnson/dr_johnson/lectures/lec8.pdf

Joseph, B. (2007). *Global kids: Best practices in using virtual worlds for education. In Second Life education workshop* (pp. 7-14). Chicago, IL: SL Community Convention, WordPress

Kontio, J., Lehtola, & Bragge, J. (2004). *Using the focus group method in software engineering: Obtaining practitioner and user experiences*. Proceedings Empirical Software Engineering, ISESE '04, pp. 271-280.

Leigh, J., Johnson, A. E., Vasilakis, C. A., & DeFanti, T. A. (1996). Multi-perspective collaborative design in persistent networked virtual environments. *In Proceedings of IEEE Virtual Reality Annual International Symposium '96, pp. 253-260.*

Market Truths. (2007). Second Life Buildings and Real Estate Market, 1st Quarter, Market truths Limited, State of the Art Market Research and Analysis.

Market Truths, (2009). Second Life Buildings and Real Estate Market, 1st Quarter, Market truths Limited, State of the Art Market Research and Analysis.

Minocha, S., & Mount, N. (2009). Design of Learning spaces in 3D Multi-user virtual environments. JISC Learning and Teaching Committee, The e-Learning Programme.

Minocha, S., & Reeves, A. J. (2009). *Interaction design and usability of learning spaces in 3D multi-user virtual worlds*. Pune, India: Human Work Interaction Design.

Nielsen, J. (1997). The use and misuse of focus groups. *IEEE Software*, 94-95.

Roussou, M., Sideris, A., Loscos, C., Dettori, A., Drettakis, G., & Lombardo, J. C. (2004). *Requirements analysis on cultural heritage—Education and urban architectural planning and design case studies* (Technical Report No. RN/04/09). London: University College.

Saleeb, N., & Dafoulas, G. (2010a). *Pedagogical immigration to 3D virtual worlds: A critical review of underlying themes and their concepts*. In Proceedings of IEEE International Conference on Information Society (i-Society 2010), London, England.

Saleeb, N., & Dafoulas, G. (2010b). Architectural propositions for enhancement of learning spaces within 3D virtual learning environments. In Proceedings of IEEE International Conference on Information Society (i-Society 2010), London, England.

Saleeb, N., & Dafoulas, G. (2010c). Relationship between Students overall satisfaction from 3D virtual learning spaces and their individual design components. *International Journal of Computer Science Issues, 7*(4), 1-9.

Saleeb, N., & Dafoulas, G. (2010d). Analogy between student perception of educational space dimensions and size perspective in 3D virtual worlds versus Physical World. *International Journal of Engineering, 4*(3), 210-218.

Saleeb, N., & Dafoulas, G. (2010e). *Investigating student satisfaction from environmental and architectural design elements of 3D educational facilities within 3D virtual worlds.* In Proceedings of the 3rd World Summit on the Knowledge Society (WSKS 2010), Corfu, Greece.

Scopes, J., & Lesley, M. (2009). *Learning archetypes as tools of cybergogy for a 3D educational landscape: A structure for eTeaching in Second Life* (Master's thesis). Southampton, England: University of Southampton, School of Education.

Sowizral, H., Angus, I. G., Bryson, S., Haas, S., Mine, M. R., & Pausch, R. (1995). *Panel session on Performing Work within Virtual Environments.* In 22nd international conference on computer graphics and interactive techniques, pp. 497-498.

Stanley, J. (2010). *What is standard deviation.* Retrieved from http://www.wisegeek.com/what-is-standard-deviation.htm

Swan, J. E., II, Gabbard, J. L., Hix, D., Schulman, R. S., & Kim, K. P. (2003). A Comparative study of user performance in a map-based virtual environment. In *IEEE Virtual Reality* (pp. 259-266). Los Angeles, CA: IEEE Computer Society.

Wilder, W. L. (2008). *Importance of color in web design.* Retrieved from http://www.cyberindian.com/web-designing/importance-of-color-in-web-design.php

PART III

STRATEGIES FOR ENGAGING THE AVATAR

CHAPTER 4

THE EYE AND HOW WE SEE

Physical and Virtual Worlds

Joseph N. Trachtman

How we see employs both cognitive and noncognitive elements in the processing of visual information. In the physical world, we certainly want to see clearly; but are also sensitive to color, brightness, contrast, depth, and perspective. Two distinct vision pathways allow us to have this sensory information. One channel is the well-known cognitive pathway from the eye to the cerebral cortex. The noncognitive, second channel is less well known, having a pathway from the retina of the eye to the hypothalamus, whose regulation includes the emotions, the immune system, brain waves, the retina of the eye, and eye focusing. Additionally, our ultimate visual image receives input from the other senses. For example, we take pleasure in seeing a plate decorated with mouth watering food, where the smells, taste, texture, and sounds add to the visual image of the meal that will be stored in our memory. Accordingly, the design of a virtual world should include all these elements of vision in the physical world: cognitive, noncognitive, and intersensory input. The challenge is to be able to display in 2D the perception of 3D. The purpose of this chapter is to describe in detail all elements to provide us with a full vision.

Engaging the Avatar: New Frontiers in Immersive Education, pp. 83–116

INTRODUCTION

How we see employs both cognitive and noncognitive element in the processing of visual information. In the physical world, we certainly want to see clearly; but are also sensitive to color, brightness, contrast, depth, and perspective. Additionally, our ultimate visual image receives input from the other senses. For example, we take pleasure in seeing a plate decorated with mouth watering food, where the smells, taste, texture, and sounds add to the visual image of the meal that will be stored in our memory. Accordingly, the design of a virtual world should include all these elements of vision in the physical world: cognitive, noncognitive, and intersensory input. The purpose of this chapter is to describe in detail all elements to provide us with a full vision.

MIND-BODY DUALISM

The origin of modern mind-body dualism is attributed to Rene Descartes (1595-1650), where visual perception was one of his areas of interest (Amoroso, 2010). The basic question raised by Descartes was "that during visual perception is it the thinking process or the sensory information primary?" Descartes' conclusion was that the thinking process was primary. Since the seventeenth century, there have been numerous theories that are both pro and con of Descartes' proposal (Wozniak, 2005).

Jumping ahead 300 years, models of perception acknowledge that vision is a combination of central (thinking) and peripheral (sensory) information. Trevarthen (1968) described the dichotomy as focal (central) and ambient (peripheral). The most contemporary understanding of vision information processing was elaborated by Livingstone and Hubel (1987, 1988), with the parvo (central) and magno (peripheral) pathways. These pathways contain separate fibers in the retina of the eye, the brain tract from the eye to the brain, and the brain itself. Answering Descrates' may lead us closer to understanding perception of visual objects in virtual worlds, and lead us to some serious design considerations.

PARALLEL PROCESSING

With the age of microcomputers, laptops, notebooks, and palmtops we have learned a new vocabulary term known as "parallel processing." Parallel processing refers to how many channels of data the computer can process simultaneously. For example, in the "dinosaur" days, \$30,000 computers processed only 8 channels, whereas, today we have grown accustomed to 64 bit machines. As is well-known, the number of channels being exponentially proportional to speed of information processing. With the vision system it is the same—the more channels we use, the faster and the more information we can process, store, and utilize. A parallel processing model relating the list will now be presented (Trachtman, 1992). With virtual worlds requiring large amounts of parallel processing to maintain the visual qualities, it is important to observe the ratio of computing power to optical power in examining the direction of virtual world acuity.

MODEL OF ATTENTION

How one pays attention in a virtual world is significant in terms of how we engage the avatar. Examining the way the brain processes information to assist attention management will give us some insights into how the physical world person maintains attention to surrounding information based on the way the eye and brain work together. How does this get carried over to the virtual world?

1. Parallel processing exists when both central and peripheral vision are processing information simultaneously. A speed reader is a great example. A speed reader must see the exact location of each word/letter (central/parvo) and at the same time be aware of all the other words/letters on the page (peripheral/magno). This state of attention also interacts synergistically with memory and other brain functions. Great academic performers can sustain this type of attention whenever they learn.
2. Alternating processing is when the central and peripheral information are processed one at a time; some can perform the alternating very quickly. The average college student will maintain this type of concentration occasionally getting into parallel processing; but not be able to maintain it on a consistent basis.
3. Central/parvo processing exists when only central information is being processed in a serial fashion—only one stream of information at a time. This is what we notice with the typical analytical person.
4. Peripheral/magno processing is when only peripheral information is being processed in a serial fashion—the student can not get focused, he is easily distracted by random thoughts. This is what we note with the typical unorganized person. There is a further twist to this type of processor. Many young students have trouble

> getting past this stage of attention as their creative-emotional processing is dominant. Their problem is that the function of this processing channel is not quite right—so the channel they process information with is not coding the information properly—this is the child with a learning problem (see Table 4.1).

If an avatar is expected to process significant amounts of information quickly in a visual environment, they likely need to be able to envision the whole environment to process it quickly. This may imply that containing information into spaces that catch the avatar's eye for rapid access to information is critical.

To better understand the extent and nature of the parallel pathways, the relationship between vision information processing and the hypothalamus, a structure in the mid-brain, will now be described.

THE HYPOTHALAMUS

It is important at this point to note that the hypothalamus is involved with the regulation of the majority of bodily functions, including the electroencephalogram (EEG), memory, learning, the emotions, the pituitary gland, the immune system, cardiac function, and vision (Lincoln, 1969;

Table 4.1. Central and Peripheral Pathways

CENTRAL	*PERIPHERAL*
analytical	experiential
cognitive	noncognitive
color sensitive	color blind
cortical	visceral
high luminance	low luminance
high spatial frequencies	low spatial frequencies
language	art
left brain	right brain
low contrast sensitivity	high contrast sensitivity
parvo	magno
respond slower	respond faster
shape	movement
sustained response	transient response
texture	depth

Swaab, 2003). In addition to the innumerable functions and its myriad connections, the hypothalamus has further wide-ranging physiological influence via the hypothalamic-pituitary-adrenal axis (HPA). The importance of the role of the HPA axis was emphasized in 1955 by Nobel Prize winners Guillemin (1977) and Schally (1977). Most importantly, what happens with one function of the hypothalamus tends to happen in other functions. For example, our emotional state affects our visual perception, and vice versa, what we see affects our emotions. Additional hypothalamic connections are made via the vision system as will be presently discussed. This implies that whatever is designed in the virtual world that affects our emotional state, may easily affect our visual perception as well. Effects such as color, lighting, enclosures, ease of movement, authenticity of the environment, all may combine to create emotional responses that can alter our visual perception and thus either increase or decrease our attention in world. Attention management in a 3D space can affect our learning outcomes.

Beginning in the 1980s, Sadun and his coworkers began publishing reports of retinohypothalamic tracts (RHT) (Sadun, Schaechter, & Smith, 1984; Sadun, Johnson, & Schaechter, 1986; Schaechter & Sadun, 1985; Kenney, Weiss, & Haywood, 2003). To date, three distinct pathways have been described. Each pathway originates at different stages of retinal processing and terminates in different locations in the hypothalamus. A recent discovery in relation to the RHT is the opsin, melanopsin. Retinal cells containing melanopsin project to the hypothalamus via the retinohypothalamic tracts (Provencio et al., 2000; Hannibal, Hindersson, Knudsen, Georg, & Fahrenkrug 2002). Melanopsin is also found in melanophores in the iris and the inner brain (Provencio, Jiang, De Grip, Hayes, & Rollag, 1998). The discovery of melanopsin greatly aids our understanding of the nonvisual aspects of light stimulation, in particular, and enhances our appreciation of the interaction between the hypothalamus and the vision system, in general.

ANATOMY OF THE HYPOTHALAMUS

The hypothalamus is located in the proximity of the geometric center of the brain. It is surrounded by several structures, most notably the pituitary gland, and the optic chiasm—where the optic nerves of each eye meet with semidecussation. (Netter, 1964; Schmidt, 1985; and Strandring, 2004). Illustrations of the hypothalamus are shown in Figures 4.1 and 4.2.

The size of the hypothalamus is approximately 4 cm^3, with the cube root of 4 cm being 1.58 cm (Swaab, 2003). The weight of the hypothalamus is

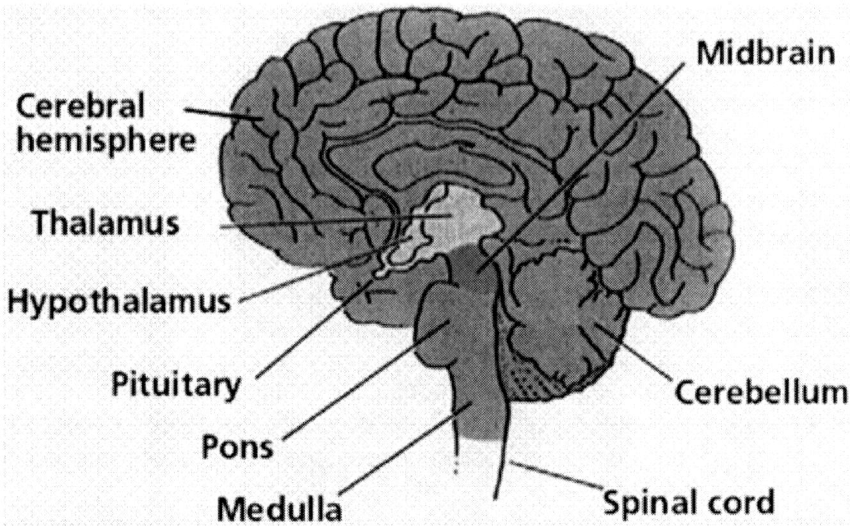

Source: http://www.emc.maricopa.edu/faculty/farabee/biobk/biobooknerv.html#The%20Brain

Figure 4.1. Anatomical location of the hypothalamus.

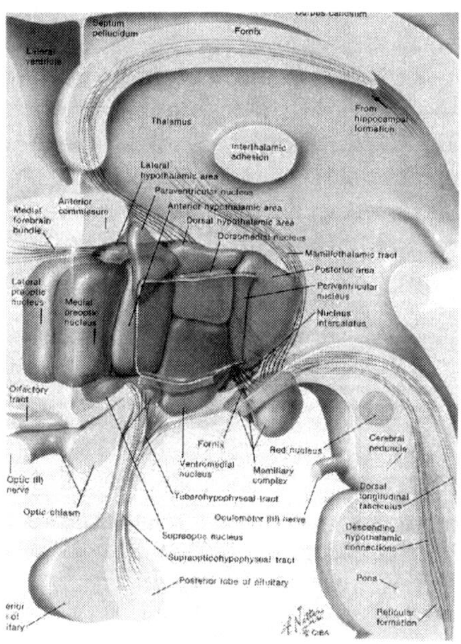

Source: www.psycheducation.org/emotion/hypothalamus.htm

Figure 4.2. The anatomy of the hypothalamus.

approximately 39 to 42 grams, which converts to 1.4 to 1.5 ounces or 195 to 210 carats, respectively (Jones & Lopez, 2006). The carat notation has been given to compare the hypothalamus to the largest jewelry diamond, the Star of Africa, which weighs 530 carats (Pulliam, 2004). For the sake of brevity rather than a very lengthy discussion on the components of the hypothalamus, Table 4.2 (Trachtman, 2010) lists the different components of the hypothalamus, with their function(s).

Table 4.2. Different Components of the Hypothalamus, With Their Function(s)

Structure	*Function*
Amygdala	depression, stress, immune system, memory, reproduction, salt intake, appetite suppression, neurotrophic properties, cerebral energy metabolism, vasodilator, cardiovascular function, respiratory functions
Anterior Cingulate Cortex	pain
Anterior Commisure	immune system
Anterior Hypothalamic Area	feeding, cardiovascular function, sleep
Arcuate Nucleus	appetite, feeding, growth, reproduction, anorexigenic, energy balance, stress, gastric acid secretion, water balance, cardiac function, blood pressure, endocrine gland regulation, pain, euphoria, exercise, neuro-muscular and -degenerative disease
Cingulate Gyrus	stress
Commissural fibers of the suprachiasmatic nucleus	epilepsy
Dorsal Hypothalamic Area	immune system
Dorsomedial Nucleus (DMN)	ANS functions, sexual functions, feeding, anorexigenic, energy balance, stress, gastric acid secretion, reproduction,

(Table continues on next page)

Table 4.2. Continues

Structure	*Function*
	cardiovascular function
Fornix	narcolepsy
Hippocampus	immune system, seizures, pain, stress, neurodegenerative disease, learning, pain, reproductive functions, neural injury and repair, salt intake, inflammation
Hypothalamus-Pituitary-Axis (H-P-A)	
Interthalamic Adhesion	sexual dimorphism
Lateral Hypothalamic Area (LHA)	sleep, appetite, drink center, reproduction, energy homeostasis
Lateral Preoptic Nucleus	sexual behavior
Mamillary Complex	temperature, memory, reproduction
Medial Preoptic Nucleus	sexual behavior
Median Eminence	immune system, sleep, neurotrophic properties, endocrine gland regulation, gastric acid secretion, depression, hepatic encephalopathy
Nucleus Basalis	cholinergic function
Nucleus Intercalatus	pain, immune system
Organum Vasculosum Lamina Terminalis	sodium regulation
Paraventricular Nucleus (PVN)	corticosteriods; NOS:injury, stress hypophyseal secretion, anorexigenic, energy balance, gastric acid secretion, water balance, cardiac function, blood pressure, endocrine gland regulation, appetite regulation, immune system
Periventricular Nucleus	growth, fluid and electrolyte homeostasis, feeding, pituitary hormone secretion, nociceptive
Pineal Gland	
Pituitary Gland	
Posterior Hypothalamus	sleep, feeding, hormone secretion
Posterior Nucleus	feeding

Premammillary nucleus	feeding
Preoptic Nucleus	sexual behavior
Preoptic-Anterior Hypothalamic Area (POAH)	temperature
Recessus opticus	endocrine regulation
Sexually dimorphic nucleus of the preoptic area	reproduction
Solitary tract nucleus	feeding, pituitary hormone secretion, nociceptive
Subfornical Organ	drinking
Suprachiasmatic Nucleus (SCN)	circadian rhythms
Supraoptic Nucleus (SON)	NOS:injury, fluid and electrolyte homeostasis, water balance
Thalamus	
Tubermamillary Nucleus	neurotransmitters
Ventromedial Nucleus (VMN)	feeding, body weight, stress, gastric acid secretion, energy balance, anorexigenic, pain, anxiety, learning, addiction, cardiorespiratory function, kidney function, urinary function, immune function, gastrointestinal function

FUNCTIONS OF THE HYPOTHALAMUS

After its initial discovery, the hypothalamus was believed to play a minor or subservient role in nervous system function and regulation. Recent studies report quite the contrary; the hypothalamus is the main component of the sympathetic nervous system, regulates much of the pituitary's activity, and produces many of the nervous system's neurotransmitters and peptides. Emphasizing the important and central role of the hypothalamus in the sympathetic nervous system (SNS), Hess (1949) summarized the many functions of the hypothalamus known at that time.

In addition to the innumerable functions and connections illustrated in Tables 4.2, the hypothalamus has further wide-ranging physiological influence via the HPA. The complete literature describing the HPA is encyclopedic and outside the realm of a brief literature review.

VISION PATHWAYS

To help understand the myriad connections between the vision system and the brain, a diagram has been configured (see Figure 4.3). It is important to note that the connections have been limited to certain structures, which have been deemed the most important and relevant to the present discussion. Structures such as the anterior cingulated cortex, prefrontal cortex, hippocampus, pulvinar, and the hypothalamic-pituitary-adrenal axis have been excluded for the sake of simplicity, and to prevent the present discussion becoming too lengthy.

As can be noted in Figure 4.3 below, fibers from the optic nerve (ON) make their connection in the lateral geniculate nucleus (LGN), and the hypothalamus (HYPO) via the retinohypothalamic tracts (RHT). Fibers from the LGN make connections with the visual cortex, and the HYPO, and then the superior cervical ganglion (SCG). (Duke-Elder & Wybar, 1961; Foxe & Schroeder, 2005). Completing the feedback and feedforward connections are the: parasympathetic nervous system (PNS) and motor input via the III Cranial Nerve, and the sympathetic nervous system (SNS): via the short and long ciliary nerves (Davson, 1990). In addition to actual neural connections, an important fact to keep in mind is that most neuropolypeptide, cytokine, and nitric oxide activities are controlled by the HYPO either directly or via its numerous interconnections (Kastin, 2006).

The purpose of the above discussion about the hypothalamus, with emphasis on the vision system, is to explain importance of our visual perception on the rest of the bodily functions. In regard to a virtual world, we are primarily concerned with the hypothalamic regulation of emotions, learning, and memory.

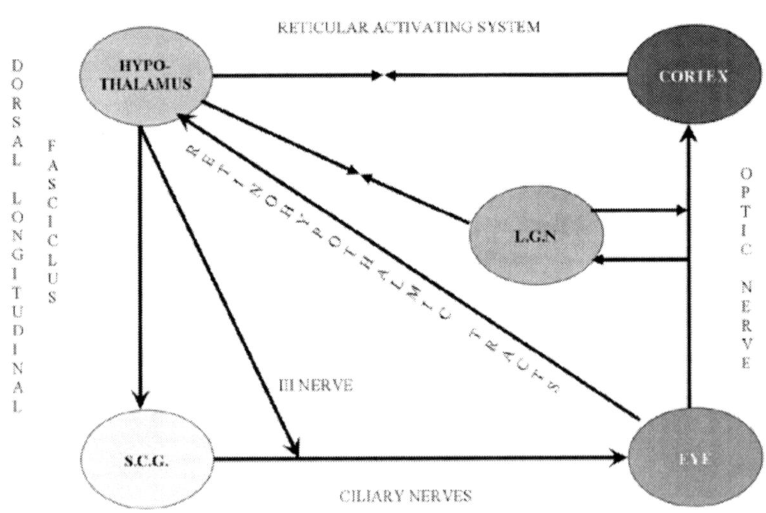

Figure 4.3. The vision pathways.

AUGMENTED REALITY

The man-machine interface is a classic topic, particularly in regard to computers (Bejczy, 1980). One area of the man-machine interface confronting a virtual world is how to relate real world images to computerized images. This challenge has been referred to as augmented reality (Wikipedia, August 9, 2010, "Augmented reality"; http://en.wikipedia.org/wiki/Augmented_reality). Some of the challenges that arise include matching real world colors of familiar objects, the use of shadows, and the affect of the viewer's movements on what is seen on the computer display. Rothfarb and Doherty (2007) note some helpful suggestions in regard to these challenges:

- Make the environment interactive—both individually and socially.
- Care must be taken when changing the scene orientation and/or perspective (Tcheung, Gilson, & Glennerster (2005).
- Use textures to add to existing perceptual cues
- Employ multisensory input; typically visual and auditory, however, smell has recently become available (O'Brien, 2010). (http://www.albertalocalnews.com/reddeeradvocate/news/local/A_virtual_world_to_see_hear__and_smell_101154339.html)
- Provide viewers with a wide variety of controls to change the visual perspective on the computer display.

Additional aspects of augmented reality have been described by:

Schalk (2008) and Straw (2008) suggest using parallel processing to augment brain-computer symbiosis, as this replicates vision and brain processing.

Gorini, Gaggioli, Vigna, and Riva (2008) suggest that interaction between the physical world and the virtual world will augment perceptual and social experiences.

Boulos and Burden (2007) recommend the use of multiple colors to augment visual perception, and give as an example Google Earth.

Zacks, Speer, Swallow, Braver, and Reynolds (2007) discuss the importance of perceptual expectations based on prior experience. For example, it is expected that objects fall at 32 feet/second, and people to walk at approximately three to four mph. and cars speed along at approximately 60 mph.

Ziemer, Plumert, Cremer, and Kearney (2009) recommend a 5 to 7 minute period of adaptation when going from the physical world to a virtual world. Inferred from their research is that it may be helpful to have viewers make distance judgements in a virtual world as they adapt to a virtual world.

Kelly, McNamara, Bodenheimer, Carr, and Rieser (2008) note that there are individual perceptual abilities, whereby some viewers will be more perceptually astute in a virtual world than others.

To better understand the application of augmented reality to a virtual world, the components of vision will now be discussed.

COMPONENTS OF VISION

In considering the design of a virtual world, it is essential to consider the various components of vision. It is not enough to build something that "looks nice"; but is it easily seen, recognized, and integrated into our learning and experience?

VISUAL ACUITY (VA)

VA is the best known and most common vision component, and is usually measured on an eye chart or as a projected image on a screen (Colenbrander, 2001). The letters on a standard eye chart are in a format known as Snellen notation (Lit, 1968). Snellen notation is based on the overall size of a letter subtending being five times that of the thickness of the lines that comprise the letter. Although VA can be used at any distance, the standard distance is six meters (approximately 20 feet) for distant vision and 35.56 cm (14 inches) for near vision.

The notation for Snellen VA is in a fraction, with the testing distance in the numerator and the proportional angle in the denominator. The generally accepted standard for "normal vision" is 20/20, the testing distance is 20 feet, where the letter subtends 5' of arc. As an example, 20/200 means that the testing distance is 20 feet, the numerator, and the letter subtends 50' of arc (200/20 =10; 10*5' of arc = 50' of arc), with 200 the denominator. Another way to understand the notation, 20/200, is that what a standard observer can see at 200 feet, the person being tested can see the same letter at only 20 feet. The definition of legal blindness, based on VA, is that someone sees less than 20/200 with their better eye with their best eyeglass or contact lens prescription.

VA is routinely measured in "normal room light" with the right eye, then the left eye, and then both eyes together. The procedure is the same for both distance and near VA testing. The classic instrument for Snellen

VA testing, using the ascending and descending method of limits, was the Clason projector, which allowed a continuous change in the letter size without having to change slides (Callaway & Thompson, 1953; and Borish, 1954), which went out of fashion in the 1960s and was relatively expensive. Currently, a variety of instruments are used to measure visual acuity ranging from computerized projectors to printed eye charts (www.calcoastophthalmic.com and www.lombartinstrument.com). The testing distance for visual acuity is 20 feet (see Figure 4.4).

COLOR VISION

We appreciate color vision by stimulation of the cone cells of the retina. There are three different types of cones, S (short wavelength for blue color), M (medium wavelength for green color), and L (long wavelength for red color). It is the interaction of the three types of cones that allow us to see a rainbow of colors. We are most sensitive to a wavelength in the M cone, a lime green color, 555 nm (Merbs & Nathans, 1992).

An illustration of the perceptual effects of the physiological bias towards M cones is the difference between Fuji 35 mm film and Kodak 35 mm film.

A most interesting aspect of the use of color perception is illustrated with the rise of the use of Fuji film and the decline in the use of Kodak film (Desmond, 1997). While price may be a factor, it will be noted in the following photographs and discussion, that the Fuji film is more perceptually appealing that is Kodak film.

Figure 4.5 is of two photographs, where both photographs were taken of the same scene just minutes apart with the same settings on a 35 mm camera (Nikon F2, Nikkor Zoom 80~200 mm set at 200 mm and focused

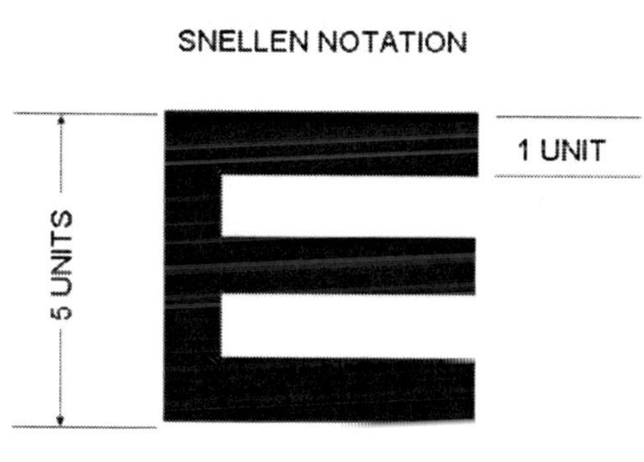

Figure 4.4. Snellen notation for visual acuity.

Figure 4.5. Kodak versus Fuji 35 mm film.

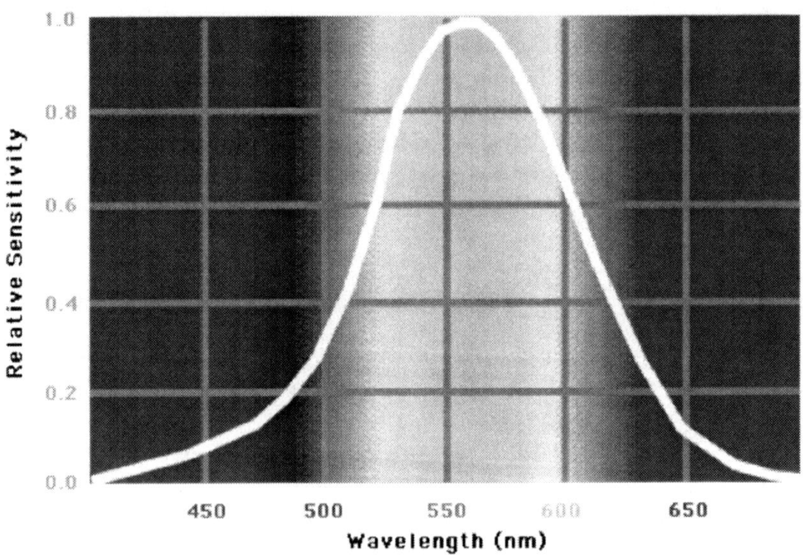

Figure 4.6. The Spectral sensitivity curve.

for ∞, f16 at 1/500 sec. The top photo was taken with Kodak 400 film (Ultramax), and the bottom photo was taken with Fuji 400 film (Superia X-tra). As can be noted the Kodak film has a bias towards reds, and the Fuji film has a bias towards to the greens. Additionally, the Fuji film has better contrast and more accurate in the blues than does the Kodak film.

Why and how is this important? An answer to this question can be found in Figure 4.7, which shows the relative sensitivity of the human eye to the visible light spectrum. It is readily noted that the human eye is most sensitive to a lime green color, wavelength 555 nm. Accordingly, the green bias in the Fuji film is more perceptually appealing to the human eye than is the Kodak film, particularly when compared to the much reduced sensitivity to red colors of the Kodak film. This concept is very important to keep in mind in designing and using colors in a virtual world.

To further aid in the use of colors, reference can be made to the CIE diagram (http://hyperphysics.phy-astr.gsu.edu/hbase/vision/cie1976.html). Looking at the intersection of the blue, red, yellow, and green colors, a white area is noted. Taking this as a reference point, we can now determine complimentary or contrasting colors. For example, if we take a blue light at 450 nm and draw an imaginary line through the reference point, we note

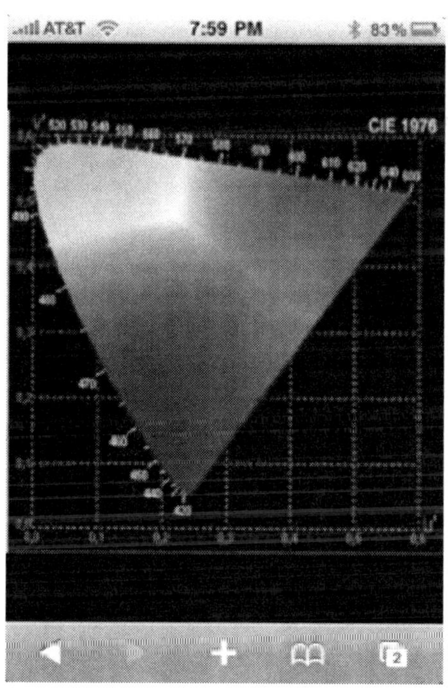

Figure 4.7. The C.I.E. diagram (1976).

that a yellow color, 570 nm, is the complimentary color and provides the most contrast. Accordingly, we can use the diagram as an aid to determine what color combinations will provide optimal or lesser contrast.

Another important consideration in the use of color is the fact that there are people who are color blind; approximately 9% of males, and 0.5% of females (Graham, 1966). The most common color vision defect is red-green discrimination, and is known as protanomalous trichromacy, when it is a red defect, deuteranomalous trichromacy, when it is a green defect (Graham, 1966; Squire, Rodriguez-Carmona, Evans, & Barbur, 2005). The "gold standard" for red-green color vision testing is the Nagel anomalosocpe (Williams, 1915; and Birch, 2008). The test consists of a circle that can range from 1° 15', 2° 10', and 3° 15' in diameter and split with a horizontal line (Lakowski, 1969). A fixed yellow color, 589 nm is on the bottom half, with the top color being a mixture of red, 670 nm, and green, 546 nm (Squire et al., 2005). The goal for the patient is to adjust the red-green mixture to match the given yellow standards. The test result is given in a range, with 45 to 73 indicating protanomalous trichromacy, and 0 to 35 indicating deuteranomalous trichromacy. A new version of the Nagel anomaloscope is a PC-based version (U.S. Patent No. 4,798,458). The test distance for the color test is between 16 to 24 in. We have recently turned the Nagel anomalosocpe into iPhone Apps, known as the ColoriTrac App, and the ColoritracB App. With this in mind, if there is extensive use of red and greens, a certain amount of the population will not be able to appreciate the colors.

CONTRAST SENSITIVITY FUNCTION (CSF)

CSF is a relatively new clinical test that became popular in the 1970s (Campbell, Hess, Watson, & Banks, 1978). CSF measures a person's ability to discriminate spatial frequencies which is important in object recognition. To accomplish this measurement spatial sine waves are used. The human eye is most sensitive to the spatial frequency of 6 cycles per degree (Watson, Barlow, & Robson, 1983). To measure CSF, charts of varying spatial frequencies are used, and a contrast between alternating black and white strips is presented. By using the ascending and descending method of limits to vary the contrast, an accurate threshold can be obtained for a given spatial frequency. Typically the range of 0.5 to 32 cycles per degree is measured, with 0.5 cycles per degree being wide stripes and 32 cycles per degree being very narrow stripes. Clinically, only five to seven spatial frequencies are measured. A decrease in perceiving higher spatial frequencies indicates refractive disorders. A decrease in perceiving lower spatial frequencies indicates visual pathway dysfunction (Regan & Neima,

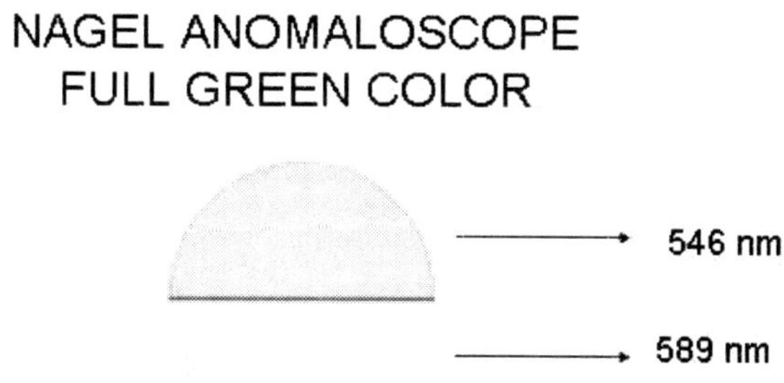

Figure 4.8. Nagel Anomalscope full green color.

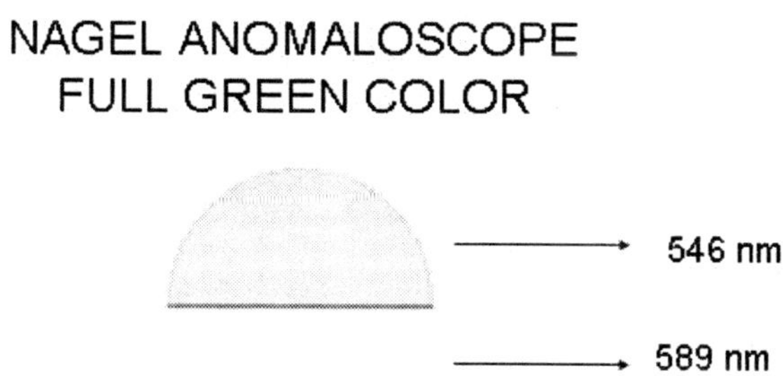

Figure 4.9. Nagel Anomaloscope full red color. A comparison of the given yellow with the full red color. The patient is to adjust the ratio of green/red to match the given yellow. An exact match indicates normal color vision. Too much red indicates a red color defect, while too much green indicated a green color defect.

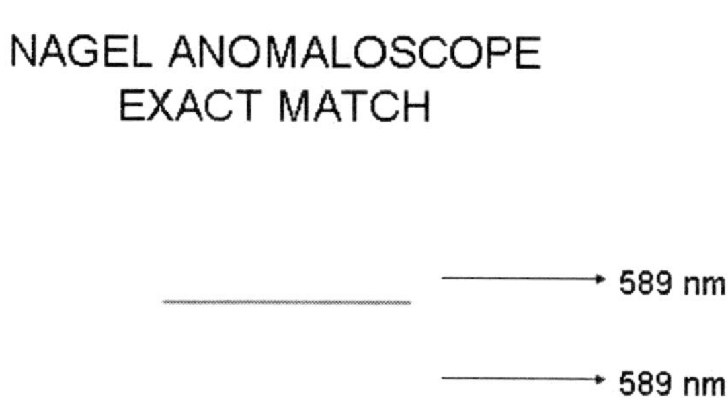

Figure 4.10. Nagel Anomaloscope exact match. A representation of a match of a given yellow (on the bottom) with a mixture of red and green (on the top)

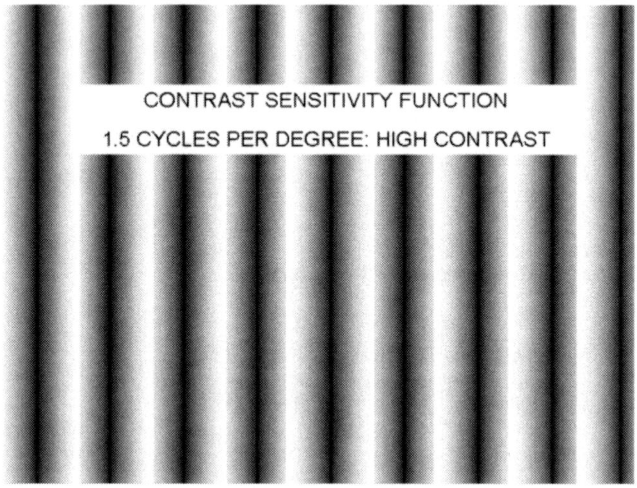

Figure 4.11. Contrast sensitivity function 1.5 cycles per degree, high contrast.

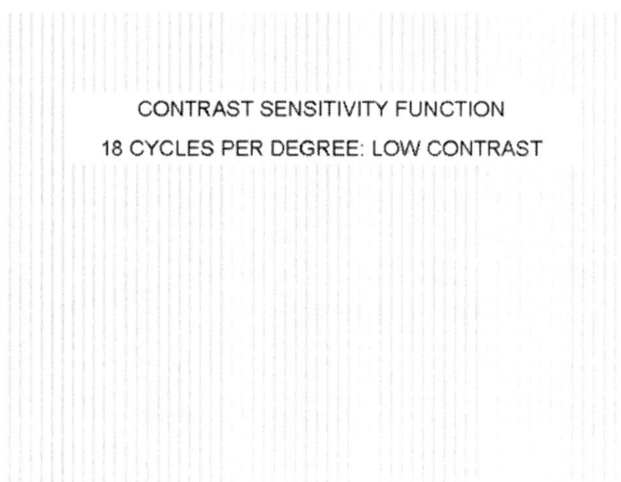

Figure 4.12. Contrast sensitivity function 18 cycles per degree, low contrast.

1984). The testing distance for CSF is 20 feet. In a design in a virtual world CSF becomes a critical factor in the use of lines or gratings, with 6 cpd being the most sensitive graphic.

DEPTH PERCEPTION

There are many cues to monocular depth perception such as parallax, superimposition, and size (Graham, 1966; and Cornsweet, 1970). Clinically, binocular depth perception is measured to determine if there is a balanced use of the two eyes. Binocular depth perception is perceived when an object is seen from a different perspective (angle) for each eye. According to recent articles, the TNO stereotest (Lameris Ootech BV, Nieuwegein, Netherlands) is the most reliable clinical test of depth perception (Garnham & Sloper, 2006; and Schmidt, Maguire, Kulp, Dobson, & Quinn, 2006). The test uses random dot stereograms (Tyler & Julesz, 1980) and red green glasses. There are a series of images when seen with both eyes that appear to be above the page. The images vary in disparity, and those with the greater disparity are seen more readily that those with less disparity. Disparity between the images is measured in seconds of arc. A standard for good binocular depth discrimination is 60 seconds of arc, and poor depth discrimination is less than 250 seconds of arc. One interesting application of stereo vision is the development of a 3D virtual library (Martín, Suarez, Orea, Rubio, & Gallego, 2009). There are several notable aspects related to depth perception and virtual worlds. One is that currently a virtual worlds displays a 3D world in 2D. Future developments will endeavor to have a 3D world displayed in 3D. Someone with little or no depth perception will have difficulty in a 3D display. The quality of depth perception is an important vision function. Accordingly, people found with poor depth perception during the testing will be informed that they should visit their eye doctor in the physical world. Poor depth perception can also lead to symptoms of eye strain, tearing, and or eye fatigue.

In the physical world, we depend primarily upon our binocular vision, and focusing mechanism to determine depth. In regard to binocular vision, the disparity of the retinal images, because of the separation of the two eyes, leads to fusion of 2D images into a single 3D image.

MONOCULAR CUES TO DEPTH PERCEPTION

The classic reference texts for information on monocular cues to depth perception are Graham (1966), and Kling and Riggs (1972). The cues are:

- Size: Familiar objects have a quality known as size constancy. For example, even though a car has a very small visual angle at 1 mile, it is still recognizable as a full size car and not a small toy.
- Shape: Familiar objects also have a constancy quality. For example, if we hold up a chair and turn it in many directions, it will still be recognized as a chair.
- Color: The color of an object helps our determination of that object. For example, if you were shown an apple that was purple it would be less recognizable than one that was red or green.
- Brightness: Objects that are closer are typically perceived as brighter than if they were far away (PercepZone, 2010). (http://library.thinkquest.org/27066/depth/nlambiguous.html)
- Contrast: The ability to discriminate among shades of gray is known as contrast sensitivity. We are most sensitive to mid-frequency contrast, that is, 6 cycles per degree, than to higher and lower frequencies. Additionally, if we want to make a distinction between two objects, there must be adequate contrast between them.
- Focus: This will be discussed in detail in the section of visual accommodation.
- Convergence: The movement of the eyes towards each other is known as convergence. As we change our focus, our eyes converge; the closer we focus, the more the eyes converge.
- Depth: In the physical world, we can use the disparity of the retinal images to determine depth; this is known as binocular vision. Because in a virtual world we loose that disparity, we should rely on the monocular cues to depth perception to provide a greater sense of space.
- Parallax: If there are two objects, one in front of the other, we can gain a sense of depth by the relative movement of the two images. The image closer will move more as we move our head from side to side than will the image farther away.
- Perspective: As is well known to artists, the use of two point perspective is an excellent cue to monocular depth perception (see Figure 4.13).
- Texture Gradient: The closer an object, the more details of its surface can be perceived.
- Superimposition: When there are two objects superimposed on each other, the object in the foreground is closer than the object that it is superimposed on.

Figure 4.13. 2 point perspective.

- Moon Illusion: Objects on the horizon appear closer than objects high in the sky. Hence, the moon appears larger on the horizon than it does as it ascends.
- Shadows: When an object casts a shadow on another object, the first object is closer to the light source than the second object.

As illustrated by the 2D drawings of my office as compared to the 3D drawings, an appreciation can be made of the importance of the components of vision in making a drawing more "realistic." The same is true in composing (constructing) a virtual world—the more simulation and stimulation of the vision system, the more it will appear that we are immersed in a physical world, rather than a purely virtual world. In order to make our virtual world like our physical world, the proper input to the vision system is required. The input from the virtual world should match the properties of the components of vision as closely as they exist in the physical world. And if possible, emotional elements are to be included. In summary, the more we make our virtual world like our physical world, the more components of vision will be stimulated.

VISUAL ACCOMMODATION

Visual accommodation is the link between the cognitive and noncognitive visual systems (Trachtman, 1992). Visual accommodation may be defined

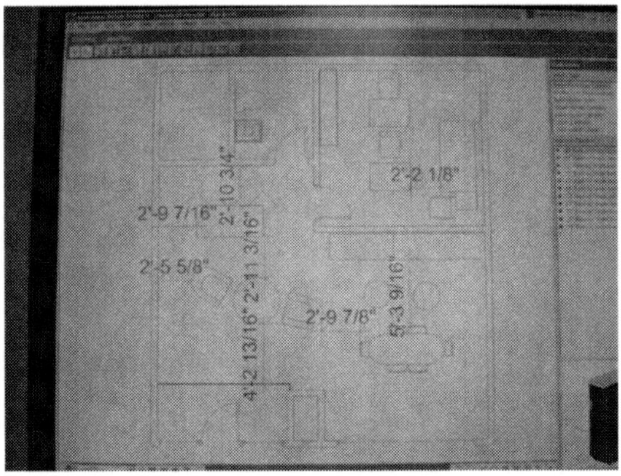

Figure 4.14. Office floor plan 2D black and white.

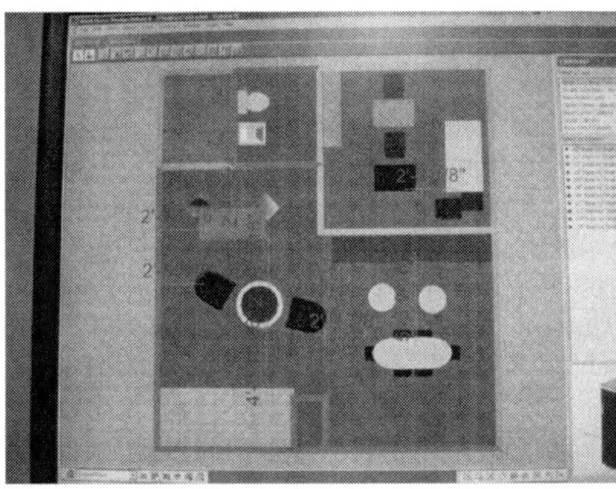

Figure 4.15. Office floor plan 2D color.

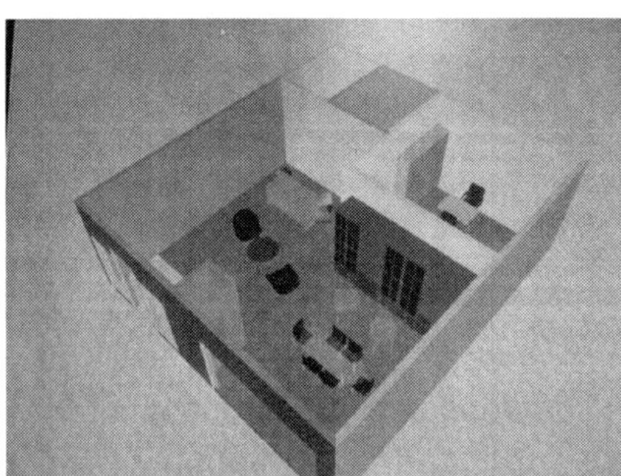

Figure 4.16. Office floor plan 3D color.

as a change in focal power of the eye by a change in the converging power of the crystalline lens due to the action of the ciliary muscle (see Figure 4.17). A full understanding of accommodation will lead to a better understanding of how to present 2D data to produce a 3D effect. Accordingly, an historical view of accommodation will now be presented.

The human visual accommodation system has been a topic of interest to vision scientists since the late nineteenth century, although the topic of accommodation had it is origin of interest much earlier. Kepler (1611) had described a function of the crystalline lens that allowed a range of vision from far to near. Young (1801) was the first to demonstrate that accommodation was due to a change in the crystalline leans and not due to a change in corneal curvature or a change in the axial length of the eyeball. The cornea was dismissed as being involved in accommodation after Young placed his eye in a water lens device, thereby eliminating the refractive effect of the cornea, and was still able to accommodate. By placing rings to the sclera circumscribing the anterior and posterior poles of the eyeball, Young showed that there was no axial length increase during accommodation. If the eyeball had increased in axial length, Young would have seen a pressure phosphene. He reported that no phosphenes were seen while looking at objects at different distances.

The measurement of the change in refraction of the crystalline lens was described by Sanson in 1837 (Southall, 1961) by observing what came to be known as the Purkinje-Sanson images. These images are catoptric images formed by a light source being imaged on the anterior and posterior surfaces of the cornea and crystalline lens. Further advances in studying the action of accommodation were made by Tscherning (1900) and Fincham (1937). Tscherning argued that the lens increases its dioptric power by increased tension on the zonules, which was a result of ciliary

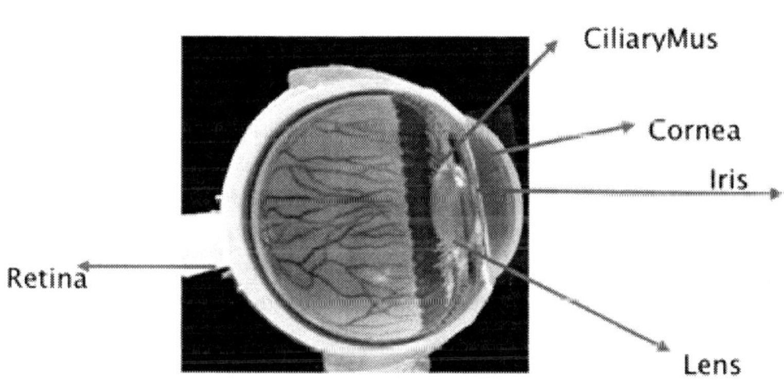

Figure 4.17. Diagram of the eye.

muscle contraction. With the increased tension, traction was put on the lens capsule causing it to flatten in the periphery and bulge in the center. According to Borish's (1970) literature review, the range of human accommodation have been reported to be from 20 diopters to one-half diopter and it inversely proportional to age.

The present mechanical model of visual accommodation in the human has been described by Toates (1972). The normal tonus of the ciliary muscle is determined by the equilibrium established between the parasympathetic and sympathetic nervous systems. For accommodation to a near object (positive accommodation) the parasympathetic nerve fibers increase their activity and the sympathetic nerve fibers decrease their activity, thereby relaxing the tension on the zonules; which in turn allow the lens to obtain an increase in dioptric value. For accommodation to a distant object (negative accommodation) the activity and the resultant dioptric change are reversed.

Stark, Takahashi, and Zames (1965) reported that the accommodative mechanism acted as a nonlinear system in response to stimuli and they proposed two models that could account for their experimental data. Model one was a description of a system with a flat, saturated response to large stimuli. The system was also nonlinear, responding with greater changes to small changes in stimulation than for large changes in stimulation. The second model, which incorporated a decrease in sensitivity to increased blur, was similarly nonlinear, has been confirmed by data reported by Toates (1972).

Another analysis of the human accommodation system was performed by Crane (1966) based on several experiments. He found that the central region of the retina, having a diameter of 30 minutes, was the relevant region for accommodative control. A control system was then described that involved feedback to the extraocular muscles, which control eye position to maintain fixation on the central retina. The latency of accommodation has been reported to be 100-300 msec and an accommodative change can occur at a velocity of six diopters per second (Randle & Murphy, 1974).

The application of control theory to accommodation has given new information on the activity of the accommodative system. In terms of control theory O'Neill (1969), the human accommodative system may be viewed as a nonlinear, closed loop servo system, which is linked with the papillary reflex pathway and the eye movement system.

In contrast with the investigations of the 1930s and the 1940s, which had been interested in the nervous control of accommodation, other investigators were primarily concerned in the characterization of the external stimuli to accommodation. The importance of the stimuli to accommodation, which led to the recent investigations of this system, is related to the overall organization of the visual system. When the retinal role in accommodation has been defined it will lead to a better understanding of the pathways as well as the physiology of these structures. In this regard, one of the major thrusts of the more recent investigators has been to identify the various stimuli to accommodation. Accommodative responses have been elicited by a variety of visual stimuli including: blur (Fincham, 1953); target size (Campbell & Westheimer, 1959); astigmatism (Allen, 1955; Campbell & Westheimer, 1959); chromatic aberration (Fincham, 1951); and stimulus intensity (Campbell, 1953). Perhaps the most commonly accepted stimulus to accommodation by researchers is blur. Smithline (1974) concluded that: "Blur is not the sole stimulus to accommodation. It is necessary, but not sufficient" (p. 1515). However, Alpern (as cited in Kling & Riggs, 1972) stated that current evidence suggested that any of the various stimuli to accommodation may be sufficient, but that no single one is necessary. Moreover, Cornsweet and Crane (1973) and Randle (1970) showed that accommodative responses could be elicited in the absence of all visual cues. Provine and Enoch (1975) suggested that a large range of voluntary control of accommodation can be demonstrated after training. for a comprehensive review of the literature on accommodation see Crane (1966), Tucker (1969), or Toates (1972). In summary, the human accommodative system is the link between cognitive and noncognitive processes, whereby an optimally accommodative system results in parallel processing.

VISUAL ACCOMMODATION IN A VIRTUAL WORLD?

With the completion of the explanation of the visual accommodative mechanism, we are left with an interesting and perplexing question: How can we take advantage of the perceptual factors involved in visual accommodation in a 3D world when the images are on a 2D surface? Hoffman, Girschick, Akeley, and Banks (2010) devised a system where 3D focus cues could be displayed on a 2D screen by making the contrast of the image dependent upon the desired perceived distance, thereby simulating blur. In order to produce such an affect the authors utilized a binocular viewing system.

Accordingly, there must be some perceptual method(s) to "compensate for the lack of visual accommodation in a virtual world." Perhaps the best methods are the monocular cues to depth perception, as listed above. Of the monocular cues, two-point perspective provides the best cue(s) for virtual world accommodation with the interaction of changing depth and size. Additional monocular cues to depth perception would only aid in producing a virtual world accommodation.

A second important factor in considering visual accommodation in a virtual world, are the connections between accommodation and the hypothalamus. Visual accommodation brings into play emotional, psychological, and physiological factors into our visual perception, which certainly can be noticed when comparing a 2D black and white photo to the same scene in 3D color.

VISUAL PERCEPTION AND COMMUNITY

How does our visual perception affect our sense of community? Before directly answering this question, an analogy with alcohol myopia will be made. Alcohol myopia can be defined as the impairment—myopia—on perception and thought (Steele & Josephs, 1990). While a virtual world is not alcohol, it does exhibit some addictive behavior (Boulos, Hetherington, & Wheeler, 2007), which like alcohol can temporarily reduce anxiety and depression. Part of this myopia is the blocking of "distractions," which can lead to possible virtual world social conflicts if someone becomes oblivious to his surroundings.

Several examples can be used to answer this question. If your virtual world neighbor is a central serial processor, he certainly will have great detail in his design; but may not be sensitive to the designs of adjacent builds. If your virtual world neighbor is a peripheral serial processor, he may not be meticulous in his design or his borders; but will be aware of his neighbor's builds around him. In other words, there should be a balance among the various perceptual factors that make a virtual world community. Colors, size, shape, and so forth, must all be carefully considered.

On a psychological basis, how we perceive affects our thinking. Accordingly, any building should take into consideration the builds on one's neighbors, in particular, and the community, in general. On the other hand, there should not be such constraints as to affect creativity and someone's desire to be an individual. However, all factors should be taken into consideration, even when it would seem unnecessary.

SUMMARY AND CONCLUSION

While many different topics have been discussed in this chapter, they are all related, and meant to be instructional. The goal: to describe the factors important for visual perception in a virtual world as compared to the physical world. On the one hand, if the above factors are incorporated into a build in a virtual world, they will simulate the physical world. On the other hand, if the desire is to design a build with a nonreal sense of visual perception, the guidelines discussed above should not be followed.

The most essential elements in the design include parallel processing of visual information, and the lack of visual accommodation and the necessary compensation for that lack.

APPENDIX: LESSON PLAN

Objective: To learn parallel processing of central and peripheral vision information.

Activity: While looking at a favorite scene in a virtual world, take note of where your vision attention is, that is, center, periphery, top left, and so on. To begin learning how to parallel process visual information, look at the center of the scene. Make a mental note of the size of your vision attention, that is, ½", 1", or 1½" diameter circle. Close your eyes and visualize the center of the scene. Gradually increase the size of the scene. Open your eyes and check to note if you accurately included what was outside the original central scene. By repeatedly rehearsing this exercise, you should be able to increase your visual information processing to include both central and peripheral information. It is important to understand that this may not be an easy exercise. For example, nearsighted people tend to dominate their vision attention in the center of the visual field, while farsighted people tend to dominate their vision attention in the periphery of the visual field. Why is parallel processing important? By parallel processing you will be able to engage not only your analytical skills; but your emotional intelligence as well.

Objective: Experience the relationship between the size of text and the desired perspective of depth.

Activity: While most people are familiar with the depth clue of two-point perspective; many may not be sensitive to applying an analogous concept to the size of text. For example, text in the foreground should be proportionally larger than text in the background. To better understand this concept, make two signs of the same size; placing one in the foreground and one in the background. Then type a sentence such as, "virtual worlds rock" on each sign using the same font style and size. Gradually change the size of the background sign with and without changing the font size. Make a note of the sign and font sizes that produce the optimal depth effect. Just one caution, make sure that the font size in the background will still be readily legible. If not, the size of the foreground font should be increased to maintain the optimal depth effect.

Objective: Experience how color affects memory and attention.

Activity: Look at the 2D picture of the hypothalamus. Report in writing what the areas of the hypothalamus are named. Now look at this picture of the hypothalamus (make sure that all of the shades of the picture are in a color or hue of red). Now report back in writing what the areas of the hypothalamus are named. Note: what did you feel when looking at the first picture? What did you feel when looking at the second picture? Did you have a harder time remembering the areas of the hypothalamus picture that was in shades of red?

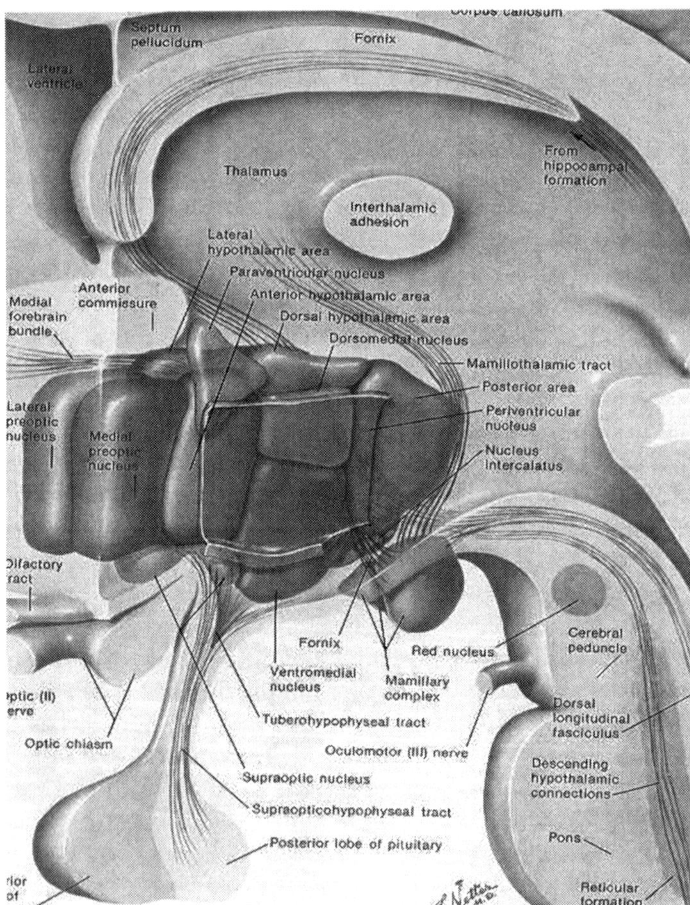

Reproduced with permission by tracht@accommotrac.com

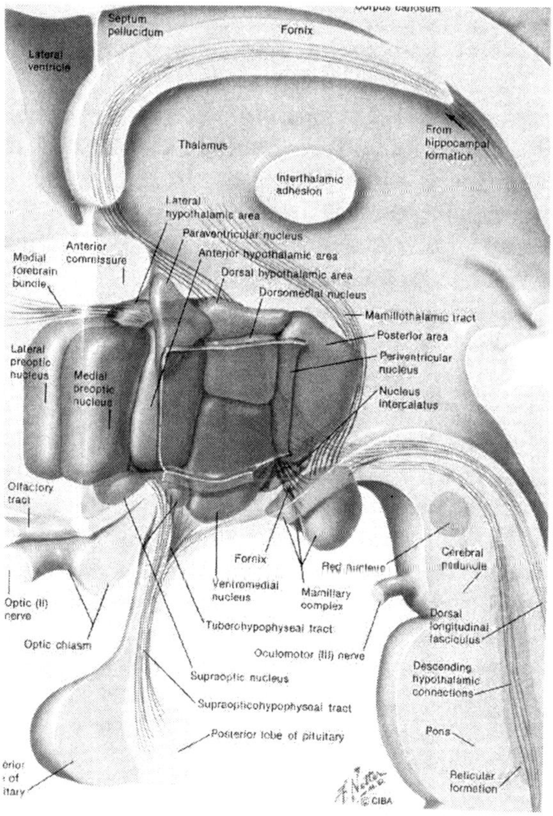

Reproduced with permission by tracht@accommotrac.com

Objective: To understand the concepts of perceptual constancy of size and shape.

Activity: We all know that if we take a chair and place it 100 feet away, we still recognize it at a standard size chair. This is known as perceptual constancy. The actual size of the image of the chair in the eye that is 100 feet away is much smaller that it was when it was within arm's reach. However, we still recognize it as a standard sized chair. We now want to apply this concept to virtual reality. Make or take a chair in a virtual world scene, and duplicate it. Place one chair in the foreground and the other in the background. Being the same size, the one in the background looks a little odd in its proportion to the rest of the virtual world scene. Gradually make the chair in the background smaller until it is perceived to be in a proper proportion to the chair in the foreground.

Objective: Apply texture to augment perceptual cues in a virtual world.

Activity: Use textures to add to existing perceptual cues, that is, contrast, depth, visibility, and so forth. Perhaps this could be the key activity that you want to begin with when exploring visual perception in a virtual world. If you were going to view an object with one set of textures, then use various choices of textures to add to existing perceptual cues. Which textures would you choose and why?

Objective: To learn the perceived distance of object by varying their brightness.

Activity: Objects that are closer to us appear brighter. We can better understand this concept by demonstrating two different scenarios. First, make or take an object in a virtual world scene, and duplicate it. Place one object in the foreground, and the other in the background. Do they both have the same degree of depth within the scene? Second, make the object in the background brighter and duller, while noting the relative sense of depth as compared to the object in the foreground.

REFERENCES

Allen M. J. (1955) The stimulus to accommodation. *American Journal of Optometry, 32*, 422-431.

Amoroso, R. L. (2010). *Complementarity of mind and body: Realizing the dream of Descartes, Einstein and Eccles*. Hauppauge, NY: Nova Science.

Bejczy, A. K. (1980). Sensors, controls, and man-machine interface for advanced teleoperation. *Science, 208*, 1327-1335.

Birch, J. (2008). Failure of concordance of the Farnsworth D15 test and the Nagel anomaloscope matching range in anomalous trichromatism. *Visual Neuroscience, 25*, 451-453.

Borish, I. M. (1954). *Clinical Refraction* (2nd ed). Chicago, IL: The Professional Press.

Borish, I. M. (1970). *Clinical refraction* (3nd ed). Chicago, IL: The Professional Press.

Boulos, M. N. K., & Burden, D. (2007). Web GIS in practice V: 3-D interactive and real-time mapping in Second Life. *International Journal of Health Geographics, 6*, 1- 6.

Boulos, M. N. K., Hetherington, & Wheeler, S. (2007). Second Life: An overview of the potential of 3-D virtual worlds in medical and health education. *Health Information and Libraries Journal, 24*, 233-245.

Callaway, E., & Thompson, S. V. (1953). Sympathetic activity and perception: An approach to the relationships between autonomic activity and personality. *Psychosomatic Medicine, 15,* 433-455.

Campbell, F. W. (1953). Twilight myopia. *Journal of the Optical Society of America, 43,* 925-926.

Campbell, F. W., & Westheimer, G. (1959). Factors influencing accommodation responses of the human eye. *Journal of the Optical Society of America, 49,* 568-571.

Campbell, F. W., Hess, R. F., Watson, P. G., & Banks, R. (1978). Preliminary results of a physiologically based treatment for amblyopia. *British Journal of Ophthalmology, 62,* 748-755.

Colenbrander, A. (2001). The historical evolution of visual acuity measurement. *Proceedings of the 2001 Meeting of the Cogan Society for Ophthalmic History, pp. 1-8.*

Cornsweet, T. N. (1970). *Visual perception.* New York, NY: Academic Press

Cornsweet, T. N., & Crane, H. D. (1973). Servo-controlled infra-red optometers. *Journal of the Optical Society of America, 60,* 548-554.

Crane, H. D. (1966). A therotical analysis of the visual accommodation system in humans. Washington, DC: NASA Report. Retrieved from http://ntrs.nasa.gov/archive/nasa/casi.ntrs.nasa.gov/19660027855_1966027855.pdf

Davson, H. (1990). *Physiology of the eye* (5th ed.). London: Macmillan.

Desmond, E. W. (1997, October 27). What's ailing Kodak? Fuji while the U.S. giant was sleeping, the Japanese film company cut prices, marketed aggressively, and now is stealing market share. *CNNMoney.* Retrieved from http://money.cnn.com/magazines/fortune/fortune_archive/1997/10/27/233297/index.htm

Duke-Elder, S., & Wybar, K. (1961). *System of ophthalmology: The anatomy of the visual system.* St. Louis, MO: C. V. Mosby.

Fincham, E. F. (1937). The mechanism of accommodation. *British Journal of Ophthalmology, 8,* 5-80.

Fincham, E. F. (1951). The accommodation reflex and its stimulus. *British Journal of Ophthalmology, 35,* 381-393.

Fincham, E. W. (1953). Controlling ocular accommodation. *British Medical Bulletin, 9,* 18-21.

Foxe, J. J., & Schroeder, C. E. (2005). The case for a feedforward component in multisensory integration mechanisms. *Neuroreport, 16,* 77-83.

Garnham, L., & Sloper, J. J. (2006). Effect of age on adult stercoacuity as measured by different types of stereotest. *British Journal of Ophthalmology, 90,* 91-95.

Gorini, A., Gaggioli, A., Vigna, C., & Riva, G. (2008). A Second Life for eHealth: Prospects for the use of 3-D virtual worlds in clinical psychology. *Journal of Medical Internet Research, 10,* e21.

Graham, C. H. (Ed.). (1966). *Vision and visual peneption.* New York, NY: John Wiley.

Guillemin, R. (1977, December 8). *Peptides in the brain. The new endocrinology of the neuron.* Nobel Lecture. Retrieved from http://nobelprize.org/nobel_prizes/medicine/laureates/1977/guillemin-lecture.html

Hannibal J., Hindersson P., Knudsen S. M., Georg B., & Fahrenkrug J. (2002). The photopigment melanopsin is exclusively present in pituitary adenylate cyclase- activating polypeptide-containing retinal ganglion cells of the retino-hypothalamic tract. *Journal of Neuroscience, 22,* 191–197.

Hess, W. (1949, December 12). *The central control of the activity of the internal organs.* Nobel Prize in Physiology or Medicine 1949, Nobel Lecture.

Hoffman, D. M., Girshick, A. R., Akeley, K., & Banks, M. S. (2010). Vergence-accommodation conflicts hinder visual performance and cause visual fatigue. *Journal of Vision, 8,* 1-30.

Jones, R. E., & Lopez, K. H. (2006). *Human reproductive biology* (3rd ed.). Burlington, MA: Elsevier.

Kastin, A. J. (Ed.). (2006). *Handbook of biologically active peptides.* Burlington, MA: Academic Press.

Kelly, J. W., McNamara, T. P., Bodenheimer, B., Carr, T. H., & Rieser, J. J. (2008). *Cognition, 109,* 281-286.

Kenney, M. J., Weiss, M. L., & Haywood, J. R. (2003). The paraventricular nucleus: An important component of the central neurocircuitry regulating sympathetic nerve outflow. *Acta Physiologica Scandinavica, 177,* 7-15.

Kepler, J. (1611). *Dioptrice.* Cambridge, England: W. Heffer and Sons.

Kling, J. W., & Riggs, L. A. (1972). *Woodworth & Schlosberg's experimental psychology* (3rd ed). New York, NY: Holt, Rinehart and Winston.

Lakowski, R. (1969). Theory and practice of colour vision testing: A review. Part 2. *British Journal of Industrial Medicine, 26,* 265-288.

Lincoln, D. W. (1969). Correlation of unit activity in the hypothalamus with EEG patterns associated with the sleep cycle. *Experimental Neurology, 24,* 1-18.

Lit, A. (1968). Visual acuity. *Annual Reviews of Psychology, 19,* 27-54.

Livingstone, M. S., & Hubel, D. H. (1987). Psychophysical evidence for separate channels for the perception of form, color, movement, and depth. *Journal of Neuroscience, 7,* 3416-3468.

Livingstone, M. S., & Hubel, D. H. (1988). Segregation of form, color, movement, and depth: Anatomy, physiology, and perception. *Science, 240,* 740-749.

Martín, S., Suarez, J., Orea, R., Rubio, R., & Gallego, R. (2009). GLSV: Graphics library stereo vision for OpenGL. *Virtual Reality, 13,* 51-57.

Merbs, S. L., & Nathans, J. (1992). Absorption spectra of human cone pigments. *Nature, 356,* 433-436.

Netter F. (1964). *The Ciba Collection of Medical Illustrations, Nervous System* (Vol. 1). New York, NY: Ciba.

O'Brien, S. (2010, August 20). A virtual world to see and smell. *Red Deer Advocate.* Retrieved from http://www.albertalocalnews.com/reddeeradvocate/news/local/A_virtual_world_to_see_hear__and_smell_101154339.html

O'Neill, W. D. (1969). An interacting control systems analysis of the human lens accommodative controller. *Automatica, 5,* 645-654.

PercepZone. (2010). *Ambiguous depth cues.* Redwood Shores, CA: Oracle Education Foundation.

Provencio, I., Jiang G., De Grip, W. J., Hayes, W. P., & Rollag, M. D. (1998). Melanopsin: An opsin in melanophores, brain, and eye. *Proceedings of the National Academy of Science, 95,* 240-245.

Provencio, I., Rodriguez, I. R., Jiang, G., Hayes, W. P., Moreira, E. F., & Rollag, M. D. (2000). A novel human opsin in the inner retina. *The Journal of Neuroscience, 20,* 600-605.

Provine, R. R., & Enoch. J. M. (1975). On voluntary ocular accommodation. *Perception & Psychophysics, 17,* 209-212.

Pulliam C. (2004, February 13). This Valentine's Day, give the woman who has everything the galaxy's largest diamond. *Harvard-Smithsonian,* Center for Astrophysics Press Release, Release Number: 04-07.

Randle, R. J. (1970). *Volitional control of visual accommodation.* Advisory Group for Aerospace Research and Development Conference Proceedings, Garmisch-Partenkirchen, Germany.

Randle, R. J., & Murphy, M. R. (1974). The dynamic response of visual accommodation over a seven-day period. *American Journal of Optometry, 51,* 530-544.

Regan, D., & Neima, D. (1984). Low-contrast letter charts in early diabetic retinopathy, ocular hypertension, glaucoma, and Parkinson's disease. *British Journal of Ophthalmology, 68,* 885-889.

Rothfarb, R. J., & Doherty, P. (2007). *Creating museum content and community in Second Life.* San Francisco, CA: Archives & Museum Informatics.

Sadun, A. A., Johnson, B. M., & Schaechter, J. (1986). Neuroanatomy of the human visual system: Part III: Three retinal projections to the hypothalamus. *Neuro- ophthalmology, 6,* 371-376.

Sadun, A. A., Schaechter, J. D., & Smith, L. E. (1984). A retinohypothalamic pathway: Light mediation of circadian rhythms. *Brain Research, 302,* 371-377.

Schaechter, J. D., & Sadun, A. A. (1985). A second hypothalamic nucleus receiving retinal input in man: The paraventricular nucleus. *Brain Research, 340,* 243-250.

Schalk, G. (2008). Brain-computer symbiosis. *Journal of Neural Engineering, 5,* 1-15.

Schally, A. V. (1977, December 8). *Aspects of hypothalamic regulation of the pituitary gland with major emphasis on its implications for the control of reproductive processes.* Nobel Lecture.

Schmidt, R. F. (Ed.). (1985). *Fundamentals of neurophysiology* (3rd ed.). Springer-New York, NY: Verlag.

Schmidt, P., Maguire, M., Kulp, M. T., Dobson, V., & Quinn, G. (2006). Random dot E stereotest: Testability and reliability in 3- to 5-year-old children. *Journal of American Association of Pediatric Ophthalmology and Strabismus, 10*(6), 507-514.

Smithline, L. M. (1974). Accommodative response to blur. *Journal of the Optical Society of America, 64,* 1512-1516.

Southall, J. P. C. (1961). *Introduction to physiological optics.* New York, NY: Dover

Squire, T. J., Rodriguez-Carmona, M., Evans, A. B. D., & Barbur, J. L. (2005). Color vision tests for aviation: Comparison of the anomaloscope and three lantern types. *Aviation, Space, and Environment Medicine, 76*(5), 421-429.

Stark, L., Takahashi, Y., & Zames, G. (1965). Non-linear servoanalysis of human lens accommodation. *IEEE Transactions,* SYST. SC. Cyber, SSC-1,1, 75-83.

Steele, C. M., & Josephs, P. R. E. (1990). Alcohol myopia: Its prized and dangerous effects. *Annals of Psychology, 45,* 921-933.

Strandring, S. (2004). *Gray's anatomy: The anatomical basis of clinical practice*. London: Churchill Livingstone.

Straw, A. (2008). Vision Egg: An open-source library for realtime visual stimulus generation. *Frontiers in Neuroinformatics, 2,* 10.

Swaab, D. F. (2003). *The human hypothalamus: Basic and clinical aspects. Nuclei of the human hypothalamus.* Amsterdam, The Netherlands: Elsevier.

Tcheung, L., Gilson, S. J., & Glennerster, A. (2005). Systemic distortions of perceptual stability investigated using immersive reality. *Vision Research, 45,* 2177-2189.

Toates, F. M. (1972). Studies on the control of accommodation and convergence. *Measurement and Control, 5,* 58-61.

Trachtman, J. N. (1992). *The etiology of vision disorders: A neuroscience model*. Santa Anna, CA: Optometric Extension Program.

Trachtman, J. N. (2010). Vision and the hypothalamus. *Optometry, 81,* 100-115.

Trevarthen, C. B. (1968). Two mechanisms of vision in primates. *Psychological Forschung, 31,* 299-337.

Tscherning, M. (1990). Pysiologic Optics (1st ed.). (Trans. C. Weiland). Philadelphia, PA: Keystone.

Tucker, J. (1969). *The construction of a recording infra-red optometers for the purpose of studying accommodation and an examination of previous work carried out on the nature of the stimulus to accommodation*. Master of Science thesis, The University of Manchester, Institute of Science and Technology.

Tyler. C. W., & Julesz, B. (1980). On the depth of the cyclopean retina. *Experimental Brain Research, 40,* 196-202.

Watson, A. B., Barlow, H. B., & Robson, J. G. (1983). What does the eye see best? *Nature, 301,* 419-422.

Wikipedia, (2010, August 9). *Augmented reality.* Retrieved from http://en .wikipedia.org/wiki/Augmented_reality

Williams, C. H. (1915). Nagels's Anomaloscope for testing color-vision. *Transactions of the American Ophthalmological Society, 14,* 161-165.

Wozniak, R. H. (2005). Mind and body: Rene Déscartes to William James. *Bryn Mawr College, Serendip.* Retrieved from http://serendip.brynmawr.edu/Mind/

Young, T. (1801). On the mechanism of the eye. *Philosophical Transactions of the Royal Society of London*, 23-88.

Zacks, J. M., Speer, N. K., Swallow, K. M., Braver, T. S., & Reynolds, R. (2007). Event perception: A mind/brain perspective. *Psychological Bulletin, 133,* 273-293.

Ziemer, C. J., Plumert, J. M., Cremer, J. F., & Kearney, J. K. (2009). Estimating distance in real and virtual environments: Does order make a difference? *Attention, Perception & Psychophysics, 71,* 1095-1106.

CHAPTER 5

STRATEGIES FOR EXPLORING INFORMATION BEHAVIOR IN SECOND LIFE

John Marino, Natascha Karlova, Peyina Lin, and Mike Eisenberg

This chapter serves as an overview of our investigation into how users seek, evaluate, and use information in a variety of contexts within Second Life (SL). Based upon the work of the Virtual Information Behavior Environments (VIBE) (http://vibe.ischool.uw.edu) project at the University of Washington's Information School (http://ischool.uw.edu), we focus on findings, insights, and recommendations that may be useful in guiding research, learning, and work practices within SL.

Of course, our interest is in the information behavior of SL residents, which is what users are called. However, we believe that our experience will benefit those seeking to use SL for a variety of purposes. Therefore, we have divided this chapter into four sections; each with strategies designed to be of practical use, and grounded in our experience with conducting information behavior research in SL.

Engaging the Avatar: New Frontiers in Immersive Education, pp. 117–150

SECTION 1.0: STRATEGIES FOR EXPLORING INFORMATION BEHAVIOR IN SL

- Macro strategies: how and why do people congregate in SL?
- Micro strategies: how do people seek, access, share, evaluate, and use information in SL?
- Special strategy: what factors into decisions people make about their SL identities?

SECTION 2.0: STRATEGIES FOR CONDUCTING RESEARCH IN SL

- Data collection methods strategies: interviewing, shadowing, and observing
- Preliminary strategies: exploration, sampling, and research design
- Ethics online: strategies for conducting ethical online research

SECTION 3.0: STRATEGIES FOR HARNESSING THE UNIQUE FEATURES OF SL

Background

Motivation Questions

> *Motivating questions: Do the capabilities and affordances of the virtual world affect the way people search for information, use information, and make decisions about information?*

The Virtual Information Behavior Environments (VIBE) project developed from previous work on credibility and trust (Fogg & Tseng, 1999) and from interest in emerging capabilities of immersive virtual environments (Bainbridge, 2007). These virtual environments, or virtual worlds (VWs), are becoming more prevalent; moreover, we expect their features to become increasingly common within other systems and tools, such as operating systems, productivity & communication tools, smartphones, etc. Therefore, we wanted to study the nature and patterns of everyday life information problem-solving behavior in immersive VWs. We chose SL for our investigation for a number of reasons. First, it is one of the most popular 3D VWs, and supports a wide variety of uses by many different communities. Second, SL is not based on a competitive win-lose paradigm (Ostrander, 2008), that is, not a gaming environment specifically, and therefore enables a broad range of activities. Last, SL benefits from participation across ped-

Figure 5.1. Wide-shot view of the Virtual Information Behavior Environments (VIBE) Research Site in Second Life. The research site used Hubs to teleport users to specific rooms and automatic avatar detection to prompt research participants with the right procedure.

agogical and disciplinary boundaries by educators and researchers and across economic sectors by government and private industry.

VIRTUAL INFORMATION BEHAVIOR ENVIRONMENTS (VIBE) PROJECT AT THE INFORMATION SCHOOL, UNIVERSITY OF WASHINGTON

Project Description and Goals

The overarching purpose of the VIBE project is to gain insights into the nature of information problem solving (IPS); specifically, information seeking, use, and sharing in immersive VWs. However, IPS also requires determinations of credibility and trust (Hertzum, 2002; Rieh, 2002), and the VIBE project attempts to understand how users understand these concepts. While we do not seek to directly compare information behavior in VWs and the physical world, we are interested in the many unique aspects, elements, or features of VWs that can facilitate or constrain IPS.

Our project mandate is to focus on the high-impact and high-stakes topic areas of health/wellness and politics/policy with the expectation that these would be likely topics for people to have questions or concerns related to their everyday lives (Levy & Strombeck, 2002; Papacharissi, 2004). While we sought to understand how SL residents address information needs related to health and politics, we expected that for many new SL residents, initial information needs will be related to the use of SL itself. Information needs related to the specific technology are characteristic of any virtual environment, and can provide the learning experiences that provide the knowledge for residents to satisfy their other information needs later in their SL use. Further, novice and experienced SL residents exhibit SL-related problem solving behaviors. Therefore, in both Phase 1 and Phase 2 of our study, we included both everyday life IPS behavior related to health and politics in addition to IPS specific to the SL environment.

Information Problem Solving

Motivating Questions

> *Do Second Life residents generally follow the linear, Big6 process or engage in frequent looping, branching, or jumping around?"*
>
> *Do Second Life residents have particular difficulty with any of the six stages of the information problem-solving process?*

We describe IPS as the process of identifying a problem or need for information, finding sources of information, assessing and determining the credibility and relevance of these sources, making decisions or finding solutions by synthesis, and reflection (Eisenberg, 2008). The model we use for describing this process is the Big6 Skills Approach to Information Problem Solving (Eisenberg & Berkowitz, 1990). People do not always complete every stage when seeking and using information to solve a problem, nor do they always proceed through the stages in a linear order. Each stage is composed of two substages in the Big6 Skills model. This model serves as the lens through which we investigate information problem solving in SL (see Table 5.2.).

Table 5.1. Questions Guiding the VIBE Research Project

What is the nature of information problem solving in Second Life?
What are the patterns of information problem-solving behavior in Second Life?
What are the unique aspects of Second Life that enable or constrain information problem solving?

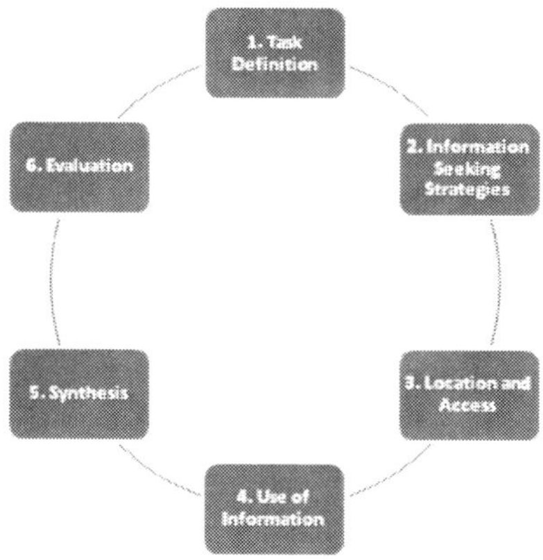

1.	Task Definition
1.1	Define the information problem
1.2	Identify information needed
2	Information Seeking Strategies
2.1	Determine all possible sources
2.2	Select the best sources
3	Location and Access
3.1	Locate sources (intellectually and physically)
3.2	Find information within sources
4.	Use of Information
4.1	Engage (e.g., read, hear, view, touch)
4.2	Extract relevant information
5.	Synthesis
5.1	Organize from multiple sources
5.2	Present the information
6.	Evaluation
6.1	Judge the product (effectiveness)
6.2	Judge the process (efficiency)

Table 5.2. The Big6 information problem-solving process.

To date, little is known about how everyday life information needs are addressed in VWs. Examples of everyday life information needs might include activities such as seeking information about alternative cancer treatments, sharing and discussing political opinions and ideas, or finding or creating an appropriate setting space for a meeting or conference whose attendees may be geographically distributed. When these activities are conducted in VWs, we would expect there to be differences given the many new capabilities VWs offer for communication, interaction, exploration, and creation. VWs are not yet typically integrated with computer applications used in a desktop environment to address information needs, such as web-based search, online libraries, personal desktop folder management systems, and so forth—but we expect they will be. By listening to and analyzing users' stories of their experiences with IPS in SL, our study can inform designers about how these experiences can be enhanced in VWs through richer interaction and experience design. To understand how people use VWs to solve their information problems and meet their information needs, we designed a 2-phase study.

VIBE Project: Phase 1

In the first phase of the VIBE project (Phase 1), we explored the nature of IPS behavior and explored the SL factors supporting these behaviors. To sensitize ourselves (Blumer, 1954) to the environment and to gain additional insight and empathy with our participants' experiences, each member of the research team created and customized their own avatar and explored many of the communities available in SL prior to and concurrently with studying users. We collected data through interviewing, shadowing, and observing SL residents in locations throughout the SL grid. We used three methods to ensure that the data was consistent across variations of the SL resident experience. For each participant and each method, appropriate permissions and informed consent were obtained beforehand.

We studied two types of residents (leaders and general residents) across two topic areas (health and politics). We use the term "leaders" to refer to SL residents who are highly involved in SL and who are main contributors to founding an SL group, coordinating events, and sustaining a particular community. With general residents, we also explored information problem-solving in general (any topic area). All data gathering took place in SL using the avatars of the research team members. In terms of setting, data collection for Phase 1 took place primarily in naturalistic settings in SL. For interviews, we used our own UW VIBE Island interview spaces, designed to be similar to a living room. Shadowing took place at the SL work space of the participant, if applicable. Examples include a library (for a librarian), a dance club (for a SL DJ), and a discussion forum (for a forum coordinator). Observation sessions took place in publicly accessible places in SL: lectures, interviews, discussion forums, dance parties, classes, and experiential simulations. Small text signage was worn above the heads of all observation team members identifying them as UW VIBE researchers.

VIBE Project: Phase 2

Phase 2 was run in June 2011; data analysis was not complete at time of printing. Whereas Phase 1 adopted an ethnographic, qualitative approach, Phase 2 employed a series of new and unique design features. These enabled researchers to focus on evaluating how specific technological and social features of SL modulate users' information behavior, including their determinations of credibility and trust. Findings from this phase of the project are forthcoming.

Summary of Findings From Phase 1

Below, we describe highlights of preliminary findings from our Phase 1 efforts. These findings will be translated to specific, concrete strategies in later sections:

Communication

SL offers a variety of communication modes (e.g., real-time text chat; real-time voice chat; note cards with text and hyperlinks; billboards, signs, and textures with text, hyperlinks, and images, etc.). Participants choose modes based on functionality, effectiveness, familiarity, and whether the parties they are interacting with are able to use the same mode. Some presentations and community forums in SL follow similar patterns of information sharing and social interaction as in "first life" (FL). However, features unique to VWs that seem to affect the nature and patterns of exchange and interaction include: distributed access, multiple channels for communication and participation, backchannel communication, anonymity, and public recording of events.

Content and Sources

Health information sources are more easily accessed outside of SL, which offers more in terms of social support. However, health information is exchanged serendipitously within groups, which serve as "information grounds" (Fisher, Durrance, & Hinton, 2004). Typically, this information is exchanged via what are called note cards in SL. Political activism in SL rarely occurs independently from related activities in FL. For example, if an election will occur soon, political activity in Second Life will increase accordingly. Political information is also shared via note cards and links to sources outside of SL. Both health and political information behaviors in SL tend to complement uses of other technologies for health and political activities in FL.

Credibility

An avatar's profile can contribute to others' assessment of the avatar's credibility and trustworthiness. Items in the profile such as FL credentials, SL and FL group memberships, synopsis of self, and type of profile picture can affect others' impressions. However, while profile characteristics can affect assessments of trustworthiness, it is not always necessary to have first life credentials. In the absence of explicit connections between one's SL avatar and other identifiers (online and offline), trustworthiness can be established through demonstrating dependable work and honest practices within a particular community.

While one would suspect that avatar appearance would be a key factor in determinations of trust, our data does not support this claim. Other studies have investigated issues related to avatar appearance (e.g., Bessiere, Ellis, & Kellog, 2009; Merola & Pena, 2010; Nowak & Rauh, 2005; Yee & Bailenson, 2007). In our exploratory investigation of avatar appearance and the determination of trust (Hussain, Nakamura, & Marino, 2011), we found that SL residents recognize that authoritative sources may be new to SL, and thus may exhibit unsophisticated avatars. Meanwhile, others may have no authority whatsoever on a given topic, yet exhibit very sophisticated avatars. Thus, experience in SL is not necessarily related to expertise in a particular domain.

Identity

Anonymity is desirable for many users seeking health or politics information to protect their identities and prevent or reduce social stigma. For example, individuals considering transgender surgery, seeking sensitive health advice, needing psychological clinical treatment or becoming involved in political activism may be concerned about protecting their FL identities to prevent social stigma. While the technology of SL can support the use of multiple avatars and disconnection from FL identifiers (including credentials), any deliberate deceit by those seeking to establish leadership is ultimately uncovered through the tightly-knit networks of social relationships in SL. Of course, many choose to remain anonymous throughout their in-world experiences, and participate as "bystanders."

Use and Users

SL is an event-oriented environment. We initially found many empty spaces in SL—even scheduled events seemed to attract only a handful of users. However, after spending more time exploring the environment, our research members learned that there were various highly-active communities. It was simply a matter of having the right social connections and understanding the dynamic nature of VWs. SL appears to be particularly relevant for scarce populations that have difficulty finding community in their local geography, such as transgendered individuals and those with special disabilities. It is also relevant for people who need to protect their FL identities from social stigma or physical abuse, such as people going through psychological adjustment related to their sexual identity, and political activists raising awareness for human rights, protection against violence, or communism. SL is also highly relevant for educators creating dynamic teaching environments and for researchers creating experimental labs. Bainbridge (2007) cites the vast number of potential participants and the user-created capabilities of SL as two of the many reasons why this is a promising environment to conduct research using methodologies such as formal experimentation, observational ethnography, and quantitative analysis, particularly of economic markets or social networks. Minocha, Tran, and Reeves (2010) suggest some practical strategies for conducting research in SL.

Figure 5.2. Many social interactions in Second Life are event-based. So, randomly selecting observation settings without deliberate, specific reasons is not the best site selection strategy.

SECTION 1.0: STRATEGIES FOR EXPLORING THE USE OF INFORMATION IN SECOND LIFE

The VIBE project explores how people go about seeking, evaluating, and using information to solve everyday life problems in immersive VWs, and evaluates the factors facilitating or constraining particular virtual information practices. We aim to inform the design of tools and services that enhance virtual user experiences in immersive 3D VWs. Of all the strategies we describe below, the most important strategy is simply to spend time in-world. Explore, observe, interact with, and learn the culture, practices, values, and activities of the VW.

MACRO STRATEGIES: HOW AND WHY DO PEOPLE CONGREGATE IN SL?

Strategy 1.1: Become Familiar With Population Patterns in SL

Some SL virtual places can be extremely crowded but do not necessarily have regular, consistent members (e.g., various "infohubs" for new members); these places seem to be frequented by residents on a regular basis, albeit with different groups of residents frequenting the places at different times. Other places can be completely empty with no sign of recent activity. In our familiarization process, we noted that health or political information behavior was seldom observed if SL places were randomly selected for observation without deliberate, specific reasons for observation there. In further evaluation, it was discovered that many social interactions in SL are event-based: people gather in a particular location for meetings, classes, workshops, musical performances, fundraisers, art shows, fashion shows, etc. These may be regular or periodic events. There are few places where people will go just to "hang out." However, infohubs are places where people will congregate; these also provide the ability to teleport to other locations throughout the SL grid.

Strategy 1.2: Use Communities and SL Features to Locate Gatherings

We suggest utilizing the search browser embedded in SL for event listings, groups, and popular places.

As of this writing, SL users currently logged in and online are represented by "green dots" on both the mini-map, a pop-up window showing the immediate surroundings, and on the world map, which shows population clusters in large swaths of SL islands. Our experience reveals, however, that it is often difficult to make sense of the purposes for gatherings discovered in such a manner (if any) and to see any actual interaction—many times we found groups of people simply "looking around."

SL provides the ability to search for events and places to add some specificity to an investigation of where to go in a virtual world. Once found, residents can teleport directly to these events and places in a single click. Once an avatar arrives at a given location, SL provides the ability to view the interests of others by clicking on their profiles, which makes commonalities immediately apparent. Discovering information about avatars is embedded directly into the avatar. Each avatar has a profile in which certain information is prepopulated. The majority of the information is entered by the user. There are several ways to access the profile: through the tool bar, hovering over the avatar and clicking, and through the menu and sidebar interfaces. Everyone can see the profile.

Communities in SL create opportunities for various information activities (sharing resources, vetting the trustworthiness of information sources), and different groups in SL are more open to sharing virtual spaces and resources than their members have experienced in FL. Other ways of finding gatherings in SL are enabled by the social nature of the interactions and community support. Community leaders are embedded in networks of social relationships that support the easy sharing of information, resources, and cross-checking on the validity of information.

Strategy 1.3: Attend to the Demographics of SL

The majority of the leaders participating in our study joined SL for their own personal entertainment, to have an alternative online medium of social interaction, or just to "check it out." However, other factors described below determined their sustained contributions in SL.

New technologies go through a period of rapid growth of user numbers during the hype created by the novelty effect (Fenn & Raskino, 2008). Once that hype reaches the "peak of inflated expectations," expectations drop sharply—as do adoption levels. Beyond this "trough of disillusionment," audiences begin to understand the technology's relevance and effective usefulness, and adoption starts to pick up again. Eventually the technology's adoption rate reaches a plateau and the technology becomes obsolete. We suspect that only a small percentage of the audience (i.e., the SL persistent users) sees the relevance of SL to their needs, and understands that SL is an event-driven environment that supports very specific user needs.

Many personal and social factors influence sustained, continuous use. For example, being home-bound provides the time necessary to maintain activity and build social relationships through a high level of activity. Often, community leaders become leaders because they are frequently requested to become involved in different groups, to found new groups, and to offer their expertise or advice to other individuals and to other groups.

MICRO STRATEGIES: HOW DO PEOPLE SEEK, ACCESS, SHARE, EVALUATE, AND USE INFORMATION IN SL?

Strategy 1.4: Recognize the Type of Information Needs People Bring to SL

SL residents do address everyday life and FL information problems in SL. Between 2007 and 2009, SL librarians in Info Island recorded approximately 150 patrons' reference questions (asking for referrals to information sources) and direction questions (asking where something is) per week (Eisenberg, Head, Lin, Marino, & Karlova, 2010).

While 35% of those questions were reference questions about SL, between 5.0-8.3% were FL reference questions (e.g., questions about where to find information about a disease) (Eisenberg et al., 2010).

Strategy 1.5: Be Alert for the Unplanned and Serendipitous Sharing of Information

One resident, a librarian, explained that SL residents tend to join SL for reasons other than health or political information. This resident then explained that SL residents are more likely to join SL because they are seeking community, and as a by-product of community activities and interaction, health information may be shared serendipitously (Eisenberg et al., 2010). Even for those residents explicitly in roles as information sources or intermediaries (e.g., community leaders, librarians, etc.), other than planned presentations, information sharing seems to happen serendipitously.

Community events are often used to share health information in a formal way, such as through presentations; however, additional information behavior frequently occurs in a backchannel. During SL presentations, residents discuss the content of the presentation in the backchannel; contradictory information or differences of opinion are readily expressed and are frequently mediated by those in attendance, leading to nuanced discourse, rather than the polarization observed in some text-only web forums and on some television shows.

Strategy 1.6: Explore How People Communicate and Seek Information

We found that people will often use a variety of communication modes simultaneously while in-world—we use the term transmedia to describe this phenomenon. We found that people will often function in and out of SL fluidly and simultaneously for communication and information, using such tools as Skype, web browsers, Web 2.0 tools linking FL and SL (e.g., Facebook).

SL offers a variety of communication modes; these can either support or constrain effective communication and sharing of information, depending on the context. Our participants typically chose a communication medium based on a combination of the medium's functionality and effectiveness in accomplishing their goals, their familiarity with it, and whether the parties they were interacting with were able to use the same medium.

In nearly every location we investigated, English was the language used in discourse and in text on objects and communications; however, "translator bots" are readily available.

Strategy 1.7: Explore How People Determine Trustworthiness

Unsurprisingly, community leaders' FL social status within a particular area of focus (e.g., health or politics) as determined by their profession and expertise contributed to their reputation and establishment of trustworthiness. Nevertheless, such attributes of individuals are mediated by social factors (such as frequent requests for their expertise and time availability), and by how other community members perceive the relevance of their contributions to their community. These factors enabled the leaders to commit to a level of involvement and sustained their participation in SL. Therefore, the leaders' background and expertise, and the community's reinforcing social factors both sustained the leaders' leadership role.

An avatar's profile can contribute to others' assessment of the avatar's credibility and trustworthiness; items in the profile such as FL credentials, SL and FL group membership, synopsis of self, and type of profile picture can affect others' impressions. In the absence of explicit connections between one's SL avatar and other identifiers (online and offline), trustworthiness can be established through demonstration of dependable work and honest practices; however, it can take a great deal of time to establish trustworthiness in this way. Volunteering and then following through to completion certain community tasks can add to trustworthiness.

Ideological fit may also affect someone's determination of another's credibility. Individuals are more likely to trust other people they perceive as being similar to them: the more similar other people are perceived to be, the more individuals will trust others; the less similar other people are perceived to be, the less individuals will trust others.

Strategy 1.8: Explore How People Solve Problems, Make Decisions, or Have Questions Answered In-World

IPS is not a well-understood term. A home-bound study participant interpreted IPS to be overcoming obstacles, such as when she discovered that SL provides access to the rich social interactions that her physical disabilities prevented her from enjoying. A political activist did not consider himself to be problem-solving as he did not perceive any explicit problems. Therefore, we learned that in order to understand how people solve information problems, we had to avoid the IPS terminology, and let the participants' descriptions of their daily activities indicate to us the appropriate terms to use.

Few information resources are stored or used exclusively within SL (unless it is in the form of a note card, and often these highlight or link to sources outside of SL). Rather, information related to health/wellness and politics/policy in particular is typically accessed outside of SL, compromising the "immersive" nature of the virtual environment, unless that outside information is located on media inside the virtual world. At the time of this writing, digital media on a 3D object in SL enabled users to access web content while not leaving the virtual world. However, this feature was client-viewer dependent and was not necessarily accessible to every user.

Strategy 1.9: Explore Ways in Which People Manage Information In-World

As of this writing, SL currently offers only folders to organize one's clothing, note cards, landmarks, and other information tools. Many common "web 2.0" features such as tagging or flagging are not available. Further research is needed to determine the extent and variety of users' practices around information management in VWs.

SPECIAL STRATEGY: WHAT FACTORS INTO DECISIONS PEOPLE MAKE ABOUT THEIR SL IDENTITIES?

Strategy 1.10: Explore Avatar Creation and the Nuances of Identity

People must make many decisions about the creation of their avatar

- How much FL information will I include?
- How similar to my actual appearance do I want my avatar to be?
- How similar to my actual behavior will my avatar act?
- What are my expectations and others' expectations for my in-world interactions?
- How extensive do I want my interactions to be?
- Is it important to me for other people to take me seriously or trust me?
- How much of my actual profession do I expect to practice in SL?

Preliminary analysis of interviews with participants in our study indicates tension between anonymity and trust in participant contexts, and suggests a framework for describing avatar transparency and trust issues, as described in Table 5.3 below (Marino, Lin, Karlova, & Eisenberg, 2010).

While interviewees claimed that avatar appearance does not affect their assessments of others' credibility, our observations suggest that avatar appearance may matter. A number of studies on avatar appearance show that SL expertise, affiliation to particular communities, and respect for

Table 5.3. Framework for Describing Avatar Transparency and Trust

Avatar	*Opaque*** (no disclosure of FL identity anonymous)*	*Transparent** (full disclosure of FL identity)*
NO established trustworthiness	"Drifter" No disclosure/claim of relevant FL work/credentials No known SL work, effort, history, affiliation Gradual/No establishment of trustworthiness/effort/history/affiliation	"Novice"* Perhaps disclosure/claim of FL work/credentials, but irrelevant No known SL work, effort, history, affiliation Gradual establishment of trustworthiness

(Table continues on next page)

Table 5.3. Continued

Avatar	*Opaque**** *(no disclosure of FL identity anonymous)*	*Transparent*** *(full disclosure of FL identity)*
Established trustworthiness	"Experienced" Perhaps disclosure/claim of relevant FL work/credentials SL work, effort, history, affiliation well-respected Rapid establishment of trustworthiness	"Professional" Disclosure/claim of relevant FL work/ credentials SL work, efforts, history, affiliation well-respected within community Immediate establishment of trustworthiness

*Avatars with known work/effort/history/affiliation that is not respected within SL ("griefers," "alts," etc.) will not establish trust.
**"Transparent" indicates that one or more conditions have been met, and the participant's FL identity can be determined.
***"Opaque" indicates that the participant's FL identity cannot be determined via these avatar descriptors. (Table reprinted with permission from the authors)

norms in a particular setting can affect the way an avatar is perceived (Merola & Pena, 2010; Nowak & Rauh, 2005; Yee & Bailenson, 2007); this an engaging avenue for research, but beyond the scope of our study on information behavior.

SECTION 2.0: STRATEGIES FOR CONDUCTING RESEARCH IN SECOND LIFE

Lessons Learned

In this section, we suggest strategies specific to conducting research in SL. These strategies stem from our experiences in setting our research in a virtual environment, and have been selected to serve the needs of education and business practitioners who may be interested in conducting research in SL.

We learned that in order to effectively create spaces for research activities in SL, we would need to either buy or rent land. We contracted with 2b3d, a Seattle-area-based VWs service provider to host our project, conduct estate management activities, and build the necessary in-world tools.

Figure 5.3. Sound proof interview rooms with restricted access only to avatars to whom entry is granted.

PRELIMINARY STRATEGIES: EXPLORATION, SAMPLING, AND RESEARCH DESIGN

Strategy 2.1: Conduct a Preliminary Exploration to Determine the Feasibility of Research

In investigating how people address everyday life information problems in SL, one of our concerns was whether we would see enough everyday life information problems in the areas of interest (health/wellness and politics/policy). We also needed to design research activities that would be feasible in SL. Therefore, we first conducted exploratory fieldwork.

Prior to designing the research, two members of the research team engaged in opportune conversations with SL residents at openly accessible SL settings. These SL residents either proactively offered help or officially identified themselves as an SL librarian or a volunteer aide. Through these conversations we were able to identify locations, events, groups and individuals relevant to understanding IPS in health/wellness and politics/policy. We utilized knowledge gleaned from this exploratory phase in the subsequent research design, recruitment of participants, and identification of observation settings.

Strategy 2.2: Target Specific Samples of the Population of Interest

We used a combination of sampling techniques in order to find specific types of participants. Participants were invited to participate in our research based on how other residents described their levels of involvement and, thereafter, by snowball sampling to identify and select leaders in the areas of health/wellness and politics/policy. We identified them as leaders to recognize that in addition to being very involved in SL activities, they founded groups in SL, organized key events for the groups, and had meaningful contributions to the sustainability of the group. Based on these initial criteria, we identified 33 residents who were the group leader or land owner of health or political groups, places, and events. Then, based on the groups'/events' sizes, levels of activity, and leaders' initial interest, we narrowed our selection to 12 leaders, or between 5 to 7 per topic area in health/wellness and politics/policy.

Strategy 2.3: Determine the Research Goals and How Best to Achieve Those Goals

Different methods will produce different pictures of the research landscape. In-depth interviews can provide personal anecdotes and stories. Broad surveys can produce large sketches. Detailed questionnaires on specific topics can create a thorough description of attitudes and beliefs.

DATA COLLECTION METHODS STRATEGIES: INTERVIEWING, SHADOWING, AND OBSERVING

Strategy 2.4: Interview With Purpose

We interviewed a total of 10 leaders: 5 in the field of health/wellness and 5 in political activism leadership positions. In a qualitative approach, it is generally recommended to determine the number of participants based on the point at which data collection efforts stop returning new data and start returning data that has already been discovered or known. For our study, we found our data approached saturation point on everyday life IP at 10 participants.

From participants, we learned how they joined SL, why they stayed in SL, and their primary activities in SL, especially pertaining to their work or information behavior context in SL. For example, our analysis of the interview data uncovered patterns of use across participants. Some of these patterns included different communication channels in SL, the types of information needs they helped address or are faced with, how they addressed those needs, how they established credibility and trust, and what they think of their SL presence in relation to their presence in other online spaces.

Leaders consented to being interviewed via private audio chat and/or private text chat in our private SL offices on UW VIBE Island. At the end of the interview, leaders were asked to recommend observation settings relevant to their topic area (health or politics). Leaders were also asked to recommend other leaders who might be interested in participating in our study (snowball sampling). Leaders were thanked with Linden dollars (the in-world currency of SL), which were, at the time of the interviews, approximately equivalent to U.S.$10 per session. The 90-120 minute interviews were later transcribed and were analyzed using a modified grounded theory approach (Glaser & Strauss, 1967).

Text chat may be more effective in terms of producing an interview transcript, and is more suitable for participants who were not comfortable in disclosing their voice (e.g., transgender individuals who had not openly come out). However, text chat is at least twice as time consuming as interviews conducted via voice chat, and responses may lack the spontaneity desired.

Strategy 2.5: Shadowing Complements a Participant's Interview

Shadowing involved following and observing the participant's avatar in-world during activities that represented a typical day-to-day work context in SL. Shadowing these leader avatars allowed us to observe the processes of health and political information behaviors as they unfolded and occurred in real time, thus complementing the memory and self-report of the participants. Since health or political information behavior may occur as a by-product of other social interactions (i.e., serendipitously, as previously described), this shadowing method was particularly useful for uncovering unplanned health or political information encounters. Six leaders (three per topic area) who participated in the interview phase were asked to participate again as shadowing participants. For each participant, there were two shadowing sessions of 1 to 2 hours each. After the shadowing sessions, we also asked participants some questions to help clarify the observations.

Strategy 2.6: Target Specific Observation Settings for the Population of Interest

The observations provided additional insight into the nature of information problem-solving in VWs. We observed types of information problems addressed in SL, the settings in which these occur, the resources used to address these information problems, and the breakdowns or barriers to solving information needs that are encountered in the process. A wide range of observation settings was identified based on popularity, topic areas, and times across the areas of health/wellness and politics/policy. Other general interest areas were also identified in order to understand the general nature of information problem-solving in SL. In all, 18 settings were selected, and two observation sessions of 1-2 hours each were conducted in each setting.

Through observation, our team learned how residents of specific communities address and recognize one another, how residents use local lingo, how to behave according to social norms and expectations, who the leaders are, and where the "cool" spots are.

ETHICS ONLINE: STRATEGIES FOR CONDUCTING ETHICAL ONLINE RESEARCH

Strategy 2.7: Consider the Ethical Implications of SL Research

There is great debate over ethical practices of research in online environments, with few generalizable solutions (e.g., Johnson, 1997). For example, the appropriateness of the presence of researchers and data collection is questioned, despite the public and anonymous characteristics of these environments.

In the process of designing our research study and obtaining Institutional Review Board (IRB) approval for research involving Human Subjects, our research team learned a great deal about the tensions among the process of informed consent, the public characteristic of openly accessible settings in SL, and about designing research which departs from the perspective of protecting the rights of human subjects versus research which is mainly concerned with data integrity.

Challenges of conducting research in SL and other VWs include:

Figure 5.4. Private interview setting in Second Life.

Blurred Boundaries Between the Public and Private

While SL is openly accessible to registered users, Linden's terms of service prohibit remotely recording conversations without permission. Many communities and events advertise themselves as "open to the public," but this really only means registered, logged in SL residents. Further, there is no public space in SL. All islands and spaces are owned either by an organization of some kind, by individuals, or by Linden. Unlike a city park which belongs to or is supported by "the general public," no island or space in SL is owned by "the general residents" in the same way as public city parks.

FL/SL Identities

Anonymity of FL identity may be a motivating factor for some residents to join SL. Many SL residents consider their SL avatar to be an extension of their FL identity, and therefore subject to the same privacy protections. There is additional debate over whether avatar names and images constitute personally identifiable information, which must be protected. When the UW VIBE team publishes avatar names and images, we use anonymized names and stock photo images in order to protect our participants' privacy to the fullest and in order to maintain good relationships with them.

Data Collection

Ethnographic-style note-taking (using a mouse and keyboard) while virtually immersed is challenging due to the need to be active and involved (typically using that same mouse and keyboard). Shifting back and forth between a text document for notes and the activities of SL can create attentional delays and a researcher can easily miss important details of the interactions being observed. We suggest conducting observations in teams of two to capture the richest, thickest descriptions possible.

Informed Consent

Obtaining informed consent from each avatar in an observation setting becomes highly disruptive, defeats the purpose of naturalistic observations, and compromises data integrity. Under IRB regulations, using any special or surreptitious equipment to record behavior makes the behavior no longer "public," and thus subject to informed consent procedures. However, it is nearly impossible and certainly impractical to obtain informed consent en masse.

Strategy 2.8: Consider Carefully the Use of Bots and Other Automated Procedures

Because we were interested in recording conversations in settings that were openly accessible to SL residents, and in settings attended by a large number of people, it would be difficult to go through private informed consent procedures without interfering with the natural flow of behaviors, thus compromising data integrity. Therefore, upon consulting with the IRB's Human Subjects Division (HSD) for ethical guidelines, and 2b3d for technical expertise, we designed an automated informed consent process through a "consent bot"—a system that automatically detects the presence of an avatar within 20 meters from the researcher's avatar position, and informs these avatars of the researcher's presence, research objectives, research activities, and enables the users to accept or decline participation on the spot (Lin, Eisenberg, & Marino, 2010).

In the process of obtaining the HSD's approval for using the consent bot, we modified our procedures to meet Human Subject ethical standards and the technological constraints of SL. Once approved by the

HSD, the consent bot strategy held great promise, but when put into practice, we encountered the following realities, described below.

Coordinators of events and groups were reluctant to agree to placing the bot in their locations (a requirement for the bot to function). The coordinators felt that for new users the SL technology could be overwhelming enough and did not want additional confusion and disruption to their planned activities.

Owners of particular settings (land owners) were unknown or unavailable to respond to requests to place the bot. In addition to meeting expected ethical research practices, asking for the sim (land) owner's permission to place objects on their territory was both a requirement of SL's Terms of Service, and a technological necessity in locations that restricted the use of automated objects.

As a result, we consulted with the HSD again, and obtained approval for a modified approach. We obtained group-level verbal consent to record textual chat conversations in group settings where this was possible (i.e., the group coordinator gives the researcher permission to go through the informed consent process with the group prior to an event). Where this was not possible, we proceeded with observations with note-taking only. Taking notes by hand maintained the "public" nature of the behavior in such openly accessible spaces as no surreptitious equipment was used to record behavior. Even so, researchers in the observation settings wore clearly visible "UW VIBE Researcher" signs.

Designing the consent bot and subsequently modifying our informed consent approach in observation settings has spurred interest in researchers investigating ethical concerns of research online. It has also inspired research questions about the effect of diverse informed consent procedures on participants' attitudes towards the research, and how natural their behavior is after different informed consent procedures. These are topics beyond our current focus, but draw from our informed consent work, and could set the ground for other research projects.

SECTION 3.0: STRATEGIES FOR HARNESSING THE UNIQUE FEATURES OF SECOND LIFE

People go to SL for the communities. Community-building is the main goal of nearly all activities in SL, as demonstrated by the consistently welcoming and friendly attitudes we encountered. While community building was beyond the scope of our research, the broad networks, complex relationships, and sheer variety and volume of communities in SL suggest that understanding the value these communities find in SL

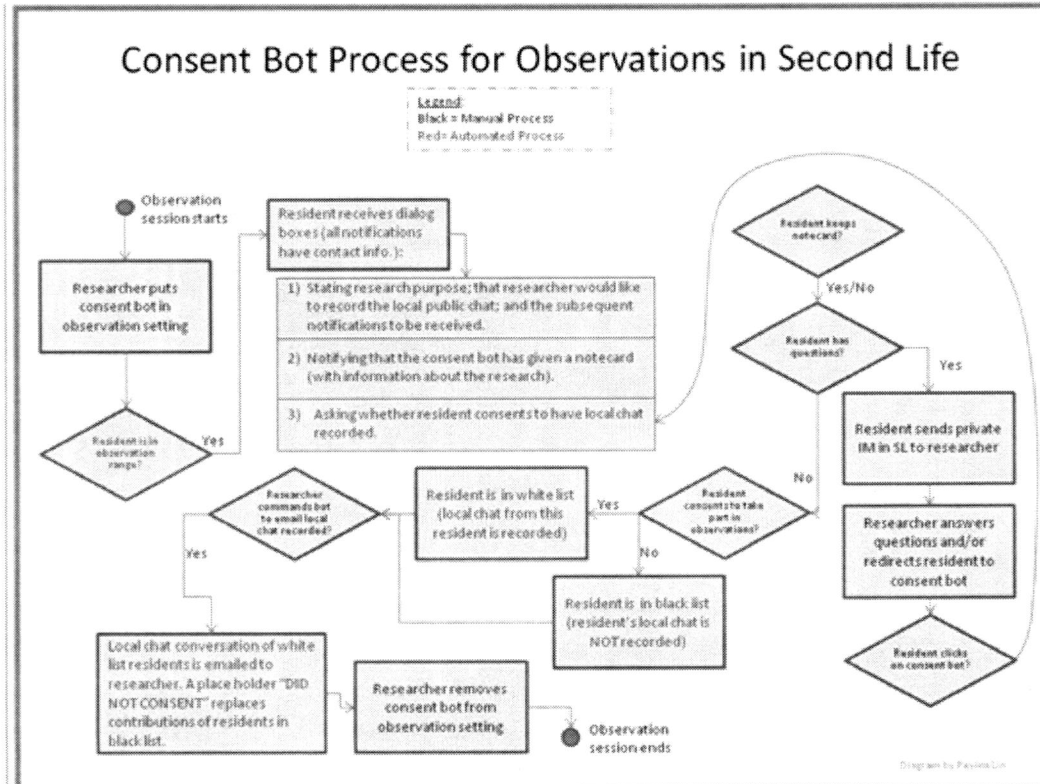

Figure 5.5. Consent bot process for observations in Second Life.

Figure 5.6. Researcher is explaining the research procedures to potential research participant.

can inform the added significance that a virtual environment like SL provides to its users. Nearly all residents have found communities with similar dis/abilities, hobbies, interests, ideologies, and self-identities.

Strategy 3.1: Engage SL's Many Unique Communities

SL also has actively involved populations that are very specific to this environment, such as Furries, Goreans, "child" avatars, Tinies, steampunk neo-victorians, merpeople, "clanks," "constructs," and other alternative cultures (Bardzell, 2006). Even in local public text chat that everyone can read, residents greet their friends with in-jokes; this is evidence of a history of involvement which contributes to cohesion and a sense of community.

Residents are concerned about showing deference to leaders and acknowledgement of leaders within a community. For example, we observed a community member publicly excuse herself from an event to change her dress when she realized that she was wearing the same dress as the event hostess. Additionally, group leaders are typically given seating on or near the stage during events, their names are suggested as sources of authoritative information during conversations, and their presence is typically announced during events. All these actions indicate active communities with strong ties among members.

Figure 5.7. Second Life objects can be highly interactive using Linden Scripting Language. The whiteboard in the back of this room enabled dynamic use by avatars in the room, thus facilitating virtual meetings.

Strategy 3.2: Become Involved in the User-Built Environment of SL

SL's most unique feature, compared to other VWs, is that it allows residents to build objects, such as houses, clothing, animals, cars, and so forth. SL offers a computer scripting language called LSL (Linden Lab Scripting Language) that residents use to craft SL to their needs. This kind of user-generated content makes SL unique among VWs. Since so many groups and communities in SL offer classes, advice, and expertise in learning LSL, resources abound. Further, learning LSL can be a great way to explore communities, build relationships, and understand the culture of SL.

Most educators or researchers will need space for their unique needs. A great way of obtaining this space is to build it or to collaborate with a resident who builds. Because SL provides residents opportunities to build the environment through scripting (e.g., buildings, signs, etc.), the environment's spaces can be highly dynamic. Rules of physical gravity, for example, need not apply and simulations are constantly undergoing change. Highly participatory environments can sometimes feel slightly disordered due to residents' enthusiasm, creativity, and deep involvement. Residents sometimes chat in the local public text chat at a lightning pace, for example, and it can sometimes be challenging to keep up.

CONCLUSION

The findings described herein were gleaned from work supported by the John D. and Catherine T. MacArthur Foundation. We also thank our colleagues and friends who helped evaluate our research instruments, all research participants, and 2b3d for their development of our research site in Second Life.

—*The Virtual Information Behavior Environments (VIBE) Project Team*

ACKNOWLEDGMENT

The findings described herein were gleaned from work supported by the John D. and Catherine T. MacArthur Foundation; we acknowledge this generous support.

GLOSSARY

Backchannel: A backchannel is real-time communication among participants through chat (public or private) or other means that occurs simultaneously as the main channel (e.g., a speaker or teacher or presentation).

IRB: The Institutional Review Board reviews research proposals to ensure the safety, privacy, and protection of participants and the ethicalness of the research.

HSD: The Human Subjects Division is a specially-designated committee within the IRB to review proposals specifically involving human participants.

Linden Dollars: Linden Dollars are the in-world currency of SL, exchangeable with currencies in FL.

LSL: Linden Scripting Language is the programming code language used to build objects in SL.

Residents: In the culture of Second Life, "residents" are users: people who create SL accounts and use SL services through their virtual representations as avatars.

Sim: A sim is a simulation. Sim can refer either to a built experience on an island or to the land itself.

Teleport: Teleportation is the fastest way of travelling throughout the SL grid. Users click on a hyperlinked object, and it takes them to a precise set of SL coordinates.

For further reference see the University of Washington Information School Virtual Information Behavior Environments (VIBE) Project: http://vibe.ischool.uw.edu, University of Washington Information School: http://ischool.uw.edu, The Big6: http://big6.com

APPENDIX: LESSON PLAN

A Suggested Lesson Plan Inspired by the VIBE Project

The potential educational uses of SL are exciting and promising. Presented here is a lesson plan based upon our experience and findings in SL, and intended for use with students in college and continuing education settings.[1]

Lesson:

The Campaign: kissing babies and knocking on doors in Second Life.

Purpose:

To apply information problem-solving (IPS) behaviors to a particular information need and engage the technological and social affordances of the virtual environment Second Life.

Project Goals:

Student recognition and practice of information and technology skills.

Effective participation in a project team.

Completion of solutions and products that will have a meaningful and positive impact on students and/or community members.

Related Skills:

From ACRL Standards[2] and Big6 Information Skills[3]:
ACRL Standard 1: The information literate student determines the nature and extent of the information needed.

- Explores general information sources to increase familiarity with the topic
- Defines or modifies the information need to achieve a manageable focus

ACRL Standard 2: The information literate student identifies a variety of types and formats of potential sources for information.

- Identifies the value and differences of potential resources in a variety of formats (e.g., multimedia, database, website, data set, audio/visual, book)
- Differentiates between primary and secondary sources, recognizing how their use and importance vary with each discipline

ACRL Standard 3: The information literate student considers the costs and benefits of acquiring the needed information.

- Determines the availability of needed information and makes decisions on broadening the information seeking process beyond local resources (e.g., interlibrary loan; using resources at other locations; obtaining images, videos, text, or sound)
- Defines a realistic overall plan and timeline to acquire the needed information

ACRL Standard 4: The information literate student reevaluates the nature and extent of the information need.

- Reviews the initial information need to clarify, revise, or refine the question
- Describes criteria used to make information decisions and choices

Big6 1.0 Task Definition

1.1 Define the information problem

1.2 Identify information needed

Big6 2.0 Information Seeking Strategies

2.1 Determine all possible sources

2.2 Select the best sources

Big6 3.0 Location and Access

3.1 Locate sources (intellectually and physically)

3.2 Find information within sources

Big6 4.0 Use of Information

4.1 Engage (e.g., read, hear, view, touch)

4.2 Extract relevant information

Big6 5.0 Synthesis

5.1 Organize from multiple sources

5.2 Present the information

Big6 6.0 Evaluation

Judge the product (effectiveness)

Judge the process (efficiency)

Learning Contexts:

Higher or continuing education settings in which information skills applied to course content are required; the task may be varied to align with course content objectives and/or to actual situations experienced by students in order to develop meaning and motivation. This plan may be adapted for students at the secondary level, but considerations for SL users between the ages of 13 and 17 must be made.[4]

Discussion:

This is a comprehensive learning activity that covers information and technology skills in a systematic way. It must be noted that not every learning activity need be organized in such a way; in fact, information and technology skills are often practiced in informal and unstructured activities without explicit instruction. Even comprehensive learning activities such as this may be organized so that these skills are practiced in an as-needed manner[5]; we encourage you to organize this learning activity in a way that best suits your goals and the needs of your instructional setting.

Task Overview:

Students will be presented with a topic related to course content; choose a topic that may be of immediate concern to students, has a strong connection to a real-world information need, and the stakes are high. In addition, the task takes place in Second Life, or relies on the use of Second Life resources. See Sample Task at the end of this Appendix.

The Assignment:

Students will address this topic by conducting a thorough search of related information, and synthesize into an objective report—they are not asked to persuade one way or another, simply to provide a comprehensive overview of the issue from multiple sources and perspectives. They may draw from primary as well as secondary resources, formal (e.g. League of Women Voters) as well as informal (e.g. a friend), online and/or in-world (SL) as well as first life (FL), and any other resources that make sense to complete the picture.

The Project Team:

Students will be assigned a project team with three others in the class, and will use SL as the environment in which they will meet, collect information, complete project work, and present the final report and product.

Deliverables:

A series of project deliverables will be due to the instructor in the timeline given in the syllabus.
Statement of Intent: includes restatement of project goal, group norms, and strategic plan for completion of each stage of project (semantic web format).

Certificate of Completion: documents successful completion of SL introductory tutorials.

Annotated bibliography: to include primary/secondary, formal/informal, in-world/online, and rationale for resource selection.

Final Report and specifications for implementation of product.

Completed simulation/support group/marketing campaign.

Presentation of project work.

The Final Report is to be presented to client; moreover, it will guide the development of an ongoing informational product within SL (a simulation, the establishment of a local grassroots organizing group, a marketing campaign, etc.)

Assessment:

We recommend informal evaluations of the set of deliverables due on a predetermined timeline to serve as formative assessments guiding student progress; a formal evaluation of the entire project at completion with self/team evaluations serve as a summative assessment.

Sample Task:

A close friend of yours is heavily involved in organizing a campaign for election to the House of Representatives. Your friend has asked several people to help look for ways to expand visibility in new media, including social networking websites, blogs, discussion forums, and virtual worlds. Your friend has asked you to further look into ways of expanding visibility in virtual worlds.

You are a fairly new user in Second Life (SL). You thought that SL would be a good virtual place to start because you've seen other political campaigns produce successful events, campaign rallies, information centers, and the like. However, you have also seen some unsuccessful efforts.

Your task is to determine how best to promote your friend's campaign within SL.

APPENDIX NOTES

1. At present, Linden Lab's Terms of Service requires that SL users be over the age of 18; Teen SL is a grid designed for users between 13 and 17 years of age, but it has been closed down as of this writing.
2. From the Association of College and Research Libraries' "Information Literacy Competency Standards for Higher Education": http://www.ala.org/ala/mgrps/divs/acrl/standards/informationliteracycompetency.cfm#stan
3. From The Big6 Skills Information Problem Solving Approach: http://www.big6.com/what-is-the-big6/
4. According to SL's Terms of Service: "By accepting this Agreement in connection with an Account, you represent that you are at least 13 years of age and you have the legal authority to enter into this Agreement. If you are at least 13 years of age and less than 18 years of age, you represent that (i) your parent or legal guardian has consented to your having an Account in Second Life, participating in the Service, and providing your personal

information in connection with the Service; and (ii) your parent or legal guardian has read and accepted this Agreement on your behalf. If you are under 13 years of age, then please do not use Second Life. There are lots of other great virtual world services for you. Talk to your parents about which services are appropriate for you." http://secondlife.com/corporate/tos.php?lang=en-US

5. For more information on constructivist learning, see Bransford, J., Brown, A. L., & Cocking, R. R. (2000). *How People Learn: Brain, Mind, Experience, and School* (expanded edition), Washington, DC: National Academies Press.

REFERENCES

Bainbridge, W. S. (2007). The scientific research potential of virtual worlds. *Science, 317,* 472-476. doi: 10.1126/science.1146930

Bardzell, S. (2006). The submissive speaks: the semiotics of visuality in virtual BDSM fantasy play. In *Proceedings of the 2006 ACM SIGGRAPH symposium on Videogames (Sandbox July '06). ACM, New York, NY, USA,* 99-102.

Bessiere, K., Ellis, J., & Kellog, W. (2009). Acquiring a professional "second life": problems and prospects for the use of virtual worlds in business. In *Proceedings of the 27th international conference extended abstracts on Human factors in computing systems (CHI EA April '09). ACM, New York, NY, USA,* 2883-2898.

Blumer, H. (1954). What is wrong with social theory? *American Sociological Review, 19*(1), 3-10.

Eisenberg, M. (2008). Information literacy: Essential skills for the information age. *Journal of Library & Information Technology, 28*(2), 39-47.

Eisenberg, M., & Berkowitz, R. E. (1990). *Information problem-solving: The Big Six Skills approach to library & information skills instruction*. Norwood, NJ: Ablex.

Eisenberg, M., Head, A., Lin, P., Marino, J., & Karlova, N. (2010). *Research on credibility and immersive virtual environments*. Seattle, WA: Annual Report for MacArthur Foundation Grant #92258-0.

Fenn, J., & Raskino, M. (2008). *Mastering the hype cycle : How to choose the right innovation at the right time*. Boston, MA: Harvard Business Press.

Fisher, K. E., Durrance, J. C., & Hinton, M. B. (2004). Information grounds and the use of need-based services by immigrants in Queens, New York: A context-based, outcome evaluation approach. *Journal of the American Society for Information Science and Technology, 55*(8), 754-766.

Fogg, B. J., & Tseng, H. (1999). The elements of computer credibility. *In Proceedings of the SIGCHI conference on Human factors in computing systems: the CHI is the limit (CHI May '99). ACM, New York, NY, USA,* 80-87.

Glaser, B. G., & Strauss, A. (1967). *The discovery of grounded theory: Strategies for qualitative research*. Piscataway, NJ: Aldine Transaction.

Hertzum, M. (2002). The importance of trust in software engineers' assessment and choice of information sources. *Information and Organization, 12*(1), 1-18.

Hussain, M., Nakamura, B., & Marino, J. (2011). Avatar appearance & information credibility in Second Life®. In *Proceedings of the 2011 iConference (iConference February '11). ACM, New York, NY, USA,* 682-683.

Johnson, D. G. (1997). Ethics online. *Communications of the ACM, 40*(1), 60-65.

Levy, J. A., & Strombeck, R. (2002). Health benefits and risks of the Internet. *Journal of Medical Systems, 26*(6), 495-510.

Lin, P., Eisenberg, M., & Marino, J. (2010). Hi! I'm Harvey, A Consent Bot": How automating the consent process in Second Life addresses challenges of research online. In *Proceedings of the 2010 iConference (iConference February '10). ACM, New York, NY, USA.*

Marino, J., Lin, P., Karlova, N., & Eisenberg, M. (2010). Avatar transparency and the establishment of trust in virtual information eco-systems. In *Proceedings of the 73rd ASIS\&T Annual Meeting on Navigating Streams in an Information Ecosystem - Volume 47 (ASIS\&T October '10), Vol. 47. American Society for Information Science, Silver Springs, MD, USA, , Article 177, 2 pages.*

Merola, N., & Pena, J. (2010). The effects of avatar appearance in virtual worlds. *Journal of Virtual Worlds Research, 2*(5), 3-12. Retrieved from http://jvwresearch.org/index.php/past-issues/25-the-metaverse-assembled

Minocha, S., Tran, M. Q., & Reeves, A. J. (2010). Conducting empirical research in 3D virtual worlds: Experiences from two projects in Second Life. *Journal of Virtual Worlds Research,* October, 2001, *3*(1), 3-21.

Nowak, K. L., & Rauh, C. (2005). The influence of the Avatar on online perceptions of anthropomorphism, androgyny, credibility, homophily, and attraction. *Journal of Computer-Mediated Communication, 11*(1), 153-178.

Ostrander, M. (2008). Talking, looking, flying, searching: Information seeking behaviour in Second Life. *Library Hi Tech, 26*(4), 512-524.

Papacharissi, Z. (2004). Democracy online: Civility, politeness, and the democratic potential of online political discussion groups. *New Media & Society, 6*(2), 259-283.

Rieh, S. Y. (2002). Judgment of information quality and cognitive authority in the Web. *Journal of the American Society for Information Science and Technology, 53*(2), 145-161.

Yee, N., & Bailenson, J. N. (2007). The Proteus Effect: The effect of transformed self-representation on behavior. *Human Communication Research, 33*(3), 271-290.

CHAPTER 6

VIRTUAL WORLDS AND WORKFORCE EDUCATION

Real Strategies for Engagement and Learning

Caroline Avey and Lyn Maize

ACS, a Xerox Company and the authors share examples of how 3D virtual worlds afford a rich, immersive and collaborative environment for both self-paced, learning activities to rich, interactive live, learning events. High-performing organizations can all be enhanced by moving learning and knowledge exchange out of the classroom and closer to the job. 3D virtual worlds provide an immersive environment for learning and knowledge collaboration on a variety of skills from on boarding to virtual lab training on technical skills. With millions of Halo users entering the workforce over the next few years, it is becoming impossible to ignore the rising expectations of younger workers for more collaborative, interactive and experience-rich distance-work environments. The authors offer a new strategic process for incorporating 3D learning that involves three phases. Analyze (define requirements), Align (develop solutions architecture) and Plan (what would it take to implement) as the final outcome.

Engaging the Avatar: New Frontiers in Immersive Education, pp. 151–185

INTRODUCTION

At this writing, there are over two dozen established 3D virtual worlds. The key in selecting the platform(s) that is right for your organization is to determine your firm's criteria in areas such as security, IT requirements, business drivers, user readiness and productivity measures. Our belief is that understanding and applying a sound instructional design process that retains the integrity of both the "content" and the "context" across a range of learning experiences, is the foundation for preparing your avatar to engage in out of the box thinking about the learning experience.

Many best-in-class organizations place a dollar value on knowledge assets or informal learning exchange. 3D worlds are excellent mechanisms to capture and share knowledge and immerse expertise throughout the organization into a digital, virtual environment. When 3D virtual worlds first entered the training scene, early adopters numbered among the avid gamers population. Second Life (SL), perhaps the most engaging of the virtual worlds, because it allowed not only consumption of experience, but enabling building your own experience, brought in new users blurring the boundaries between the physical world and the virtual world. Users found all sorts of purpose to create in this environment. People began conducting business, building libraries, teaching and creating science communities. On the social side, people dated, danced, got married, made movies, conducted fashion shows, and played interactive shooter games. Though we have been following online game technology to determine their potential for learning, we wondered if Second Life could offer more to learning than a game environment because of its ability to engage the enterprise worker in work business activities such as meetings, conferences, and activity based learning. One of our clients started asking questions about Second Life to the pointed we decided to take a closer look.

We began exploring 3D virtual worlds as viable applications for learning when our instructional designers and technology aficionados became enamored with the promise of designing and hosting real-time, synchronous learning events using avatars and corporate environments recreated in virtual worlds. The first overtures emerged as virtual worlds replicating workshop-type experiences as cool alternatives to presentation driven webcasts and conference calls. Amazingly, our early attempts were quite successful in transforming what had become simple (yes, even dull) learning events into high-engagement experiences. During these early proof-of-concepts and pilots we realized two things about engaging avatars. First, that virtual worlds offered robust environments that required us to dig deeper into our design skills and knowledge of the technology to exploit their potential. Second, we quickly realized that virtual world environments required an intensive understanding of the technology enablers and constraints, and how these affected the learning process, kept the learner motive and engaged, and delivered on our learning objectives. In short, we learned that we needed to develop an overall strategy for using these new 3D virtual worlds that spoke to both our business stakeholders and corporate learning leaders.

We know now that 3D virtual worlds will not replace webcasting, teleconferences, much less formal eLearning or face-to-face events. But as learning and performance support activities become more closely fused with daily work activities, many organizations are shifting their training focus from assuring completion and compliance to influencing performance outcomes. One of those outcomes is the ability to be immersed in a co-presence state with other digital learners. The web does not provide that sense of togetherness and we found avatars in virtual worlds that occupied the same spaces as other avatars, felt that the increased presence of others created the sense of being there.

Virtual worlds, with their ability to provide a high-fidelity user experience, can offer experiences that mirror physical-world situations and interactions. At a distance, they offer us powerful new opportunities for teaming and interaction that could change the way we work and learn in corporate environments. As yet, we have not found that any one single platform or provider meets all of the needs of our clients. Additionally, our clients struggle with issues such as ability to conduct all facilitated physical activities in a distance environment. Similarly we found that dealing with certain technical issues obstructed adoption because of required plug-ins and downloads which are not being acceptable to internal IT. Additionally, we observed that the constraint of scalable VOIP (voice over IP) issues inhibited larger numbers of workers to be functional in this environment. They struggle with the hearing or listening issues with the technology, since there are so many ways to manage voice, including the operating system, the headset and the virtual world's settings. When voice fails in the learning environment, students dwindle very quickly from the technology. This makes the technology inappropriate for large events—such as all hands meetings or product shows. Another barrier is the server load and the beefy graphic card requirement. Although the high fidelity and near photorealism increase the intensity of interaction, the companies may waive with this constraint the requirement to maintain strong multipoint VOIP capability. We have found that it is important to match the requirements of the organization with the capabilities, constraints and functionality of individual platforms and providers as described later in this chapter. Despite these limitations, we are optimistic that virtual worlds have their uses for certain types of learner engagement and will evolve to enable a higher-uptake for use in learning.

In this chapter we offer perspectives and practical examples of how multidimensional, virtual worlds can be used for adult learning in corporate and other work related environments. We will share our experiences in building and facilitating 3D virtual worlds and learning programs, and discuss why these immersive learning environments are particularly powerful in impacting performance and proficiency. We note that virtual worlds also facilitate the capture of the "informal" or collaborative knowledge exchanges that occur every day in our organizations. We will begin by looking at 3D virtual worlds as part of a strategy for a high-performing organization.

Virtual Worlds Can Enhance Organizational Performance

High-performing organizations by definition are those that implement strategies or instill cultures that enhance individual and organizational performance. These high-performing organizations foster teams and collaboration, creating a "generative learning" environment that enhances their capacity to create and address future challenges that face the business (Senge, 2006). High-performing organizations also ensure that alignment, efficiency, effectiveness and impact can all be enhanced by moving "learning" and knowledge exchange out of the classroom and closer to the job. According to research on the value of collaborative learning technologies conducted by a team at North Carolina State University, "Research on (eleven) team's performance using collaborative technology has shown that the high-performing teams incorporate capabilities of the Web 2.0 and Second Life (virtual world) into their collaboration activities. Conversely, the lower performing teams relied more heavily on traditional Web 1.0 tools, particularly e-mail and traditional document repositories. (Massey, Montoya, & Bartelt, 2011, p. 5).

At the same time, we must continue to apply solid instructional design strategies which have been adapted for the new medium to take advantage of the strengths of this new channel. "The process of learning on an individual level itself is not changing. Adults will continue to learn as they always have. What is changing dramatically is the organization's ability to enable, extend and enhance learning on an enterprise level" (Wisniewski & McMahon, 2005, p. 1). So in designing virtual worlds, we first ensure that the efforts are in alignment with the organization by evaluating the virtual world against the business requirements of the various stakeholders.

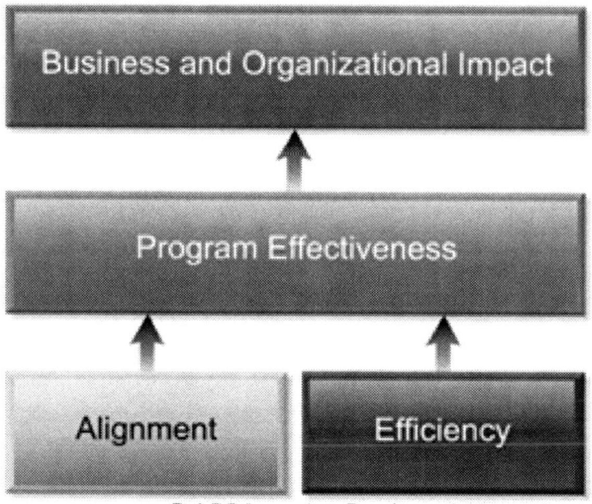

Figure 6.1. High-Performing organization strategy alignment.

Making the Business Case for Virtual Worlds

Globalization, a challenging economic environment and an insatiable demand for faster and faster access to skills and knowledge, has driven many of our organizations to be much more open to strategies that help them improve performance and increase the productivity of their people. The NCSU research showed similar results. Massey et al. (2011) state, "Our findings suggest that Web 2.0 and VW (virtual world) collaborative tools may begin to address the missing component of human interaction that has plagued collaborative learning in distance education" (p. 7)

With Xbox sales topping 30 million, iPhone apps exceeding 200,000 and millions of Halo users entering the workforce over the next few years, it is also becoming impossible to ignore the rising expectations of younger workers for more collaborative, interactive and fun work environments. When these factors are combined with the growing requirement for workers to interact with colleagues in multiple locations and time zones, this convergence creates favorable conditions for creating learning and knowledge exchanges in 3D virtual worlds. According to Wexler et al. (2008), "Younger workers will demand we provide e-Learning 2.0 approaches to performance support" (p. 8)

If and 3D virtual worlds might be most effective in an organization, however, is somewhat uncertain until we gain more evidence. Guiding the decision with a business strategy will result in much better learning outcomes. We provide a list of the most common business drivers attributed to the use of 3D virtual worlds in enterprise learning. These drivers were discovered in our explorations of virtual worlds as a learning space. Detailed examples will follow with discussion on how business value is being measured to support the list.

- Need to reduce travel costs for global employees to minimize face-to-face learning or enable interaction from a globally dispersed audience.
- Requirement to practice a skill that is not feasible or safe for the learner in the real world.
- Access to expertise that is not normally accessible, such as interacting with global experts or with resources not readily available without going to the physical location such as art galleries.
- Need to create an immersive understanding of a location or situation that is not accessible to the learner.
- Ability to enable just-in-time engaging, collaborative learning on a learner's desktop that normally would require travel to a location.
- Establishment of a venue that creates "presence"; easily capturing informal knowledge sharing and exchanges that occur among team members, coaches or mentors without actually being in the same place.

MAKING THE LEARNING CASE FOR VIRTUAL WORLDS

In order to ensure that the 3D virtual world initiative, or any emerging technology enabled learning initiative, is aligned with the goals of the organization or stakeholder, we follow (and recommend) a strategic process at ACS that involves three phases: Analyze, Align and Plan, outlined in Figure 6.2.

During the Analyze phase, we define the organizational requirements. This definition includes the technology used, the stakeholder interest, the user's capability and a content owner's interest or goal. The Analyze phase may include reviewing previous documentation or experiences, talking with stakeholder groups regarding the proposed outcomes of the initiative and conducting assessments to document the current and desired state. With virtual worlds, the Analyze phase helps to determine the current state and desired state but also identifies the cultural "tolerance" for this new immersive technology within the organization. The current state and desired state needs assessment is reviewed with stakeholders for affirmation and confirmation. Then the specific requirements are documented in a

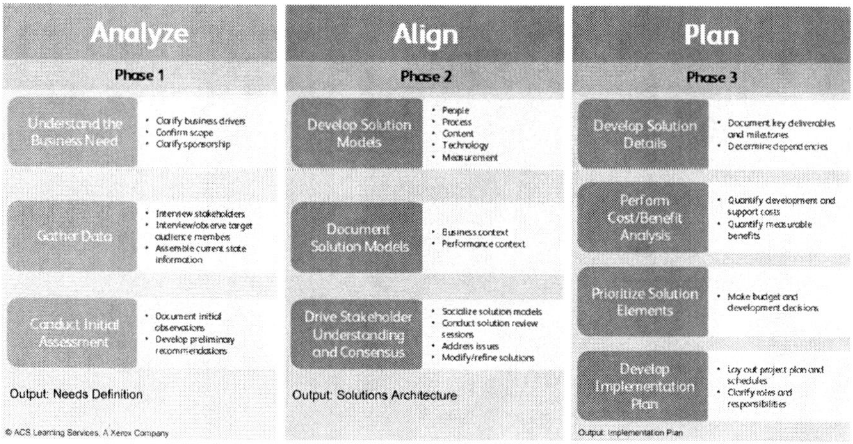

Figure 6.2. Technology strategy assessment.

needs requirement statement that includes cultural, technical and behavioral issues that will also need to be addressed.

During the Align phase, we explore potential solutions and assess them against the known requirements from the Analyze phase. We choose from dozens of effective 3D virtual worlds that include thick client solutions. Thick clients are those client / server solutions which offer rich functionality on the client, that do not have to depend so heavily on accessing the server. The richer the functionality of the application, the thicker the client solutions is. Thicker clients required high bandwidth capability and significant desktop computing capability. When the client is thick, it depends less on organizational storage, so it can be an institutional advantage. Lighter thin-client solutions exist in virtual worlds but they are not as mature with functionality, but tend to minimize technology constraints.

In our model, we weigh various potential alternative solutions against the requirements in selecting a virtual world technology architecture (hardware and software) that best meets the stakeholder's needs. This is different for each client, so a thorough analysis is always required. Just remember, the thicker client solutions may require obtaining computers and software components that need to be incorporated into your IT infrastructure. The thinner clients may require you to run a browser, open some of the Internet ports, and enable certain security measures.

Often in this stage of the process when selecting a tool, people forget to double check that they have addressed all five areas of the Human Capital Framework in Figure 6.3. This framework includes five factors for strategic learning initiatives that include People, Process, Content, Technology and Measurement.

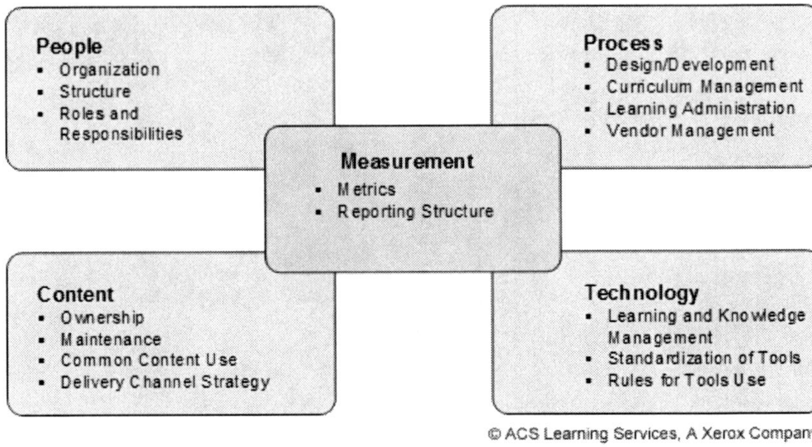

Figure 6.3. Human capital framework.

In a 3D virtual world initiative, we suggest looking at this framework as a check and balance to the analyze stage. Address each factor and know how it relates to the proposed virtual world. For example, the people factor could refer to the learner and how they will become comfortable in the new environment. But people could also refer to facilitators or subject matter experts (SMEs) and their needs relative to uploading or changing content in the environment. These are normally questions that are considered in the Assess Phase, but the framework can help trigger factors that should be considered in all three stages.

During the Plan phase, we take the solution outcome from the Align phase and consider all elements for a successful implementation. Elements to consider might include: investment analysis; testing of software; exploration of change management for course designers; facilitators and learners; determination of metrics and how they will be measured, and establishing details to ensure that the virtual world initiative is successful. The end result of the Plan phase focuses on an implementation plan for the virtual world that has been reviewed by peers and endorsed by your stakeholders.

Launching Your First 3D Virtual World Project

Not surprisingly, we quickly learned that understanding a corporate culture and addressing the concerns or questions of our various stakeholders was essential to the long-term success of our project. One bad experience for a group of stakeholders (or misunderstanding as to what they were supposed to get out of the experience) will likely shelve further exploration and development in virtual worlds. As with the use of any new technology or idea, it is important to follow a systematic process for evaluating 3D worlds in ways that increase the engagement of the avatar. It is also important to establish assessment tactics for the technology review and address how the various expectations (or misperceptions) of stakeholder groups are defined in the Assess, Align and Plan phase.

A process we found to be successful in managing expectations and "immersing" our end-users and stakeholders in this new technology includes a series of steps: Demonstration, Proof of Concept (POC), Pilot, and Implementation. First, we conduct comprehensive demonstrations of the 3D world or tool in conjunction with the vendor, explaining and exploring various functions and technical considerations in context to the business case. We also make sure to showcase the aspects that make these environments powerful, such as presence, engagement, access to experts, collaboration or fun that were addressed earlier in our business considerations. Should the virtual world meet expectations, and the recommendation is approved, then the next step is to conduct a proof-of-concept (POC) within the tool (see Figure 6.4).

Not unlike the experience of test driving a car before purchase, we take the potential 3D world tool for a "test drive" within a structured POC. We deploy the software in the organization's environment and create a "mock scenario" relevant to the organization. Then we invite various testers (stakeholders, content owners, course designers, technicians, HR, facilitators and sample learners)—to utilize the tool within a facilitated and controlled experiment. We found that this POC step quickly exposes issues and expectations that can make or break a project if they are not addressed early on Issues we often encounter include: difficulty or complexity of downloading software, bandwidth or computing constraints, open network port or firewall issues, security considerations, usability issues, and audio preferences, such as whether to use Voice Over Internet

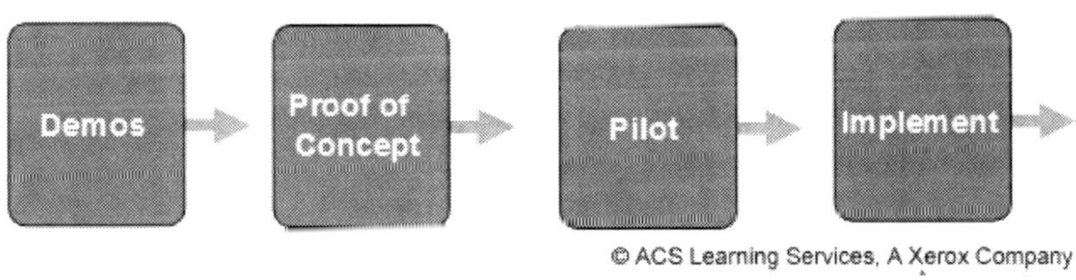

Figure 6.4. New technology exploration process.

Protocol, online chat, conference chatting in groups, accessing sound through the computer speakers or headsets.

Taking the time to resolve these issues during the POC—or determining that the chosen tool really isn't a good fit for the organization—is a very critical step. This step allows you to make changes and adjustments without jeopardizing the entire project. Your POC should be designed to elicit enthusiasm and excitement for the new medium which can build into your future marketing and communication plans. 3D virtual worlds are unlike most 2D software. 2D software tends to be text based and reside in a browser. The familiarity of this model is not an obstacle for people when learning how software accomplishes a test. With 3D visualization, non-gamer users find the 3D interface more complex to learn. The natural way to navigate, communicate and interact with the interface is quite new. Users, for example, have to use an avatar and a camera to view content in a 360 degree format. Also, working with an avatar interacting with 3D objects which observe physics is a new model for people who are not used to it. Once overcome, however, the rewards tend to be great. Here are a few comments from our learners we heard at the POC stage:

- "This provides the interactive element without having to hold classroom training."
- "This 3D training goes one step further than typical webcast/conference call training."
- "... Required more attention to participate than other forms of learning."
- "I felt like I was there!"

Assuming your 3D world initiative "passes" the POC stage, the next step is to design a controlled pilot. This is where you create an actual facilitated learning experience with learning objectives and anticipated learner outcomes. These pilots are quite similar to other new courseware pilots, but have the added complexity of also assessing if the learning was effective in a 3D virtual environment. The difference is that we measure not only the effectiveness of the learning, but also the user's feedback on their experiencing in the virtual world. If live facilitation is involved, we also assess the facilitator's experience. Once a pilot is deemed "successful", then we move to the final step of planning for implementation of the 3D virtual world for enterprise use.

A critical component for implementing engaging 3D virtual worlds is communication. Does the organization understand that this is a new technology with unique features and benefits? Do we have a plan in place to set expectations for various groups, and remedy the various issues encountered during the POC or pilot? Maintaining a group of realistic, supportive stakeholders is critically important to your success. Dr. Judith Bayliss of the Defense Acquisition University made an insightful comment recently after a Pilot of their Nexus 3D virtual world encountered load test issues. "This was frustrating and exciting at the same time. It is like seeing the rollercoaster, but not being able to ride. I can hardly wait until Nexus works and we can use it!" Dr Bayliss' team promptly set about a course of action to address the load test issues and communicate the changes.

Even more importantly, learning outcomes have been extremely successful in these worlds. One financial services organization utilized Second Life to prepare junior auditors for conducting inventories in a warehouse. Assessments indicated that the learners participating in the simulated 3D virtual world program reported a higher tendency to "feel prepared" to conduct an inventory audit versus those learners who only participated in the classroom training (Second Life with 2B3D were the developers) (Kapp & O'Driscoll, 2010). This project was developed by Seattle based 2b3D and is chronicled in a case study in Kapp and O'Driscoll's *Learning in 3D. Adding a New Dimension to Enterprise Learning and Collaboration*.

Another example that we experienced was a pilot designed in the Forterra (now SAIC) OLIVE Platform for a group of practice leader in a professional services firm. The course was an investigative inquiry skills program, where learners participated in live, online role plays with their avatars while senior partners were in attendance. Course reviews showed learners scored 10% higher on their postcourse assessments and managers reported a higher level of skill demonstrated by those learners who participated in the 3D course compared to the traditional one-day workshop.

Selecting the 3D Virtual World that is Right for You

At this writing, there are over two dozen established 3D virtual worlds. Some virtual worlds are designed for gaming, others for business application. Second Life, a virtual world design and operated by Linden Labs is the most well known virtual world, but we think less likely used for organizational learning in large enterprises. In this chapter, we will focus on those platforms with unique qualities and functionality appropriate to learning, collaboration and knowledge sharing, and offer some criteria for evaluating which might be most appropriate to your organization

Once your team has decided that 3D virtual worlds or platforms are a viable business choice for learning programs within your organization, we strongly encourage you to educate yourself to speak knowledgably about the basic features and functionality your organization requires and how they vary across the options. One criterion to consider when evaluating 3D virtual worlds for enterprise learning is to understand at a high level how they differ technically and from an end user perspective. Here are a few of the differences:

- 3D virtual worlds are offered in either a thick-client application (large software download), or a thin client solution that runs in a browser,
- 3D virtual worlds exist as open access, gaming/public environments or as private, proprietary business-to-business applications,
- Many 3D virtual worlds are sold in a per-user subscription model, or on a proprietary platform where multiple users can access a private application,
- Applications vary in development capabilities, from a "build your own" development model versus the come as you are environments requiring minimal development, and
- Finally there are a number of "quasi" 3D applications (web portals with the appearance of 3D, but the end-user only clicks through the environment) versus the fully immersive 3D environments (such as Second Life) where the end-user manipulates and maneuvers through the environment as an avatar.

Another way to identify which tools or applications might be the best fit for your initiative is to understand a bit about the tool and the history of its development, as this is often a clue to the tool's strengths and weakness. Second Life began as a tool for the gaming world, yet organizations like ours quickly saw its power and application for enterprise learning. Second Life allows an organization to purchase virtual real estate, and "construct" an environment. In our case we built a training center and park that served as a "set" for various synchronous and asynchronous learning scenarios. We created rooms with chairs and screens to project content. We created various other spaces within our environment that could be used for facilitated role playing, skills development and team projects. For one organization we built a complex set that hosted a time management course, sending learners on a facilitated path of activity check points through which they achieved competency in time management and practices. Xerox built an insider's view to document processing with a walk through virtual copy machine. Second Life has also been used for on-boarding, role playing, coaching, new hire training, complex skills based training programs.

Other enterprise ready 3D platforms grew out of a single purpose use and evolved into a broader business application. SAIC's OLIVE platform, for example, is quite robust and primarily known for its "in-world" simulated activity that first appealed to military organizations until enterprises like ours saw an application for traditional business training. Proton Media's Protoshere product started out as a training tool alternative for other live learning delivery solutions. But because of its functionality it has evolved as a robust collaboration space used for real-time training for sophisticated products in industries like medical devices, pharmaceuticals and telecommunications. Teleplace, a robust collaboration space for data intensive projects, is a great application for learning but requires much less interaction on the part of the end user. These applications are used for virtual meetings, classes, events, or facilitated group settings where content is shared with many virtual end users. Finally, IBM's Sametime 3D is a robust brainstorming tool and also works well for training or meetings. These enterprise tools come with preexisting environments and functionality already built in. Instead of making a development investment, an organization typically pays software license fees (either on an enterprise basis or per user basis) for access to the environment.

Please note that all of the aforementioned tools have been in use for a few years and continue to add functionality and scope. Their one constraint is that they are thick-client tools meaning that learners must download the software (called a client). The software is robust as the 3D images are dynamically rendered on a constant basis, putting a strain on the bandwidth and graphics capabilities of most corporate configured computers to effectively render the 3D world on the user's screen. Some tools require Internet access ports to be opened that some organization's security measures prohibit. As a result, most of these applications are exploring a thin client solution that run in the browser. Browser based virtual worlds, especially the newer ones that are emerging, provide applications that are in terms of software download, easier to render, utilize generally accepted Internet ports, and tend to perform well across a diverse range machines. Thin-client solutions are fairly new, so while they tend to have less technology constraints, they are not quite as mature in their functionality as the thick-client solutions.

Already, there are close to a dozen thin-client providers. These 3D virtual worlds each have unique offerings for learning. Reaction Grid's Unity Jibe is a 3D world following the model of Second Life in that it is an open-source environment where organizations buy virtual real estate. Similar to Second Life, organizations must engage solution providers to design the sets and functionality for their environment. VenueGen is on the other end of the spectrum with a thin-client tool focused on business application. This tool has prebuilt sets for organizations to select which range from conference rooms and board rooms to casual coffee shop booths and presentation halls. VenueGen offers pricing plans from a monthly enterprise use fee to a per-user fee. Another emerging 3D immersive environment is offered by Nexus. It is currently in development with the Defense Acquisition University and was designed to be a virtual classroom alternative. This tool promises to be a thin-client solution with robust functionality for facilitating live learning events. Other promising thin-client tools for learning are Open Wonderland, Assemblive, ActiveWorlds, Web.Alive and Canvas.

One last differentiator is that all of these tools are fully immersive 3D virtual worlds. This means the learner uses an avatar to maneuver through the content either by moving the avatar through the environment with the tool (continually rendering a new perspective with each move) or by the user manipulating content such as expanding or opening an object in the world. There is a new category of virtual worlds that we call quasi-3D. These environments are web-based portals built with a rich media tools such as Flash. These tools provide the feeling of presence but the learner does not actually move through the environment. The authors have had good success with the quasi-3D environment Unisfair as a learning portal to host traditional collateral, run webcasting type sessions and link to traditional courses via a deep link to the Learning Management System.

Our advice to learning leaders and early adopters is to avoid becoming enamored with the technology or software, but to choose the right technology or platform that serves the learning needs of the participant/end user while addressing the business needs of the company and works with the company's it and infrastructure.

Here is a quick rating tool we have used to help organizations select the criteria most important to them in a 3D virtual world. We ask individuals to rank each feature and each section as low, medium or high, and then analyze the results when making recommendations (see Figure 6.5).

EXTENDING THE BLEND— INSTRUCTIONAL DESIGN IN VIRTUAL WORLDS

3D technology has been used by the military and other industries for years to conduct simulated hands-on training required to fly sophisticated aircraft or to learn to operate and maintain complex equipment. Today, we can use this capability to address varying learning styles, create interactions and design for learning outcomes unobtainable with traditional courseware. 3D technology offers the ability to broadly reach dispersed audiences efficiently and effectively, creating high-fidelity learning experiences

Environment	Navigation and Functionality	Social Networking	Business Applications	IT Services	Costs	Implementation and Support
Avatar indentify/Behavior	Teleport between regions	User profile	Live meetings	Security	Vendor pricing model	Implementation less than 90 days
Structure available (sets)	WASD navigation	User search	Live training	Access	Environment build	Technical support
Graphics	Camera perspective	Contact list	Online training and education	Customization	Solution deployment	Creative support
Audio	Zooming capability	Social platforms, Documents and Links	Simulated learning activities	Deployment	Integration with systems	Training support
Scripting	Communication Tools	Integration with learning portals	Just-in-time content	Reliability	Ongoing support	Technical skills requirements
Scalability	Voice	Integration with user computer	Branding		Training deployment	
508 Compliance	Text	Upload and download documents	Virtual lab			
Simulation build, tracking and measurement	Video		Virtual simulations			
Link to external technologies or machines	Telephone /Skype integration		On line brainstorming and action planning			
	Email integration and Productivity Tools					
	Presenter tools					
	User tools					

Figure 6.5. Virtual world criteria.

online that leverage our classroom experiences. According to a recent American Society for Training and Development study, 10% of the respondents reported current use or were in development of virtual worlds. Another 30% are considering it for the future (American Society for Training & Development, 2010).

Some organizations are moving to a "networked learning model" where learning is a blend of traditional assets with nontraditional assets. Courseware starts out with traditional online learning, and then adds experiences in 3D. This evolves to learners adding tweets, posts, or links to other sites from within the virtual world. We predict 3D learning is a part of the large MOOC (multiopen online courses) which are emerging.

A common issue among learning leaders that we have worked with is choosing to let the technology or functionality of 3D drive their learning design decisions. In order to help learning leaders visualize design considerations related to the various learning tools or modalities, we have created a decision matrix that helps map plausible options based upon known requirements such as the need for the learning to be synchronous or asynchronous and whether it needs to involve technology or minimize technology. In the chart depicted in Figure 6.6, 3D virtual worlds appear twice to include either individual or group experiences, both with high uses of technology. We have found that 3D virtual classrooms are highly effective learning environments. Using 3D for individual skills practice in a simulated environment is also effective.

The chart examines learning environments from two axes: sophistication of the technology (low tech versus high tech), and the amount of collaboration (asynchronous or synchronous learning). Following, we include a few key 3D experience attributes, where the contextual, collaborative capabilities of 3D can help generate discussion among your team as you visualize how 3D virtual worlds could become a rich part of your learning blend:

- Low risk, high collaboration. Deployment of facilitated experiential/simulated learning or group activities in a secure or semi-secure space. An example of this type of 3D learning is the I95 corridor project (http://www.i95coalition.org/i95/Default.aspx), where multiple emergency response teams practiced collaboration for emergency situations demanding cross jurisdiction support.
- Multiple Learning Styles. The capability to address varying learning styles, interactions and outcomes in a unique way. Learners can experiment with physical assets such as petroleum engineers reconfiguring pipelines to determine efficiency and safety. 3D environments are by nature a highly visual environment as learners explore

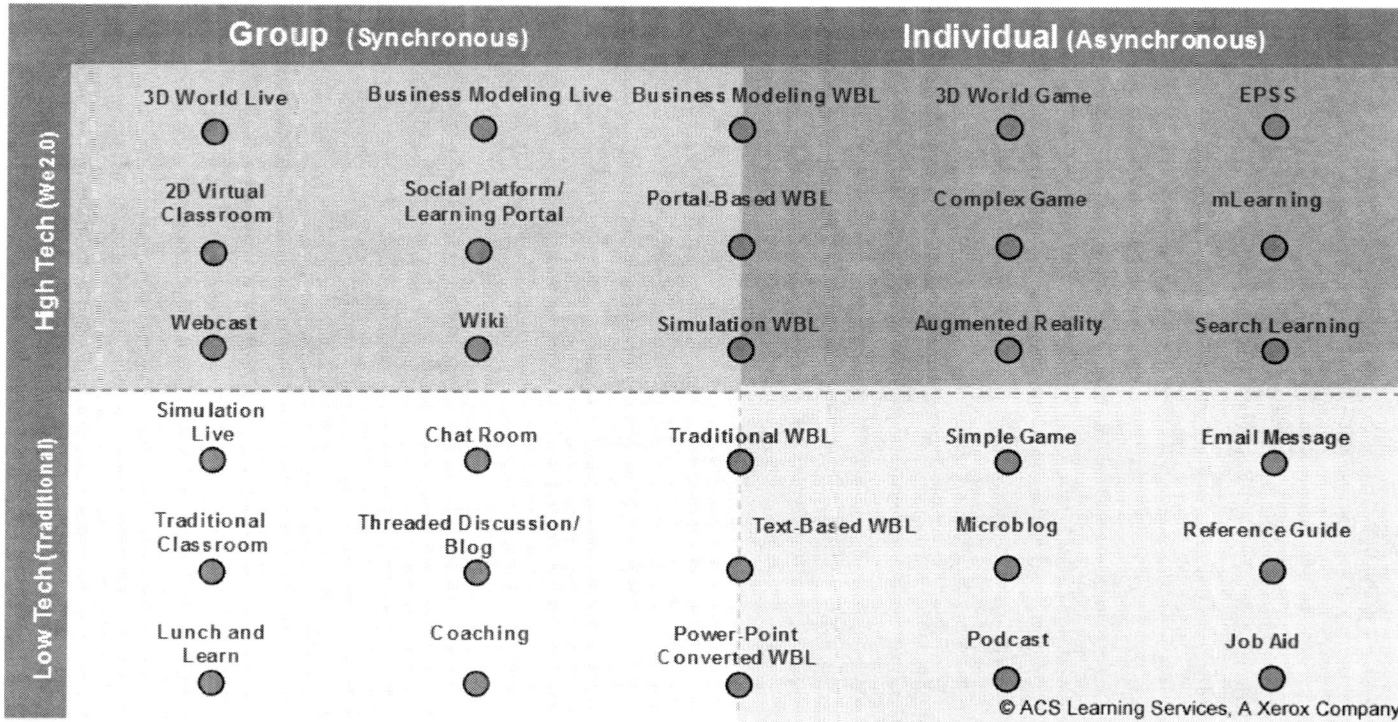

Figure 6.6. Learning delivery decision matrix.

through a graphically rich, visual domain. Built-in speakers and headsets enable the audient learner to participate with sound effects or discussions available in environments with voice over internet protocol (VOIP). Signs, tags and downloadable documents support the text learner.

- Context. 3D virtual worlds allow for the complete range of the learning taxonomy. 3D offers ways to adapt the environment and enrich the learning experience that allow for varied levels of learning. These include cognitive and affective levels and skills or competency-based learning. For knowledge and comprehension levels, asynchronous activities can be guided by making "on demand" knowledge documents available or other contextual prompts associated with objects in the world.
- Engagement. Our early experiences showed that virtual world applications are extremely effective at engaging adult learners, particularly when exposed to high-value learning exchanges. Not only did they exhibit the same sense of "presence" and collaboration as single site events, but we were amazed that cultural protocols and deference to "seniority" carried through to the virtual environment.
- Speed. A simple virtue, but some 3D virtual worlds offers significant business advantages in cost, time and the ability reach dispersed audiences efficiently and effectively (we will talk much more about this when we address measuring success and ROI).
- Versatile. These environments lend themselves to incorporating other technology solutions and content used by the organization. These 3D scenarios can enable functionality your environment uses such as networking capability, document collaboration and/or use of "sets" in the environment.
- Blended Solution. 3D learning is very easy to integrate with other learning modalities. Use of online learning, webcasts and even job aids can be blended with 3D to provide a robust learner experience and achieve the learning objectives and business requirements. An example is an on boarding program for a Human Resources firm. A month-long residence program is replaced with a blended solution of traditional online learning, webcasting, on-the-job assignments and a 3D virtual world. The virtual world is used for collaboration sessions with the new hires, as well as discussion sessions in the virtual world. The graphic below represents a consultative skills course that occurs in a 3D environment, but also blends group, individual and team activities.

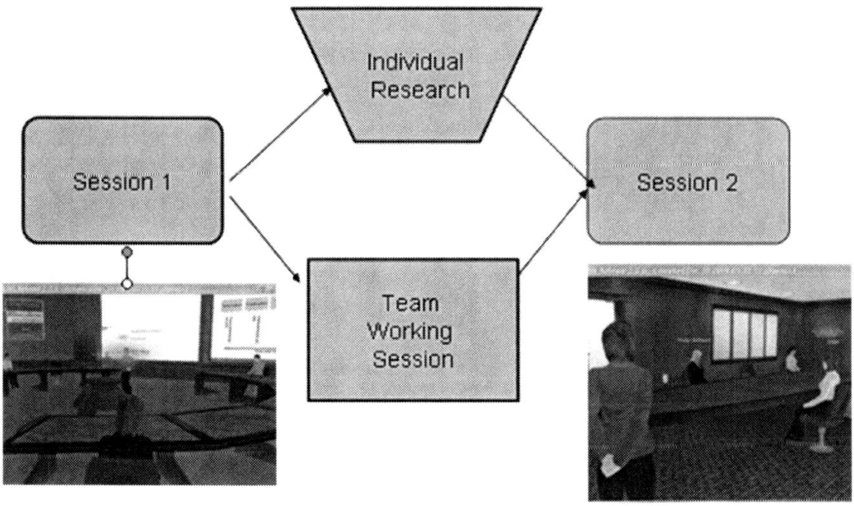

Figure 6.7. Example of a blended solution of synchronous to asynchronous virtual learning.

THE ADDIE DESIGN PROCESS WITH A TWIST

There have been more than a few debates on following a traditional instructional design process, such as ADDIE (Analyze, Design, Develop, Implement, and Evaluate), in the development of 3D virtual learning initiatives. Our belief is that understanding and applying a sound instructional design process that retains the integrity of both the "content" and the "context" across a range of learning experiences is the key to achieving the success and sustainability of any learning program. With virtual worlds, we believe that the learner experience is paramount. A recent ASTD study comes to similar conclusions about the learner. The basic premise of an e-learning 2.0 strategy involves a shift from an infrastructure that is content-based to an infrastructure that is built around people. E-learning 2.0 is a shift from stand-alone systems to a combination of many systems, processes, and environments in support of the learner. E-learning 2.0 is a change in pedagogy—how we instruct and how people learn (American Society for Training & Development, 2010).

Our e-learning stages model is a simple spin off of the traditional design model, but is adapted to incorporate an expanded mix of decision factors, tools, design considerations and group facilitation strategies specific to the use immersive learning environments. Once these design fac-

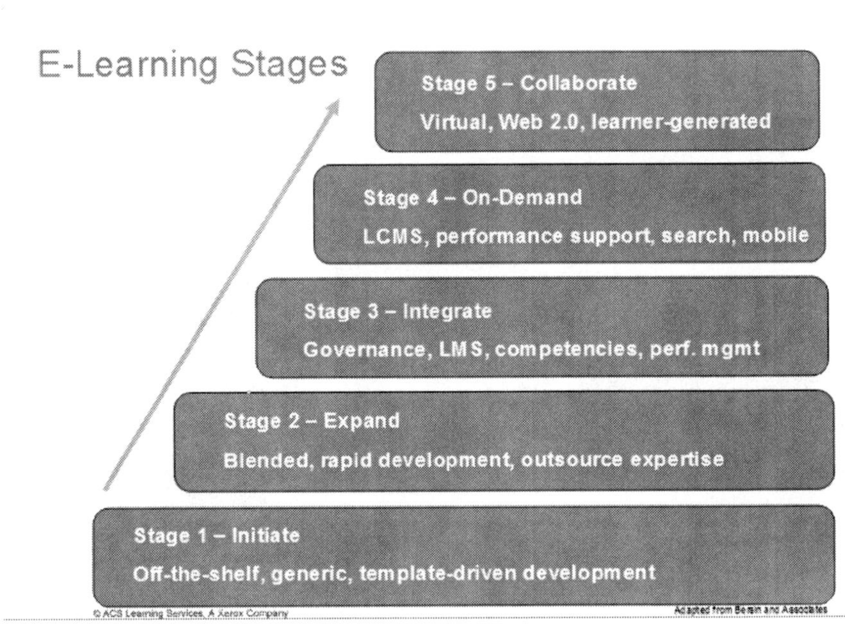

Figure 6.8. E-learning stages.

tors have been determined, the ADDIE model still offers a great approach to the instructional design process; only with a slight twist focusing on learner performance and facilitated interactivity not always found in a traditional content development process (see Figure 6.8 above).

Analyze: As in any "analyze" phase of learning design, we work with learning leaders to first consider what business issues are to be addressed and what resulting performance gaps are to be addressed by the proposed training. The twist to designing in 3D virtual worlds is to ensure learner performance of skills and knowledge in the world to close performance gaps and create contextual situations for transferring skills back to the job.

Since 3D lends itself to application-level learning, we help clients envision activities that can be simulated in 3D to help close a performance or skill gap. Then we write the instructional strategies or learning objectives accordingly. 3D environments also allow for an array of "verbs" to be included in learning objectives such as: manipulate, maneuver, observe, explore, reconfigure, build, deconstruct, and interact. These words are powerful directives when guiding the learner's paths or actions. An example of an objective from a program designed for a global energy company, states "identify issues with structural performance then reconfigure the asset to remedy the performance in a simulated environment." In this case, the learner must look for issues (awareness), and then they must make the desired changes by manipulating the components of an engineering asset (application level learning).

We conclude the Analysis phase with a scoping document detailing the business need, the performance need, the audience and the objectives. The twist here is also incorporating the specific business rationale for 3D, and how the company plans to assimilate the learners into the medium as part of the project scope.

Design: As is the case for all courseware, the design plan first depends on creating good learning objectives and then it depends largely upon the ability of the designer to utilize the capabilities of the environment to accomplish the 3D aspect. A great example of both is found in the design plan for a virtual technology lab by MTS Communications in Canada. The objective was to configure a set of servers and switches for telecommunications technicians. The design was to use the ability of the 3D world to link directly to an active computer running configuration software. The learner was directed to manipulate a computer "in world", that reacted with the real world computer which displayed changes in configuration back to the learner via the virtual world. Then the learner walked over to a virtual rack, selected the appropriate telecommunications components and places them on the virtual rack. The virtual world components were actually linked to physical world components which gave feedback to the learner when the system was configured correctly.

Other design examples that maximize 3D capabilities for greater engagement involved animating entries in a spreadsheet to dynamically populate a 3D pie chart. Another example used to generate team input and brainstorming asked participants to grab, build and create writable 3D objects to share their ideas. Still another example used a stop sign shaped object to render "barriers" and a green sign for "enablers" during an executive planning discussion. Another consideration is the ability to address multiple learning styles in 3D. These include the use of audio associated with objects for the audient learner, text on graphics for the visual learner or manipulating objects for tactile learners. Think about the possibilities for on the-job transfer of knowledge in the design of exercises that mirror real-world scenarios (such as walking around the oil rig to spot safety issues). Imagine being able to reinforce the learning with downloadable job aids or links to other sites.

Virtual worlds are graphically intensive environments. Look to the tool you have chosen and think creatively about the environment you can create to enhance the learning experience. For environments with existing scenes or "sets," what functionality will you need to create, brand or enhance to achieve your objectives? If you are building from the ground up, then there is a parallel process that involves designing the look, feel, sound and functionality that will be required to achieve the objectives. The actual set graphics can convey subtle corporate messaging. A firm that values innovation will want a high-tech, futuristic looks to their virtual world. A financial firm will probably want a more traditional campus look and feel to impart the values of the firm. Either way, having good graphic designers and project managers as part of your team is a necessity.

Finally, we have talked a great deal about the sense of "presence" in virtual worlds. This is where context comes into play in designing environments that foster exploration, connection and comprehension. Consider cutting back the time for lecture or prescriptive instruction and instead create scenarios or role-playing exercises that enable the learner to discover on their own. Think about positioning learners on teams in virtual breakout areas for role-play or discussion. Or send them on an online treasure hunt to gather, share and process information. For self-paced simulation activities, think about how you are going to move the learner through the environment. Where will they go? What will they experience?

We conclude the design phase with a traditional design document that includes a project scope description, audience profile, learning objectives and then a description of how those objectives will be met, measured and communicated. For 3D programs we also address what changes will need to be made to the 3D environment and why those changes should be made. Another aspect of design is describing the "experience" that the learner will have during the learning. First, we use a "windowpane" or agenda to map out the timeline and the experience. We do this for both synchronous and asynchronous learning. The windowpane is a table that describes the learner's experience for each event with the time to conduct the activity, the activity mapped to the corresponding objective and the resources needed in the world to achieve the objective.

Development: "Development" of learning scenarios in 3D virtual worlds is dependent on two major criteria. First, the inherent capabilities of the virtual world you are using and second the complexity of the program design. If you are using worlds that already have location sets, features and functionality, then your development activities will be limited to the enhancements you wish to make to your program. On the other hand, if you are building your own sets, objects and need advanced features and functionality then you will need to create specifications for each aspect, then make a build or "buy" decision.

Another aspect of developing learning programs in 3D is to incorporate environmental experiences that help learners increase their comfort in using avatars or other navigation tools. Examples are providing asynchronous challenges to the learner such as picking up objects for note cards, or moving through the 3D world to observe and participate in very specific experiences prior to "class." For instructor-led 3D events, we created online "icebreakers" to encourage participants to use their avatars to interact. In one Proof of Concept trial, our learners connected pre-event through a link to their personal profile on their LinkedIn page. As avatars greeted each other and talked via VOIP, they also clicked on their fellow avatars to review their information stored in LinkedIn. By designing these subtle interventions, we have found that as learners progressed through a learning program, their capability and confidence increased with each module. Finally, encouraging participants to meet in informal team sessions, outside of the formal learning events helped them explore and practice in a casual learning experience. (We created a sandbox and waterslides for this purpose on our Second Life Island).

Earlier, we mentioned the importance of the user experience and working closely with stakeholders. It is often difficult to convey what that experience will be. We have found it useful to take the intangible idea of the experience and make it tangle through the use of "storyboarding"—a series of graphic renderings depicting what the space will look like and a simulation of the learner's experience. This helps our clients respond to the "look and feel of the environment." Utilizing storyboards also helps to gain a sense for the interactions that will take place during the learning experience. This helps to ensure that a learner's experience will address the performance issues identified within the Analyze phase.

We also use prototyping as a step in our development process where we will develop one set of activities in the world and then have our clients react to the activity prior to developing the remaining learning activities.

Implementation: Since these virtual environments are very dependent upon technology (the tool, the Internet and the user's computer), it is very important to test and then to test again. Internally, we perform Alpha tests for functionality, usability, accuracy, achievement of objectives and load tests for numbers of participants. We Beta test for the same criteria on the client's site with client machines and connections. For facilitator-led courses, we conduct train-the trainer sessions for both facilitating in the environment and facilitating the course. We pilot at least once, if not twice, with facilitators and technical support personnel. We also consider a communication and change management plans for engaging learners, facilitators and supervisors who are new to virtual world environments.

Accommodations must be made for the support of any download activity, avatar creation or enrollment and registration considerations. Additionally, computer configuration changes and/or obtaining headsets for audio must also be addressed in advance.

Evaluation: Even though measurement occurs after the course is deployed, we take into consideration everything we will want to evaluate during the Analyze phase. This carries over to evaluation of specific considerations such as usability, the experience or application attributes specific to the 3D environment.

We typically measure for Kirkpatrick's Levels 1-3, and we have done some advanced program analysis at Level 4. For Level 1 measurement, we often provide a link directly from the 3D world to a survey tool—a virtual smile sheet. A twist related to 3D development is to extend this level 1 measurement survey to presenters, facilitators and others if this was their first experience in this medium.

We use traditional evaluations that measure the learner's perspective on learning effectiveness, transfer to the job, etc. But we also have created a questionnaire specific to the environment, eliciting information about the learner experience. Here are a few of the sample attributes we measure:

- I was able to maneuver freely in the environment.
- I was comfortable using my avatar to express myself.
- I could hear the discussions and media in the environment.
- I could communicate effectively in the environment.
- I could manipulate the objects in the environment.
- Overall, I found the virtual world to be a rich learning environment for this course.

For Level 2 evaluations we have used two approaches. The first approach is the traditional knowledge assessment where we link to an online quiz from the 3D environment. This is fairly quick and easy to deploy, but only measures the knowledge gained not necessarily the application of that knowledge. For other situations, we have designed experiences that test the learner's ability to apply the knowledge. An example is for a learning activity to "demonstrate the use of effective questioning by using closed and open-ended questions in a simulated client discussion." Here the learner participates in a role-play where they are assessed against the use of correct questioning procedures. For a learning objective to correctly configure the telecommunications server, the learner participates in an asynchronous configuration activity where the 3D world tracks the sequence of the objects selected as a measurement.

For Level 3, we use standard approaches to assess on-the-job performance or behavioral change. Once we have sufficient history to draw upon, we will begin measuring data pertaining to performance comparing traditional versus 3D learners in compatible curricula.

VIRTUAL WORLDS MAKE REAL BUSINESS SENSE (CENTS?)

I hope by this point you have become as fascinated and enamored with the learning potential of 3D virtual worlds as we have. We have explored the power of this environment for discovery and experimentation and offered ideas for designing new dimensions into traditional blended programs. The broad range of scenarios we can create for learning are only limited by our imaginations. But as learning executives, we understand the real world where budgets, spending freezes and attention to staff size are the realities of a challenging business environment. As you begin your discussions with stakeholders and business leaders, remember to have your facts at hand and always speak to the business results and cost considerations if you want to make 3D enabled learning a reality in your organization.

Cost Considerations for 3D Learning

The simplest way to evaluate or show the value of virtual worlds is to create a simple cost comparison between what you are doing now (classroom?) and specific 3D programs. It's not the case we prefer to make given the power of the technology, but one most business leaders can easily understand.

If your organization is like many, you still offer up to 70% or more your curriculum in live, instructor-led formats. A good first step when introducing 3D virtual worlds to your organization is to create a simple cost/value comparison between the travel expenses of live classroom or group events and the much lower cost of 3D environments. However, you will also have to have proof in hand that virtual 3D learning retains the rich interactions and sense of presence experienced in face-to-face programs, and can deliver measurable results in terms of outcomes. We will talk more about this step later in this section.

Figure 6.9 offers a simple visual for comparing the cost side of delivery modalities, not unlike those we utilized when first introducing eLearning as an enterprise solution. Do not forget to factor in all the soft costs or intangibles of face-to-face or teleconferencing versus creating virtual worlds in order to make effective comparison.

Once you have "level set" your business audience with some high-level comparative value (plus and minuses; features and functions) of the three modalities, you can "run the numbers." As you can see in Figure 6.10, this simple comparison becomes really powerful when you increase the quantity, frequency or distance of learners. Obviously, the T&E calculation is much higher for live events, but learner volume

	Face to Face Event	Teleconference (on a per learner basis)	3D Virtual World (on a per learner basis)
# of Disperse Learners	200	200	200
Cost of facility or space	Variable	0	0
Cost of technology	Low	Medium	Low – Medium
Cost of travel	High	0	0
Access to Expert	Medium	Low	High
Learner Engagement	Medium	Low	High
Immersive	High	Low	High
Collaboration	Medium	Low	High
Ability to Measure Job Performance Outcome	Low (without extra $$)	Low	High

Figure 6.9. Learning delivery cost considerations.

becomes a significant metric when comparing 3D virtual worlds to teleconferencing/telepresence software when calculated on a per user/per event basis.

Quantifying the Value of the Outcome

Like any other good training ROI calculation, next we must identify the outcome you wish to measure and attribute a value to it. As learning professionals we often attribute the value of learning in terms of degree of proficiency, performance change or mastery. In this section we will look at how to align value to our learning outcomes using business data that has "value" to business leaders.

Figure 6.11 looks at a few of the "best practice" outcomes that work great in 3D environments and the benefits normally associated with them. The trick in this step is to find the data source or correlation to quantify the value of change. Was productivity increased? Attrition reduced? Customer service metrics increased? Much of this data can be found in your organization's ERP, HRIS, CRM or other BI system. Sometimes it can be found in your organizations a balanced score card.

Here are a couple use cases that explain the cost versus value proposition:

Scenario One: Mentoring, Coaching and Team Projects. 3D virtual worlds allow for both passive and active learning, as well as synchronous

Software or Platform	Annual Cost Per User (100-500 unique users)	Comments
Audio Conferencing	$500 - $1,000	Benefits: familiar, easy to use Challenges: more expensive than perception, no visual display of data, limited engagement
Web Conferencing	$500 - $2,000 In some cases additional per event charges	Benefits: familiar, easy to use, synchronized visual and audio Challenges: Costs vary depending on # or hours of use, presentation design, license type must be purchased ahead of time, and costs are variable based on audio preference, duration, and number of people and webcasts. Limited engagement
Telepresence Solutions (Software solutions such as Cisco TelePresence, HP, Unisfair and others)	$20K-$300K installation fees; variable costs per event or user	Benefits: Similar ease of use to webcasting, but more engaging, immersive experience Challenges: High upfront investment, often network and equipment fees, tech support, users required to visit installed sites
3d Virtual Worlds (Second Life, Protosphere, Teleplace, and Avaya Web Alive	$100 - $250K Varies based on platform and custom development costs	Benefits: Immersive face to face experience, multi media file access, full range of learning taxonomy, scalable and varied modalities for real time communication Challenges: Some client side download, variable development costs, tech support, small learning curve for users to navigate environment
3D Virtual Worlds Pay-per-view SameTime 3D VenueGen	20-50K per year Varies upon either the number of users or the number of servers	Benefits: Immersive and rich functionality, but tend to have defined functionality and "look" Challenges: VenueGen a newer product so still evolving. SameTime 3D uses SecondLife viewer and is bandwidth intensive.
Quasi-3D (Unisfair)	50K per year plus media and collateral objects Varies on the number of sets and number of live events	Benefits: Solid history, large numbers, quick to deploy, good measurement capabilities Challenges: Lack of presence, lack of peer to peer or to tutor interaction. Click content instead of move through content. Interactions limited to clickable assets and webcast type events.

*Cost and pricing analysis used in this table reflect information available at the time of submission for this publication.

Figure 6.10. Learning delivery comparison.

Desired Learning Outcome	Aligning to Core Business Metrics
Speed to proficiency or skill	Percentage of utilization Productivity Time from workplace reduction Increase in sales
Compliance based learning	Value of certification or compliance versus lack of Reduction of job site errors or accident rates (Correlated to data is fine – hard to validate) Quality control or other quality metrics Risk avoidance / loss prevention/reduction reduction in law suits
Customer care or other customer facing proficiencies	Sales (before and after product training) Customer service satisfaction or quality metrics Attrition or retention of key customer care employees

Figure 6.11. Learning best practice outcomes.

and asynchronous learning. As in many companies, at our company many of us work at a distance as coaches to our firm's future leadership and resources.

Example: A planning team met in Sametime3D for two sessions. This team of eight would normally travel to meet for day long planning session. Instead, this team met for two 2-hour sessions in the 3D virtual world. During the first session, the team provided updates on their activities. During the second session, the team was able to brainstorm 60 ideas, sort the ideas, prioritize the ideas and then post the ideas onto a virtual action planning wall. Approximate cost of year license and hardware is approximately $50,000. The cost of travel for the team was $15,000 (not allowing for cost of time away from the job). Since this Sametime3D world allows for four simultaneous sessions, this tool could pay for itself in four events similar to this planning session. The key outcome was that both the director and the participants felt that the session was "highly productive." The results of the session were exported to the director via a spreadsheet.

Scenario Two: On-boarding, new hire training. 3D virtual worlds allow for both passive and active learning, as well as synchronous and asynchronous learning. 3D virtual worlds for new hires, on boarding, or other immersive experiences allow learners to explore, discover and engage with other learners as well as master material or new skills.

Example: A Global Business Consulting firm spent $100,000 annually in travel costs and entertainment for multiple on-boarding sessions for new hires. A redesign of the on-boarding approach included webcasts, on-the-job experiences and mentoring, plus facilitated learning activities in the virtual world. Using VenueGen, redesign costs were half of the face-to-face sessions. The key outcome was that the new hires retained the same level of information as previous programs, yet learned to make connections and collaborate virtually. This mirrored their physical-life work requirements for distance networking and team projects. One group reported a very high satisfaction with conducting one-on-one coaching with learners while seated on the virtual sailboat set in VenueGen. A last metric to be discovered over time was to look at attrition of new hires at year one, in comparison to other on-boarding classes.

Performance improvement, speed to competency or proficiency are key metrics for any organization who look at training interventions. Studies have shown that adult learners require engaging interactions in order to improve retention. Our trials with audiences as diverse as groups of professional services consultants or engineers have shown that learning in a 3D environment allowed the enhanced interaction with both the content and other individuals in a unique way due to the "advanced presence"—or sense of feeling like you are really there. Since our 3D environment is electronically generated, the platform allowed for creation of interactions that are both unique and meaningful for the learner, "tactile" or "hands on" experience with tools, equipment or environments and can achieve the desired outcome more quickly because of the immersive nature of the experiences.

On the ACS Island robots or facilitator avatars act in a role, in a staged event, to allow for simulated learning such as modeling a sales call or reinforcing a business process. 3D animated objects, when clicked, reveal layers of information such as the inside workings of a machine. 3D formal classrooms with video screens provide streaming video, electronic presentations and/or software application sharing. Informal discussion areas, such as tables in a garden and a tiki bar on the beach, encourage conversations between avatars for informal learning and knowledge-sharing. The key outcome was that these environments have been able to more quickly allow for recording of the experiences. These environments also allow the training to be "archived" for future users or they can be replayed immediately for assessment by the learner.

Learning reinforced by repetition or role playing. For the life sciences, having the ability to train workers in simulated environments without exposure to risk or danger of mistakes can be invaluable. One of the early Virtual Worlds providers has had great success in training and preparing clinical teams to conduct site visits for clinical studies. The company was able to create repetitive, auditable role-play practice sessions simulating clinical environments to assess competency. The key outcome was that with an average of $250,000 study costs and new product pipeline at risk, ensuring that these professionals meet proficiency and protocol standards was invaluable. This online simulated scenario cost savings does not even take into consideration of the absence of face-to-face training costs.

One last example is our work in the quasi-3D environment Unisfair. We launched an innovation program for 15,000 employees. The program dictated a need for real-time interaction, peer-to-peer interaction as well as discovery and exploration with content and applications were used as part of the program. The quasi-3D environment was quick to deploy, allowed for live and self paced interactions, and enabled access for hundreds of users to "attend" the event at the same time. The environment deployed for under $30,000 and was perfectly suited to achieving the awareness-level learning objectives. More importantly it is still "live" a year later for users to go back in and peruse the virtual conference area.

According to an industry study from former ThinkBalm,

> More than 40% of those surveyed (26 of 66) saw a positive total economic benefit from investments in immersive technologies in 2008 and 1Q 2009. More than 50% of respondents (34 of 65) expect to obtain a positive total economic benefit in 2009. The number of respondents who expect to obtain economic benefit of $25,000 USD or more in 2009 is more than double the number who indicated they achieved this level for 2008/1Q 2009. (Driver & Driver, 2009, n.p.)

One last thought on metrics, Knowledge Exchange. Many best-in-class organizations place a dollar value on knowledge assets or informal learning exchange. 3D virtual worlds are excellent mechanisms to capture and share knowledge and expertise throughout the organization. 3D world also help to track and capture "innovation" as a process and an asset.

WHAT'S NEXT IF 3D IS IN YOUR FUTURE?

3D virtual world learning environments provide exciting and challenging learning environments. The industry is working to address issues for both access to the environment and ease-of-use for the user. Remember to evaluate business objectives against the need for virtual world environments. Do an effective analysis of learning outcomes using some of the models we have provided in this chapter. Consider foremost how to engage everyone in the process—the stakeholders, the instructional designers, the users, and the people they serve as well. Not everyone in the stakeholder list has an avatar today. But, as virtual worlds emerge into the toolkit for learning in the enterprise, they may find themselves designing one soon enough. As the next generation of workers arrives in their digital suits, expect to engage them in new and exciting ways.

In our last thoughts, we offer a few trends that we are excited about, that you may want to follow as well: Thin-client capability and demand will increase. Most of our clients have issues with the thick-client solutions either in terms of performance on company machines or security with ports and software download. The thin-client solutions promise to minimize/mitigate the issues offering a good solution for both IT departments and learning organizations.

Improved Functionality. Thick client solutions will continue to add improvements such as integration of documents in and out of the 3D world; writable and manipulated objects by the learner; video capture for role play replays; and more linking of 3D assets to real world assets for software and hardware training. Meanwhile, thin-client solutions will pursue basic functionality such as: white boarding, application sharing, presenting documents on the fly, effective VOIP and social networking profiling.

Learner Generated Objects. Open source tools like Second Life allow users to generate objects. Most business-to-business tools do not. As learners become more sophisticated in 3D, they will demand to be able to create and manipulate objects. Facilitators and subject matter experts will also want to change the world to represent what they are trying to convey. Virtual sandboxes and/or simplified user object creation tools will come into play.

Cost Model Changes. Most tools are designed for enterprise usage. Since organizations are slow to make significant investments, some worlds are going to a pay-per-use model. As public platform spaces mature (like Reaction Grid's Unity Jibe) more developers will be proficient in building sets and features which will help reduce the investment into building out spaces.

3D on the Go. Mobile technology continues to evolve with some phones allowing for 3D video. For quick reinforcement sessions, coaching sessions, model demonstrations, expert interactions, we might soon be seeing 3D virtual worlds on our mobile devices.

Generic Courseware in 3D. As the tools strive to become more universally accessible (i.e., no bandwidth, port, download or computer processing issues), then off-the-shelf courseware providers will venture into providing 3D learning on a pay-per-view basis for both asynchronous and synchronous learning.

APPENDIX:
LESSON PLAN

Instructor	Date
Kelly Rider	TBD
Course Title Consultative Skills for Account Development	Lesson Numbers 1-3

Instructional Goals

At the end of this course, students should be able to demonstrate asking questions effectively to engage a client, open a dialogue and explore prospective needs for the client in a simulated situation.

Performance Objective

- Provide examples of three types of questioning skills
- Describe personality profiling and how to adjust probing questions to the profile
- Work together in a team to provide a demonstration of an effective questioning interaction
- Assess performance of a team's simulated client interaction in terms of effectiveness for building client confidence.

(Appendix continues on next page)

Rationale

In order to prepare senior managers for partner responsibilities, managers will need to gain skills to lead pursuits of new business by effective inquiry into client needs. This course on consultative skills provides the learners with the opportunity to both learn and practice effective questioning.

Lesson Content

- Effective questioning techniques
- Open-end questions and closed-ended questions
- Business questions and personal questions
- Blending effective questions
- Personality profiling and effective questions

Instructional Procedures

- Opening event: demonstration of a poor client interaction
- Teaching procedures: brief lecture, investigative inquiry, team discussion and demonstration, assessment of demonstration
- Formative check: polling, short quiz at conclusion of first session. Evaluation report at end of course.
- Student Participation: student lead team discussion, self-exploration of additional content, team demonstrations and student assessment of team's performance
- Closure: Summarize key observations from team demonstrations an discuss next steps to transfer skills to use on-the-job

Evaluation Procedures

- Level 1: postcourse evaluation
- Level 2: skills demonstration and postcourse assessment
- Level 3: postcourse supervisor observations

Materials and Aids

Students will need access to the virtual world platform via their personal computer. Learners will need headsets for VOIP online discussions.

Prework

Time (in CT)	*Activity*	*Purpose*
9:00-10:00	• Avatar newbie training. • Location OLIVE desert set • Test and retest sound/headsets • New to the environment learners to practice communicating, movement, and gestures	• Enable new participants to effectively participate in the virtual world. Allow everyone to test audio via computer headsets

Lesson 1: Day 1—Large Group and Teams

Time (in CT)	*Activity*	*Purpose*
10:00 to 10:14	• Social Networking in 3D warm up activity • Learners associate LinkedIn URL with personal avatar. • Learners mingle with two other people and click on their avatar's URL	Ice breaker activity
10:15 to 10:25	• Video Observation • Critique consultative style through polling	Awareness for the importance of effective questioning in client interactions
10:25 to 10:40	• Consultative Skills Presentation, group discussion and polling • Continuum • Sneaker List • Building Trust	Awareness of effective questioning

(Lesson 1 continues on next page)

Lesson 1: Day 1—Large Group and Teams Continued

Time (in CT)	*Activity*	*Purpose*
10:40 to 10:45	• Team Assignments • Link to asset with task description • Link to backpack with effective questioning information objects • Link to Ning (blog) for team input	Prepare teams for assignments
10:45 to 11:45	• Self-organizing team work session—use of breakout • Teams explore questioning techniques on own • Teams organize for team readout of content and plan for demonstration	Collaborative learning and exploratory learning of key content topics for effective questioning
11:45 to 12:00	• Discussion of key content • Confirmation of team assignments • Link to short quiz on question content	Verify teams on track during self-discovery and team sessions

Lesson 2: Day 2—On Own and Teams

Time (in CT)	*Activity*	*Purpose*
2 hours	• Self-discovery of content objects in world • Team preparation for report and demonstration • Links to Backpack of knowledge objects • Updates to Blog • Self-organizing	Collaborative learning

Lesson 3: Day 3—Large Group

Time (in CT)	*Activity*	*Purpose*
10:00 to 10:10	• Recap of prior session discussion	Reground learners from first session content and highlight quiz results
10:10 to 10:30	• Team • Team demonstrations	Application of questioning skills in practice client situations
10:30-11:00	• Replay team demonstrations for debrief and feedback • Use polling	Assessment of questioning skills during simulated client interaction

11:00 to 11:30	• Discuss use of questioning skills back to job and specific client prospects	Transfer learning to on-the-job
11:30 to 11:45	• Summarize key points and wrap up	Clarification and validation of key learning points
10:45 to 11:00	• Evaluation and Assessment • Link to LMS for postcourse assessment • Link to Survey Monkey for postcourse evaluation	Evaluation

Six weeks out

Postcourse evaluation of question skills performance sent to supervisors. Feedback and reminders sent to learners.

REFERENCES

American Society for Training & Development. (2010). *Instructional systems design: Today and in the future*. Alexandria, VA: American Society for Training & Development.

Driver, E., & Driver, S. (2009, May 26). *Immersive Internet Business Value Study* [Web log post]. Retrieved from http://www.thinkbalm.com/2009/05/26/thinkbalm-publishes-business-value-study/

Kapp, K., & O'Driscoll, T. (2010). *Learning in 3D: Adding a new dimension to enterprise learning and collaboration*. San Francisco, CA: Pfeiffer.

Massey, A. P., Montoya, M., & Bartelt, V. (2011). Cross-university collaborative learning: extending the classroom via virtual worlds. In G. Vincenti & J. Braman (Eds.), *Multi-user virtual environments for the classroom: Practical approaches to teaching in virtual worlds* (pp. 333-347). Hershey, PA: IGI-Global.

Senge, P. (2006). *The fifth discipline: The art & practice of the learning organization* (Rev. ed.). New York, NY: Doubleday.

Wexler, S., Hart, J., Karrer, T., Martin, M., Oehlert, M., Parker, S., Schlenker, B., & Thalheimer, W. (2008). *E-learning 2.0: Learning in a Web 2.0 world*. Santa Rosa, CA: E-Learning Guild.

Wisniewski, B., & McMahon, K. (2005, March 29). Formalizing informal learning. *Chief Learning Officer*. Retrieved from http://dns2.mediatecpub.com/articles/view/3673

PART IV

IMMERSING THE AVATAR

CHAPTER 7

SALT MARSH DYNAMICS ~ A PROBLEM-BASED LEARNING SCENARIO

Heidi Trotta and Marian Glenn

Multiuser, persistent virtual environments foster community and situated learning by supporting immersive, collaborative, extended experiences through simulations similar to those found in real world contexts. This project developed a problem-based learning environment using the online virtual world of Second Life, aimed at undergraduate environmental science courses. Both process and content promote collaborative learning, advanced problem solving and the application of research and scientific thought as students gather, interpret and analyze data to solve an ill-defined real-world problem. Specific scaffolding, modeling of process, and frequent reflection helped students advance from reliance on miniscripts acquired through past content-driven courses, to search for new information and to observe characteristics of the scenario beyond what they expected, based on superficial acquaintance with real world sites.

The Challenge

Effective education helps students to *think*. The ability to think critically, reason in a variety of ways and solve ill-structured problems creatively has become an essential skill needed to succeed in today's world.

Engaging the Avatar: New Frontiers in Immersive Education, pp. 189–215

Models of critical thinking provide a means for evaluating problem-solving maturity. Susan K. Wolcott (2005) describes the development of critical thinking with a series of shorthand phrases, remarking that half of the students entering college are "confused fact-finders" lacking the skills to put facts together to solve problems. At the end of college, the majority have advanced to the next stage, "biased jumper." They jump to a conclusion, based upon previously-held biases and then look for evidence to support it. Few master the next stages, "perpetual analyzer," which involves the critical analysis of data and assumptions, and "pragmatic performer," making well-founded decisions based on objective analysis. Even fewer reach the higher order thinking skills of the "strategic re-visioner." The challenge to educators and instructional designers is to create opportunities for students to mature in their critical thinking skills.

The Theoretical Perspective

The "knowing" of concepts, definitions and facts has dominated traditional undergraduate education. However, educational theorists for years have been questioning the very definition of what it means to know something and how knowledge is actually acquired. John Dewey in his 1916 book, *Democracy and Education,* remarked that understanding is not obtained by "telling" and "being told," but instead is the result of an active constructive process. Brown, Collins, and Duguid (1989) argued that knowledge is a "tool," understood only through its use. Knowledge is situated through experience, by "doing," instead of rote memorization of facts and concepts in isolation. In the process of "doing," students come to understand the world through their own models and representations obtained through personal experience and reflection (Duffy & Jonassen, 1992). They actively construct their knowledge rather than passively receiving it from others.

In recent years these situational theories have been moving from learning being a solitary endeavor to one where knowledge is acquired through shared experiences and social contexts (Lave & Wenger, 1991). With this anthropological approach, learning takes place in a participation framework where it is distributed among the participants similar to the reciprocal nature found in apprenticeship between the apprentice, the master and his or her workshop. In the process the learner adopts the identities and "habits of mind" of that practice.

Learning theory is important because what one believes about learning and acquiring knowledge will influence the type and nature of the learning environments he or she designs. A well-designed problem-based learning environment models real life with contextual narratives that situate concepts in practice (Brown, Collins, & Duguid, 1989). It asks students to articulate the data they use to solve the problem, and to develop methods to test their proposed solutions. The learning environment guides students to analyze theory in practice, develop "ways of knowing" and sharpen critical thinking skills.

The use of problem-based learning has been widespread in the field of medicine for many years (Williams, 1992). Characterized by small cooperative learning groups, learners are responsible for developing their own position on an issue or solution to a problem as opposed to case-based learning where one studies solutions others have identified. In problem-based learning, the problem comes first. This is in sharp contrast to the traditional strategy of teaching concepts and facts first to be later followed by a lab or "end-of-the-chapter" problems for application. In problem-based learning students must identify what they know, what they do not know and what they need to know to solve a problem that has no one "correct" answer. Through the process of identifying the problem and recommending solutions, students are exposed to a variety of viewpoints, acquiring new perspectives and increasing their understanding of the complexity that surrounds seeking solutions to real-world problems. A seemingly simple issue becomes complex as decisions impact stakeholders in different ways, both positively and negatively. In this method, the instructors become facilitators of learning instead of presenters of knowledge, as they shift from teaching content to engaging the learner in authentic tasks through which they are most likely to acquire the desired concepts and skills (Barab & Duffy, 1990).

Problem-Based Learning and Virtual Worlds

Advances in technology have opened up new forms of instructional delivery while providing unlimited access to content. Many of the popular online multiuser games have their foundation in problem-based learning as players gain skills and resources to engage in quests and various challenges with other players (Gee, 2003). These venues allow the participant to become situated within the play space, taking on the identity of a virtual character. It is through participation that the player becomes part of the world itself in which he gains knowledge though actions and consequences. Key to virtual online games is their persistent physical and social nature that evolves over time in much the same way as any learning community (Squire, 2006).

Unfortunately, these applications are very expensive to produce and beyond the skills of most educators and designers. However, virtual worlds such as Second Life offer educators the tools to build graphically rich, interactive problem-based learning environments, similar to those found in online gaming. Using effective design principles, one can provide clues to encourage students to intelligently proceed through a problem based on the layout of the material environment and the objects in it. Through this embodied experience, students discover that objects, artifacts and the design of the environment can store knowledge. A well-constructed environment provides subtle clues and bits of evidence that offer students the opportunity to practice skills in inductive reasoning. They become more proficient in "reading" a space. Because the virtual world supports a variety of media, meaning and "ways of knowing" can be built through multiple modalities such as image, action, and sound, impossible with the traditional "paper and pencil" case. This supports a variety of learning styles and challenges students to use observational skills related to all their senses to interpret information.

Applicability to the Ecology Classroom

A key objective in science education at the college level is to engage students in real world scientific practice (i.e., careful, subtle observation, data analysis, hypothesis generation and controlled testing, all carried out within the context of a socioscientific community). Scientific practice begins with an appreciation for the nature of complex systems and the fact that real-world problems have complex causes and solutions. These system-level dynamics function across disciplines often involving a balance of ethical, economic, political and scientific factors and interests.

Multidisciplinary problem solving is a necessary skill, yet is seldom practiced in discipline-based courses. Science courses focus on understanding scientific knowledge and perspectives, whereas parameters relevant to other disciplines, such as political, social or economic factors may be ignored or dealt with superficially. In addition, students have many pre-assumptions, based on their experience in the world. When faced with problems to solve, they often jump quickly to a conclusion, by interpreting their limited observations, such as what they can physically see or have heard through mass media. For example, in searching for a cause for ecosystem decline, rather than critically analyzing all possible causes, both direct and indirect, they see evidence of litter or medical waste and presume them to be causative agents. As a result, their solutions to environmental problems tend to be simplistic, usually lacking depth of knowledge, appreciation of different perspectives as well as creativity in integrating multiple causes.

To bridge the gap between theory, lab exercises and real world practice, the course may include field trips so that students may learn from nature in their own setting. Scheduling constraints make these short, intensely guided experiences. Students tour an area, learn about vegetation and animal life, discuss effects of such events as global warming and weather on the ecosystem, demonstrate simple tests and field measurements, and write up some "reflection questions" to tie together theory and field observations. The quality of these experiences tends to vary greatly from student to student, although most students find field work enjoyable and the experience leaves a more lasting impression than a classroom lecture. However, students rarely get the opportunity to solve an actual problem that involves working collaboratively with a multidisciplinary team of scientists, as well as dealing with the specific concerns of a stakeholder group of citizens. Using problem-based learning, students are guided to work as a team of investigators, finding their own strengths, as they learn from each other to investigate, and to engage in practices that are consistent with those of real world practitioners.

Using a virtual world such as Second Life for creating a problem-based scenario for the science classroom, offers rich collaborative inquiry-based explorations through which students can apply principles and content covered in class and during the field trip. Educators simulate real-world problem situations that would normally not be possible or accessible to students (Barab, Sadler, Heiselt, Hickey, & Zuicker, 2006; Dede, 2005). By taking advantage of persistent and social affordances in virtual spaces, students develop skills in team building, creative problem solving and critical thinking with enough time to acquire the "habits of mind" that are the basis for authentic learning.

The Salt Marsh Dynamics Environment

This project took place in an online 3-D, multiuser virtual environment known as Second Life, created by Linden Labs in June, 2003. Residents, by creating a virtual representation of themselves called an avatar, explore the environment, meet others, attend special events and create and sell virtual goods and services. To enter Second Life, one downloads the free application viewer to their computer and launches the application.

Everything in Second Life has been created by residents through three-dimensional modeling tools based around simple geometric shapes built into the software. Functionality can be added to objects within the environment by embedding special scripts within them. For example, by interacting with an object, an avatar can receive an informational note card, be directed to a website or play streaming media. More complex elements and interactions can be created in outside programs to be imported into the environment. Unlike traditional online computer games, Second Life does not run on an underlying game engine, being instead, more of a social online *space* known for its creative potential.

Educators have been exploring the potential of Second Life as a learning space since its inception. Well over 200 colleges, universities, libraries and museums have a presence within the environment (Harrison, 2009). Educational subjects such as Chemistry, English, Language Instruction, and the Arts have been taught in-world, and entire continents are devoted to science and technology. For the first time, educators have the tools to actually build graphically rich, interactive environments similar to those of online gaming. In December of 2008, the United States Air Force launched MyBase, a Second Life island devoted to Air Education and Simulation Training.

Project History

In 2006, Seton Hall University partnered with the New Media Consortium to develop plans for the creation of virtual environmental learning spaces inspired by actual endangered ecosystems within Second Life. Professors Marian Glenn and Martha Schoene in the College of Arts and Sciences, along with the instructional design team, identified flora and fauna and developed learning activities with the intent to pilot the effectiveness of these spaces and experiences on teaching and learning in Biology, Earth Science and Ecology courses on the Seton Hall campus. Design emphasis centered on promoting collaborative learning, advanced problem solving and the application of research and scientific thought though authentic real world contexts. In 2008 the team was awarded a New Media Consortium Virtual Learning Prize, thus starting the development of *Salt Marsh Dynamics,* on Pirate Island, virtual property owned by Seton Hall University. Professors Marian Glenn and Martha Schoene served as content experts, providing resources, writing narrative and composing assessment questions. The instructional designer served as project manager, Second Life content developer, multimedia creator, and inspirational coach.

The Design

Designing the Salt Marsh Dynamics learning environment was similar to building a house from the ground up. The project goals and outcomes were identified early in the process and served as a guide to determine what creative possibilities to include and what to set aside. This provided an initial structure and framework for design of the scenario, environment and user experience.

The following project goals and key learning outcomes (objectives) were identified.

PROJECT GOALS

- Provide students with a problem that is meaningful, interesting, relevant, engaging and ill-defined with no right-wrong answers.
- Deconstruction of students' pre-assumptions to allow for the reconstruction of new, more complex notions on the workings of an ecosystem.
- Focus on the process of learning (collaboration, scientific inquiry, inductive reasoning) over learning content.
- Provide opportunities for the acquisition of century skills such as knowledge gathering, collaborative team building, establishing knowledge communities and creative problem solving.
- Encourage diversity of thought and multiple perspectives.
- Determine and provide appropriate scaffolding.
- Present information in a variety of formats (multimedia, print, online resources).
- Integrate into existing curriculum, supporting other learning activities such as annual field trips.
- Apply successful online multiuser virtual gaming and epistemic game theory to virtual world environment where appropriate.
- Provide continual student assessment, explore group assessment.
- Establish best practice.
- Determine project assessment early to set up possible comparative studies.
- Publish findings.

Key Learning Outcomes

At the completion of the case-based activity in Second Life, learners would be able to …

- Gather, analyze, manage, and evaluate data from numerous sources to include historical and provided data, environmental observations, and outside resources to substantiate and support arguments.

- Apply scientific principles and concepts to real-world contexts.
- Work collaboratively to exchange ideas, information and perspectives.
- Debate, discuss and arrive at a group consensus, identifying and evaluating possible solutions.
- Create and present a plan of action to alleviate/solve present environmental impact and restore area to its natural state, including methods to assess and modify the plans scientifically.
- Demonstrate the acquisition of "ways of thinking" found in scientific and research communities.

Problem-based instructional models based in other online virtual worlds such as *Quest Atlantis* (http://atlantis.crlt.indiana.edu/) from Indiana University, *River City* from Harvard University (http://muve.gse.harvard.edu/rivercityproject/) and *PowerUp* by IBM and the New York Hall of Science (http://www.powerupthegame.org/) were explored to get an overview of their organizational and educational framework, design, scaffolding and embedded learning activities. These educational learning environments served as inspiration and examples of best practice. Also examined and played were the commercial online games of *Nancy Drew Mysteries* (http://www.herinteractive.com/) and *Myst* (http://www.mystworlds.com/us/), looking at user interaction with objects and overall experience.

Knowing that the project was to be instructionally grounded as a problem-based scenario, a storyline was written to support (1) the concepts of eutrophication, water quality and system dynamics (2) build skills in hypothesis generation, scientific inquiry, water quality/data analysis, and socioscientific reasoning (3) encourage the development of "reading" an environmental space. It was also important that the narrative be compelling enough to draw the students into the experience, motivate their interest and provide a realistic problem-based scenario of an ecological issue in our geographic area. Through character development within the story, students were expected to develop an appreciation for the complexities involved in drawing scientific conclusions, applying scientific knowledge to management decisions and public policy and balancing economic, social, and political factors with the conclusions of scientific investigations.

The storyline begins with a mysterious fish kill in the virtual marsh on Pirate Island. Local officials have called in a team of experts to perform follow up tests in the bay, ocean, inlet and marsh with volunteers needed to help in the ongoing investigation. Further details are provided as students navigate and click on in-world, nonplayer characters, four scientific experts and four stakeholder-citizens, each with a particular set of information and clues about the possible causes of the fish kill. In creating the story plot, the developers listed all the information desired in the environment, "mapping" the information onto the eight characters, including some repetition. When there was overlap, there was also some discrepancy in the results that students would be challenged to deal with. In this way the storyline formed a complex web that also served as a "watertight" information package. In creating the mystery this way the developers actually modeled how a team of experts operates in solving a mystery. The environment was embedded with artifacts (i.e., field guides, graphs, charts, scientific journals, and newspaper articles) some that directly related to the scenario and some with irrelevant information and red herrings.

Once the storyline was drafted, project tasks were identified which fell into three main design categories: (1) scenario (2) natural environment itself and (3) learning experience. Specifics ranged from designing the graphics (i.e., logo and banner) to brand the project, creating characters and their narratives, building the learning environment in Second Life to finding subject content with supplementary resources. Challenging, but essential, was the interrelationship between these components, requiring the team to continually review and reassess process, components and material to further support and define the student experience. For example, once the storyline and character narratives were defined, a "blueprint" was created for the Second Life environment that included land topography, watershed boundaries, buildings and learning areas to support the narrative, hold content, and provide easy avatar movement. This environment was then changed and adapted as objects were added and revisions were made to the overall experience.

Participants and Setting

Students were added to the project in the fall 2009 undergraduate course BIOL2341AA- Ecology taught by Dr. Marian Glenn, Professor of Biological Sciences in the College of Arts and Sciences at Seton Hall University. Dr. Glenn was a main content contributor and co-author on the project. The class met on Tuesday afternoons for 75 minutes, and on Thursdays for 4 hours. The class was comprised of 18 junior and senior Biology majors who had previously done other project-based, collaborative, inquiry based work in core courses in the major, however this project was their only ecology course and their first experience in Second Life.

This face-to-face course was held in a workshop format. Discussions and class activities covered assigned readings. There were four field trips to nearby natural areas including Horseshoe Cove Salt Marsh at Sandy Hook

National Recreation Area. Class assignments included keeping a nature journal for reflection along with special individual and group projects. Assigned readings included *Life and Death of the Salt Marsh*, by John and Mildred Teal (1969), a classic natural history text focused on salt marsh ecology. This book was also used by the instructional designer of the project as a reference for creation of the Second Life marsh environment.

BIOL2341AA course goals included:

1. Cultivate and apply the skills of a biospheric scientist: analysis, imagination, compassion.
2. Practice observing nature with a "deliberate gaze."
3. Apply quantitative analysis to ecosystem problems.
4. Apply knowledge of salt marsh ecology to management issues.
5. Evaluate multidisciplinary synthesis applied to biospheric ecology.

Project Integration Into the Curriculum

Along with the discussion of *Life and Death of the Salt Marsh*, students had a guided field experience at the actual salt marsh, led by an educator from the New Jersey Marine Sciences Consortium. Students conducted water quality testing and practiced identifying local plant and marine species, as well as learning general information about estuary and salt marsh ecology. This took place in mid- September, prior to the virtual marsh learning activity. Following the field trip, students were sent an e-mail that described a mysterious fish kill that had occurred in the virtual salt marsh at Pirate Bay. It concluded with an invitation to participate in the investigation (i.e., the scenario) currently underway by the virtual professionals and scientists. Students were provided two initial contacts within the virtual environment to get started on their inquiry. They were also informed of an orientation to the Second Life environment that was to be held during a lab period. At the orientation students created their avatar and practiced basic Second Life skills needed for participation in the problem-based learning scenario.

Working in groups of 4, students were given 2 weeks to explore the virtual salt marsh and develop their initial hypothesis and supporting arguments about the causes of the dead fish, to be presented to the class. After each group presented their findings, the instructor facilitated a general discussion. Each group then had the opportunity to revise and refine their original hypothesis. A few weeks following, groups submitted a final report which included their final hypothesis, justifications and reasoning, along with plans to restore the environment to a healthy state and prevent future fish kills. Groups presented their ideas orally with classmates having the opportunity to question each other's conclusions and offer alternatives. To wrap things up, the faculty member revealed the rationale on which the case was based, including the many bits of evidence, both relevant and irrelevant placed into the scenario for them to discover. This "answer key" revealed the complexity of the cause-and effect relationships in this ecosystem, as well as explaining the scientific consensus about the true causes of fish kills, rationalizing the mysterious observations presented in the case.

Assessment

An analysis of the outcomes of the project was made from student's written and oral presentations, the instructional designer's anecdotal field notes, student focus group meetings conducted by the instructional designer, and a survey of student opinions, with the intent to evaluate the learner experience and to document the effectiveness of the virtual environment in teaching critical thinking skills. Faculty and instructional design staff reviewed student feedback and reflected on the activity, identifying processes that went well and those needing revision. Following the design-based research model approach (Reigeluth & Frick, 1999), we looked at the data collected in respect to four goals: (1) recording an account of the experience (i.e., in terms of aspects that went well or needed improvement) (2) identifying possible solutions to enhance the learning outcomes (3) devising a means to assess the effects of the changes (i.e., "next steps") and (4) making recommendations to others considering creating similar environments.

Results

The input and observations are grouped into three categories: (1) the design (2) the student experience and (3) evidence of learning. There is some overlap and interdependency however these divisions helped to organize the findings.

The Design refers to the actual layout of the Salt Marsh Dynamics learning environment within Second Life and the functional artifacts contained within. The focus here is whether educators, with limited resources and experience can build authentic problem-based scenarios within this venue. The results of this project showed that it is indeed possible to create this type of learning environment, with some scripting and building, on a modest budget.

The largest expense in Salt Marsh Dynamics is the rental or purchase of space within Second Life. Since Seton Hall was renting a full "sim" (256 x 256 meters) prior to Salt Marsh Dynamics, and space was available, this was not an issue. However, for others considering such as project, Salt Marsh Dynamics was built on a Quarter Sim (128 x 128, 3750 prims) renting for U.S.$1500/yr (February, 2010), from New Media Consortium (NMC) Virtual Worlds. For those on very limited funding, free land rentals do exist from other educational institutions and much smaller parcels are available from the NMC starting at $100/yr. The $5000 grant from the NMC Virtual Learning Prize was spent on scripting for the clue giver boxes which were designed to help students identify important evidence in the case. Total purchases of objects for Salt Marsh Dynamics totaled U.S.$400 with many objects being obtained for free or little cost. Some editing was done to customize objects for the environment with only a few made from scratch (i.e., heron nesting structure).

All objects worked as expected and students were able to view pdfs and website material, however they struggled with the clue giver and guided question boxes with over half finding them difficult to use even though directions were provided. These objects were meant to help students focus on important clues and evidence provided either in the environment, character narratives, or artifacts. The clue giver, indicated by the letter "I" for information, worked on individual avatar proximity and provided a different clue in the chat window each time the avatar approached the object (Figure 7.1). It was possible for each member of a group of students exploring the area together to get a different clue that could be shared with other team members.

The guided question boxes provided a set of multiple choice questions that the student answered through chat. These were designed to help students see the relationships between the artifacts and scientific theory. Feedback was provided for each question, including the correct answer and reason that the preferred choice was the best (Figure 7.2). The limitation to this box was only one avatar could interact with the guided question box at a time and once they moved away from the box, the question period ended. Also, it was difficult to send student results to the faculty member in any readable form making it difficult to see if participants were making necessary connections in the environment.

With the most recent upgrade to the Second Life viewer, viewer 2, initially released in February 2010, there is now the ability to display third-party software tools on a Second Life object (i.e., question box). This upgrade would eliminate the scripting problems and limitations experienced, making them easier for student use and capture. The clue givers and guided question boxes could also be used during the Second

Figure 7.1. Clue Giver object provides key information.

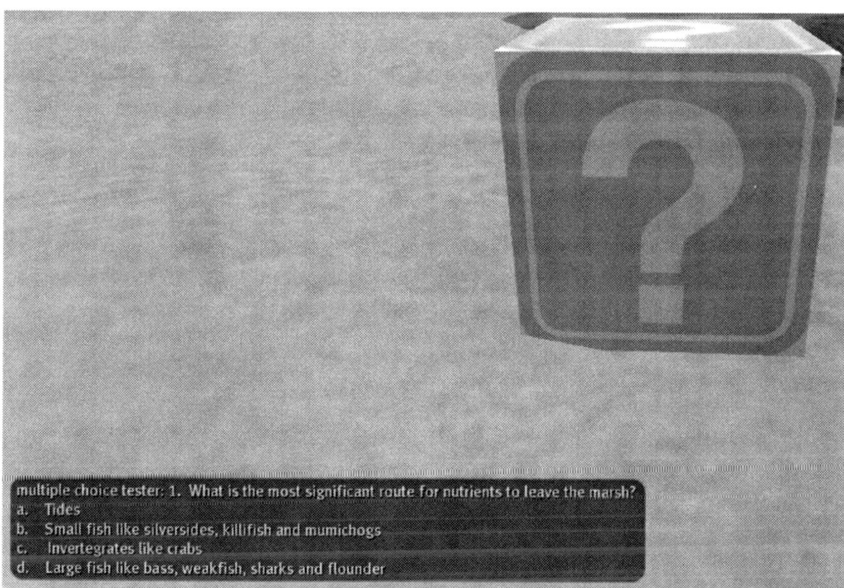

Figure 7.2. Guided Question Boxes helps students understand theoretical concepts.

Life orientation so that students get used to them, and understand their purpose and centrality to the experience.

The topography of the Second Life salt marsh area was inspired by photographs of Horseshoe Cove salt marsh, Sandy Hook, New Jersey, taken by Professor Martha Schoene. The layout and design of the virtual environment was built by the instructional designer. Although time-consuming to plan and build, the virtual environment looked surprisingly similar to its real counterpart. Students overwhelmingly felt they could make a cognitive connection between the two and enjoyed that aspect of the experience.

As expected, students still experienced a delay in objects appearing within the environment known as "rezzing in," associated with using Second Life. They complained about the poor graphic resolution on their university laptops which influenced their ability to visually view the environment and interact with it. When asked what would be one thing they would change in the virtual environments, responses included better graphics. This issue has always provided a barrier to using virtual worlds in the classroom. In the *NMC Two Minute Survey on What's Happening in Virtual Worlds* (February, 2010) results indicated university infrastructure needed to support the application is still an issue. Virtual world environments perform best on a hard-wired Internet connection however the processing power and graphic card capability of the computer is also a contributing factor to user experience.

Figure 7.3. Salt Marsh on Seton Hall University Island in Second Life.

When looking at the Student Experience, the ability to navigate the environment and interact with objects within it was definitely a contributing factor to satisfaction. Some students felt the orientation was not enough to pick up the Second Life skills they needed stating that navigating their avatar was one of the aspects of the activity they found the most difficult. Since the majority of the students indicated they explored the environment individually, they were not able to help each other with basic Second Life functions throughout the semester. Perhaps adding a scavenger hunt to the orientation within the virtual marsh would help students practice navigational skills and provide a more focused, structured introduction to the activity itself. Exercises could also be built that would encourage students to meet in-world together to perform specific tasks and then share outcomes or methods with each other.

From results on the survey, the majority of students seemed to feel their group worked well together and they enjoyed this process with an added social benefit. However, the groups also struggled in knowing exactly what they were supposed to do once they had interviewed their two characters, if they could find them. It was difficult for them to "decide where to go and what to do." Confused, the majority of students needed the scaffolding provided in the face-to-face sessions to organize themselves. This difficulty with the lack of structure is typical of that experienced by students in other problem based scenarios as reported by Williams (1992). Roger Schank (1996) proposed that we learn from our past experiences by forming scripts that we take into the next situation. These scripts are composed of a number of microscripts that can be quite unconscious. One could say in our project that some of the students here were lazy learners. However an argument could be made that it was not entirely their fault. They were using microscripts formed from more linear educational experiences. Regardless, students felt they needed more background on the scenario and what they needed to do.

Project background could be easily provided by adding an anchoring introductory movie which would set the scene, explain the challenge and provide suggestions as to "next steps." Anchoring movies invite students to engage in the problem and provide a "macro context" (Cognition and Technology Group and Vanderbilt [CTGV], 1993a). They help learners to see the problem as one worth solving while providing important background context and direction. Anchors equalize the preparation of the students for the project and help them set goals within the context of the problem, analogous to seeing the "whole task" rather than only pieces (Barab & Duffy, 2000).

The ultimate success of the project is assessed by evidence of learning. Did the students achieve the learning outcomes identified for the project even though they may have struggled through the process? The initial and final group presentations offered the best opportunity for confirmation. It was during this time that teams presented their hypothesis and supporting arguments along with making recommendations to restore the marsh to a healthy state. Students had the opportunity to compare the differences in their conclusions and receive input from others regarding their hypothesis. This resulted in a rich, scientific discourse, ending with the students experiencing their "ah-ha" moment as the complex set of causes of the mysterious happenings was revealed by the faculty member. During the focus group session, students commented that during that time, everything fell into place for them.

With that said, it was clear during the presentations that students did not distinguish well between relevant and irrelevant data nor were they sure how to organize information or evaluate its applicability to their hypothesis. Students also did not pick up clues placed in the environment as well as originally envisioned. Most did not understand the significance of important environmental clues and failed to pick up on small details. For example, few noticed the abundance of jelly fish or that the microbiologist was unfamiliar with the marsh habitat, necessitating the need for his observations to be confirmed. Others fell for the "red herrings" embedded in the case such as 20-year-old storm data, and reports of horses on the beach. Student comments on the survey revealed they felt they should have been told to evaluate information provided and to search for more clues.

This presents an interesting challenge to the developers. Students relied heavily on the miniscripts that they had picked up through past content-driven courses. They implicitly thought that if the information was provided, it must be important and relevant. Instead of looking critically at the environment, they noticed the things they expected to find, and failed to appreciate the significance of unexpected observations. The guided questions and clue givers scattered throughout the environment were designed to help them focus on information important to the case. However students failed to evaluate their hypothesis in light of that information, relying instead on the scripts told to them by the character narratives. They were more attuned to the "lecture" method than the "discovery" method, to being told by a so-called "resident expert," than to investigating for themselves.

In redesigning the experience, this would be a perfect opportunity to teach inductive reasoning, the value of assembling small bits of data to form a hypothesis, which is an important step in creating a scientific theory. Inductive logic looks at the relationship between a set of facts to explain an observation or experience, stating some degree of probability for the conclusion. For Salt Marsh Dynamics, students would gather as much information as they could (i.e., gather widely), and only then put it together to form a hypothesis. This would provide students with a reasoning process to use to solve the problem. Inductive reasoning could either be covered in class by the faculty member or inserted in-world though a virtual agent (i.e., Bot Buddy). In the form of a Scientist, the Bot could serve as a "guide on the side" to help students think through an effective method to solve the problem. Uses of virtual agents are based on the cognitive apprenticeship framework which emphasizes learning at the elbow of an expert who is present to coach and model the cognitive activity (Collins, Brown, & Newman, 1989). The new Second Life Viewer 2 opens additional options for creating the agent in third-party software and importing in-world perhaps in the form of small videoclips.

Also needed are better opportunities to capture what students are thinking as they work through the scenario to gauge their progress, provide "just in time" teaching if needed and identify areas for future development in the environment. This could be done with an online field journal or workbook and a series of reflective assignments. Also helpful would have been a grading rubric for group presentations. This would serve to connect the project objectives, reasoning process and case scenario, helping students see the salt marsh virtual activity more holistically.

Conclusions

Salt Marsh Dynamics demonstrates a real-world, problem-based scenario in a virtual world (Second Life). The scenario is both engaging and instructional. With a little money, elementary scripting and building experience, and a sprinkling of creativity, the virtual environment can look similar enough to an actual location visited on a field trip to expand the experience and create a learning community within a classroom. However, building an information-studded virtual environment makes various activities possible but does not guarantee that students will pursue them (CTGV, 1993b). In addition to the venue, scaffolding has to be provided to identify and model the processes needed to help students develop team building and creative problem-solving skills. Real world instructors may provide such scaffolding, but another approach is to build it into the virtual scenario.

Students need to be guided to develop new expectations for learning that involve deeper critical thinking about information, information sources, and greater reliance on testing inductively-generated hypotheses from evidence they collect. They need to be guided to seek out data beyond what is presented or what they can physically see. The detail-oriented ability to "read an environment" is impeded by students' microscripted learning styles, and revising these presumptions requires guidance from the instructors and/or scaffolding within the scenario. Salt Marsh Dynamics would better facilitate new learning styles with more scaffolding to integrate data collection and theory-generation, facilitate creative collaboration, and guide effective reflection. Opportunities to practice skill development need to be carefully embedded with the expectation that the learner does not currently have them.

The students in the fall 2009 pilot noticeably wanted to be "told" about the nature and design of the instruction. However one might argue that this clearly shows a need for more problem-based learning in the classroom. Understanding the complexity of real world problems requires a higher level of critical thinking. Not all information from sources is valid or appropriate. Students become accustomed to linear content-driven courses which provide little experience in physical-world problem solving. To develop into confident problem solvers key proficiencies include: observation and data-gathering, creative yet skeptical inquiry, openness to teamwork and collaborative solutions. We found that these higher-order skills need time and reinforcement to develop. A project such as this can assess where students are in this continuum and provide opportunities for further development. However it is important for students to explore these issues in greater depth as their problem-solving skills mature and they become more knowledgeable about the complexities of ecology.

Salt Marsh Dynamics proved to be an excellent tool in both supporting situated learning and assessing student expertise in solving real-world problems in undergraduate science. In the next iteration, we plan to overlay our well developed content layer with an improved process layer to help students look more critically at a problem using inductive reasoning. We intend to also provide better anchoring devices to enable students to see the "big picture," capture student thinking for "just in time" instruction and develop a grading rubric for student presentations.

In closing, Salt Marsh Dynamics demonstrates the value of Second Life as a setting for a virtual world scenario that simulates real world problem solving. It is hoped that others will look beyond the science model presented here to consider designing scenarios in other contexts. A discipline is "practiced" not in isolation but in relation to other contributing agendas (i.e., social, political, and economic) which is how content is experienced in the real world, supporting the situated learning theory presented in the paper. Technology has made it possible for us to bring this real-world into the classroom through virtual worlds; however, students are also likely to bring their physical-world presumptions into the virtual world, so technology alone does not ensure development of new, more critically-attuned learning styles. Learning environments need to be carefully constructed to allow for the deconstruction of students' pre-assumptions to allow for the reconstruction of new, more complex notions on how things work.

ACKNOWLEDGMENTS

The authors of this chapter gratefully acknowledge the financial support of the New Media Consortium and the research expertise and collaboration of Martha Schoene, Physics Department, Seton Hall University.

APPENDIX: SALT MARSH ECOLOGY LESSON PLAN

Learning Objectives

At the completion of Salt-Marsh Dynamics ~ A Case-Based Scenario, students will be able to:

- Navigate the virtual world of Second Life using their own avatar
- Collect data from multiple sources and compare results
- Evaluate the relevance of their data as evidence for explaining the causes of a complex, ill-defined environmental problem (i.e. fish kill in a salt marsh).
- Develop an evidence-based hypothesis about the cause of the fish kill
- Engage in group discussion to write a consensus report
- Present conclusions and recommendations for discussion and critique by classmates
- Evaluate and revise conclusions and recommendations after class discussion
- Hold informed discussions about human activities that affect the healthy function of a salt marsh

Introduction to the Case

Students need some familiarity with the ecology of a salt marsh before engaging in the Virtual Salt Marsh exercise. If they have not had a field trip, they can read about salt marsh ecology, or do some research on their own. For example, *Life and Death of the Salt Marsh*, by John and Mildred Teal (1969) or *Salt Marshes: A Natural and Unnatural History* by Judith Weis and Carol Butler (2009). There are also many websites with information about specific marshes and the challenges they are facing. For example, many of the ecological issues occurring in the salt marshes of Barnegat Bay in New

Jersey have been built into the virtual marsh (see: http://www.littoralsociety.org/Protecting_Barnegat_Bay.aspx).

In Class Exercise—Introduce the Problem

Ask each student to write down a list of all the possible factors they can think of that could have led to a fish kill (large number of fish dying at the same time). Each student can contribute one possibility that could be listed on the board to elicit something similar to the list below.

- Detergent suds from sewage treatment plant overflow
- Overflow of raw sewage from sewage treatment plant
- Phosphates or nitrates in the sewage treatment plant effluent
- Farm animal waste in storm water runoff
- Litter from storm drain runoff
- Fertilizers-Pesticides-Fungicides in storm water runoff
- Pesticides sprayed for mosquito control
- Other poisons that got into the marsh
- Highway salt
- Introduction of an alien or nonnative species
- Severe weather event
- Silt or soil erosion
- Thermal stress from circulation of cooling water at the power plant
- Visitor litter

Explain to students there are eight informants (i.e., residents, scientists, town folks) to be found in various places in the virtual salt marsh scenario. The informant is represented by a cell phone or a folded newspaper, or a camera, or some other object that carries information. Each group will be assigned to find and gather information from at least two specific characters. There is also a worksheet for each student to complete that guides the student in collecting data. The worksheet is the basis for the groups to prepare a report that is presented to the class.

Suggested informants for students working in four groups:

- Group 1: Fish Biologist and Citizen Activist
- Group 2: Microbiologist and Reporter
- Group 3: Environmental Scientist and Park Interpreter
- Group 4: Hydrologist and Grandpa

Orientation to the Virtual Marsh

Hold an orientation to the Second Life Environment (1½ hours recommended) for students to create their Second Life avatar and practice basic Second Life skills. (Students may customize their avatar on their own time or at the end of the orientation.) Topics should include:

- Second Life Toolbars and Menus
- Avatar Movement
- Camera Controls
- Communication Among Avatars
- Interacting with Second Life Objects
- Inventory—Avatar's Collection of Stored Items

Class Discussion

After each small group presents their findings, the instructor facilitates a general discussion for the students to collectively evaluate the likelihood of the various hypotheses and the appropriateness of the recommendations. Each group then reconsiders the issue and may revise their conclusions and recommendations, submitting a final report. To wrap things up, the faculty member may explain the rationale on which the case was based, including the many bits of evidence, both relevant and irrelevant placed into the scenario for students to discover and evaluate. This confirms the complexity of the cause-and effect relationships in this ecosystem, as well as explaining the scientific consensus about the various causes of fish kills, and the importance of continuing to collect data to monitor the ecological functioning of the marsh, making possible an adaptive management process.

Sample Schedule

- Introduction to Salt Marshes
- Salt Marsh Discussion, Chapter 1—Life and Death of the Salt Marsh
- Salt Marsh Discussion, Chapter 2—Life and Death of the Salt Marsh
- Salt Marsh Discussion, Chapter 3—Life and Death of the Salt Marsh

- Sandy Hook Field Trip
- Salt Marsh Discussion, Chapter 4—Life and Death of the Salt Marsh
- How Marshes Die
- Salt Marsh Discussion, Chapter 5 and 6: Life and Death of the Salt Marsh and each group reports on marsh of a particular area
- Second Life Orientation
- Salt Marsh Discussion, Chapter 7 and 8: Life and Death of the Salt Marsh and Marsh Plants
- Salt Marsh Discussion, Chapter 9-11: Life and Death of the Salt Marsh and Marsh Animals, Season, Productivity
- Student reports on management issues in particular marshes
- Group Presentations on Initial Hypothesis in Salt Marsh Dynamics
- Final Student Salt Marsh Dynamics Presentations and Classroom Discussion

Key for Story Characters

Maddie Ball, Environmental Scientist

Found inside at the Environmental Center. Works for the Second Life Environmental Council on ecological restoration project in Pirate Bay. Cool lady!

Focus: Euthrophication, native and invasive marsh plant species, formation of the marsh, how nutrients are generated, importance of balance, rainwater effects.

Sunny Cheeky, Citizen Activist

Found outside the Environmental Center. President of local Save the Bay chapter.

Focus: Litter on the beach, thermal pollution, raw sewage in the bay.

Jamie Meltzer, Grandpa

Found down by the beach, old-time resident fisherman.

Focus: Historical perspective, weather changes, sea nettles, trash on the beach.

TLTC Indigo, Park Interpreter
Found in the Visitor Center. Young, enthusiastic, has a border collie. Somewhat new to the job.

Focus: Provides historical information regarding storm damage at Pirate Bay, draws student attention to nearby sheep and geese, discusses salt concentration on the growth of native grasses.

Dawn Vale, Reporter
Located by the dead fish in the marsh. Local college student and part time reporter for the paper. Always looking for a good story.

Focus: New condo development on Pirate Island, past newsworthy events on Pirate Island, horses on the beach leaving behind fecal material.

Ashleigh Hartmann, Hydrologist
Found by the Water Quality Testing building. Hired by the condominium developer to work on marsh restoration.

Focus: Water Quality testing as a method to determine the ecological status of the marsh, role of detoxification and tides, historical on marsh formation and size.

Holly Meltzer, Microbiologist
Located in the Research Center. Called in by the Second Life Marine Sciences Consortium to investigate the fish kill before the reporters get the story.

Focus: Determines if the beach needs to be closed down. Reports findings to the Second Life Environmental Commission.

Jenny Steeplechase, Fish Biologist
Found by the seining area outside the Visitor Center. Works for the Department of Fish and Wildlife.

Focus: Marsh macro species, food web, water quality and ph.

E-mail Letter

This morning employees and residents of Pirate Bay were astounded to find the shores of the Salt Marsh littered with dead fish. Officials estimate that between 150,000 and 200,000 fish died mysteriously during the night. Water Quality testing showed no evidence of toxins or low oxygen content in the Marsh water.

Local officials have called in a team of experts to perform follow up tests in the bay, ocean, inlet and marsh. There is a need for volunteers to help in the ongoing investigation. You have been assigned to assist the (Hydrologist and the Park Interpreter) in order to determine the cause of the fish kill and to propose a solution for this event. Please join us at Pirate Bay on the Seton Hall Island (aka Pirate Island), in the virtual world of Second Life, as soon as possible.

Make sure to first attend a special Second Life Orientation where you will create your Second Life account and get your "feet wet" in this virtual world environment to be held: State a date and a time, in the Foreign Language Lab, second floor, Fahy. Step-by-step directions are attached along with a map of the area.

Once you arrive on Pirate Island, you will want to head southeast to find the Visitor's Center at Pirate Bay. Obtain important directions from the board outside the Center labeled "Begin Here".

Yours,

The Mayor of Pirate Island

The Hidden Secret of the Marsh

This environmental science problem-based scenario is based on the topic of nonpoint source pollution (NPS). Unlike pollution easily traced to a particular source, nonpoint source pollution comes from many diffuse sources. For example, melting snow or rainfall washes material from the land into lakes, rivers, wetlands, and coastal waters causing pollution that affects the communities of plants and animals in these areas.

In the case of the Salt Marsh Dynamics, a storm event has occurred in watershed. The storm water has washed lawn fertilizers and animal waste into the rivers that flow into the salt marsh. In addition, the excess water has caused the sewage treatment plants along the river to overflow, releasing sewage into the salt marsh. This nutrient rich run-off has caused eutrophication (over-feeding) in the marsh observable by:

- An increase in concentration of nitrates and phosphates, and a large fecal coli form count.
- Excess growth of algae (phytoplankton) and influx of jellyfish into the area.

- Anoxia (low dissolved oxygen) causing death of sensitive fish species such as the common Atlantic silverside.

Virtual Salt Marsh—Fish Kill Data Sheet

Now that you have oriented yourself both to the issues around Salt Marshes, and you have been oriented to Second Life, use the virtual salt marsh fish kill data sheet to find out what happened on the island and in the virtual marsh land that killed the fish. There are clues everywhere, and people in the environment to help you along. Be sure to observe what is going on and take the time to annotate your observations. Note: this activity is best performed when you are working synchronously with other class members.

LESSON PLAN WORK SHEET

Your Name____Date_________

Virtual Salt Marsh—Fish Kill Data Sheet

Objective

1. Gather general and specific information from the links, objects, character narratives, water quality tests and information contained in the Virtual Salt Marsh (*Salt Marsh Dynamics*) to develop a hypothesis as to the cause of the fish kill.
2. Use evidence to make recommendations for restoring the health of the marsh, and preventing future fish kills.
3. In groups: Discuss each student's hypothesis and recommendations, and prepare a group report to present to the class for discussion and critique.

The following water quality parameters are often measured to assess the health of the marsh ecosystem. Each person, independently, should collect data for these items, and indicate the source of the data.

- Clarity/Turbidity
- Dissolved Oxygen
- Fecal Coliforms
- Nitrogen/Nitrates/Nitrites
- pH
- Phosphorous/Phosphates
- Salinity/Specific Gravity
- Water Temperature
- Other Parameters

On a separate sheet, each student should:

1. Describe any other observations and facts that provide evidence for the cause of this fish kill. For example, weather reports, historic information, newspaper articles, outside websites, lab data, field data, things that you learned in the interviews. Remember that confirming a hypothesis requires interpreting data from diverse sources and perspectives. Explain how you can evaluate the relevance or reliability of the different sources of data.
2. Explain which data you consider the "best" evidence for possible causes of the fish kill.

Small Group Discussion

1. Compare results for each parameter, and discuss any differences you find among the sources of data.
2. Each group should prepare a written report explaining their final hypothesis, justifications and reasoning, along with plans to restore the environment to a healthy state and prevent future fish kills.

REFERENCES

Barab, S. A., & Duffy, T. (2000). From practice fields to communities of practice. In D. Jonassen, & S. M. Land (Eds.), *Theoretical foundations of learning environments* (pp. 25-56). Mahwah, NJ: Lawrence Erlbaum Associates.

Barab, S. A., Sadler, T. D., Heiselt, C., Hickey, D., & Zuiker, S. (2006). Relating narrative, inquiry, and inscriptions: Supporting consequential play. *Journal of Science Education and Technology, 16*(1), 59-82.

Brown, J.S., Collins, A., & Duguid, P. (1989). Situated cognition and the culture of learning. *Educational Researcher, 18*(1), 32-41.

Cognition and Technology Group at Vanderbilt. (1993a). Anchored instruction and situated cognition revisited. *Educational Technology, 33*(3), 52-70.

Cognition and Technology Group at Vanderbilt. (1993b). Toward integrated curricula: Possibilities from anchored instruction. In M. Rabinowitz (Ed.), *Cognitive science foundations of instruction* (pp. 33-55). Hillsdale, NJ: Lawrence Erlbaum Associates.

Collins, A., Brown, J. S., & Newman, S.E. (1989). Cognitive apprenticeship: Teaching the craft of reading, writing and mathematics. In L. B. Resnick (Ed.), *Knowing, learning and instruction: Essays in honor of Robert Glaser* (pp. 453-494). Hillsdale, NJ: Erlbaum.

Dede, C. (2005). Planning for neomillennial learning styles: Implications for investments in faculty and technology. In D. Oblinger & J. Oblinger (Eds.), *Educating the Net generation* (15.1–15.22). Boulder, CO: EDUCAUSE. Retrieved from http://www.educause.edu/educatingthenetgen

Dewey, J. (1916). *Democracy and education: An introduction to the philosophy of education*. New York, NY: Macmillan.

Duffy, T. M., & Jonassen, D.H. (1992). Constructivism: New implications for instructional technology. In T. M. Duffy & D. H. Jonassen (Eds.), *Constructivism and the Technology of Instruction*, Hillsdale, NJ: Lawrence Erlbaum Associates.

Gee, J. P. (2003). *What video games have to teach us about learning*. New York, NY: Palgrave Macmillan.

Harrison, D. (2009, February 18). Real life teaching in a virtual world. *Campus Technology Online.* Retrieved from http://campustechnology.com/Articles/2009/02/18/Real-Life-Teaching-in-a-Virtual-World.aspx?Page=1

Lave, J., & Wenger, E. (1991). *Situated learning: Legitimate peripheral participation*. New York: Cambridge University Press.

New Media Consortium. (2010, February). *NMC Two Minute Survey on What's Happening in Virtual Worlds.* Retrieved from http://www.nmc.org//2minute-survey/virtual-worlds

Reigeluth, C. M., & Frick, F. W. (1999). Formative research: A methodology for creating and improving design theories. In C. M. Reigeluth (Ed.), *Instructional-design theories and models* (pp. 633-651). Mahwah, NJ: Lawrence Erlbaum.

Schank, R. C. (1996). Goal-based scenarios: Case-based reasoning meets learning by doing. In D. Leake (Ed.), *Case-based reasoning: Experiences, lessons & future directions* (pp. 295-347). Cambridge, MA: The MIT Press. Retrieved from CBRMeetsLBD_for_Leake.pdf

Squire, K. (2006). From content to context: Videogames as designed experiences. *Educational Researcher, 35*(8), 19-29.

Teal, J., & Teal, M., & Fish, R. G. (1969). *Life and death of the Salt Marsh.* New York, NY: Random House.

Weis, J., & Butler, C. (2009). *Salt Marshes: A natural and unnatural history.* Piscataway, NJ: Rutgers University Press.

Williams, S, M. (1992). Putting case-based instruction into context: Examples from legal and medical education. *Journal of the Learning Sciences, 2*(4), 367-427. Retrieved from putting_casebased_into_context.pdf

Wolcott, S. K. (2005, December). *What are they thinking?* Presented at the Washington Society of CPAs Consulting Services Committee Meeting, Bellevue, WA. Retrieved from http://www.wolcottlynch.com/Downloadable_Files/WSCPA_Handout_051214.pdf

CHAPTER 8

STUTTERING SUPPORT AND NURSING EDUCATION

Two Case Studies in Second Life

Grant Meredith, Charlynn Miller, and Greg Simmons

This chapter makes a novel and valuable addition to research in online learning in virtual worlds, focusing specifically on the use of 3D Multi-User Virtual Environments (MUVEs) for simulation-based teaching and learning. The chapter applies an informed approach to detailing developmental features and examining contemporary initiatives, supported by internationally received literature in the areas of simulation learning, constructivism and online learning. Two specific case studies are described, one outlining how the technology can enhance the lives of people who stutter, the other describing how 3D MUVEs can be used to improve the training of nurses. The chapter draws upon 30 years worth of research and integrates it with contemporary developments and examples that have come about since the emergence, expansion and continuing evolution of the "3D Web." The chapter concludes with a discussion of future initiatives from a point of view of best practice in MUVE-based simulation.

Engaging the Avatar: New Frontiers in Immersive Education, pp. 217–254

INTRODUCTION

Over the past 20 years, the Internet has had a profound effect on the teaching and learning practices in higher education programs. The integration of technology with constructivist methods in simulation training has resulted in computer-based simulations that can assist in developing both technical and nontechnical skills which can then be effectively transferred into the physical world (Alinier, Harwood, Gordon, & Hunt, 2006; Koutantji et al., 2008; Johnson, Zerwic, & Theis, 1999). Studies suggest that there is a strong movement towards providing online learning as students appreciate the flexibility and enjoyment of learning in this environment (Carbonaro et al., 2008; Kenny, 2002).

Three-dimensional Multi-User Virtual Environments (3D MUVEs), including virtual worlds such as Second Life and Active Worlds, as well as popular Massively Multiplayer Online Role-Playing Games (MMORPGs) such as The Sims Online, are receiving significant attention and interest from educators as a means for providing their students with engaging, immersive, multimodal online learning experiences. These environments hold great potential for enriching higher education teaching and learning by presenting opportunities to bridge the gap between students' experiences with technology in formal education settings and those that occur in other aspects of their lives (Prensky, 2001, 2006; de Freitas, 2008; Wolburg & Pokrywczynski, 2001). Excited at the possibilities, universities and colleges across the globe are building models of their campuses in and using a variety of virtual world platforms to deliver in-world lectures, tutorials, seminars and collaborative exercises. Such initiatives are also allowing education institutions to reach new audiences. Indeed the tyranny of distance faced by many students, either due to campus disparity, living conditions, or time, provides a compelling argument for the use of 3D MUVEs as a means of fostering a learners' sense of presence, which is widely acknowledged as a key constituent of effective online learning (Bronack, Riedl, & Tashner, 2005).

In addition to replicating on-campus or classroom-based teaching activities, 3D MUVE technology can be used as a basis for simulations, to deliver authentic learning experiences that are difficult, impractical, or impossible to achieve in the real world (Saunders, 2007). Frequently described as "non-linear exploratory environments" (Aldrich, 2004, 2005), simulations have been used as part of computer-assisted learning (CAL) materials for at least three decades, with SimCity (Wright, 1989) being one of the earliest and most popular examples. Until recently, however, the prohibitive costs of equipment and computational power required to enable high-fidelity 3D simulations, confined them to specialized laboratory environments and medical/military training facilities. Today, the ubiquity of the multimedia-capable, Internet-connected PC has led to a resurgence of interest in web-based virtual reality (VR). The new wave of multiuser online games and virtual worlds is increasingly being seen as offering viable alternatives to visiting a physical place or performing a physical activity. 3D MUVE-based simulations can allow the learner to practice vocational, professional, and other skills in a real-time, safe, collaborative and cost effective virtual environment with the dynamic complexity of the real world.

Virtual worlds such as Second Life are technologies that can provide educators with unique teaching and learning strategies, assisting in meeting the educational and personal needs of today's students. Second Life is an online computer-based simulation of a physical-world environment where the user is given the impression of being in another (real) place through the replications of physical world objects. Second Life enables learners to interact with and manipulate information and representations of an environment and synchronously communicate with other people via a digital representation known as an "avatar."

The benefit of Second Life in education is the lack of a specific location and the ability to create an artificial social structure where problem-based scenarios can be created. This allows students to actively co-construct mental models of technical and interpersonal skills through experiencing human interaction in problematic environments (Dickey, 2005). Unlike traditional multiuser virtual environment technologies, Second Life can incorporate a variety of learning materials with effective educational gaming strategies (Gee, 2003). This provides an ideal setting for proactively engaging students in the construction of their own knowledge through online simulated role play activities (Haycock & Kemp, 2008). This establishment of educational simulations in Second Life is at the early stages of development, however and research conducted to investigate the success of virtual worlds as a teaching platform is small, but growing. Its pedagogical affordance regarding its quality as a teaching and learning tool, in providing meaningful learning experiences to students has also not been established.

This chapter begins with a critical exploration of the rationale/motivation for using 3D MUVE-based simulations in stuttering and nursing education teaching, with a specific emphasis on literature from the nursing education and stuttering support domains. Theoretical and pedagogical models and frameworks that underlie current and proposed uses of 3D MUVE-based simulations in stuttering and nursing education are reviewed, before reporting on current initiatives that represent the major issues surrounding research and practice in this area. The aim is to distill and illuminate the key developments/milestones and challenges in order to provide an overall "snapshot" of the field, to help guide and inform scholars and professionals.

The chapter also examines the issue of transfer of learning from the virtual environment of a 3D MUVE-based simulation to the corresponding real environment, such as an actual operational or job environment (Higgins & Champion, 2000; Orlansky et al., 1994), which is central to the debate about the efficacy of simulation-based approaches to learning, training, and skills development. That is, while learning transfer relies upon a degree of fidelity to the real phenomenon or task, deciding on the degree or level of physical and cognitive fidelity (Hays & Singer, 1989) required for different learners, learning objectives, and learning scenarios is contentious and problematic (Andrews & Bell, 2000; Fadde, 2004). For this reason, implications for design and efficacy of MUVE's for simulation based learning are discussed.

The chapter concludes with a discussion of the future of MUVE-based simulations in stuttering and nursing education, including future research and practitioner opportunities.

BACKGROUND

In presenting a review of closely related literature in this chapter, key ideas and concepts are examined to provide an insight into the nature and purpose of this chapter. The first section deals with the advancements made in online learning and specifically the technology of virtual worlds, presenting research on the ability of Second Life (SL) to provide effective and realistic replicas of real world problems and its implications of providing simulated experience. The following section examines the use of Second Life and other virtual worlds in educational environments, such as stuttering support and nursing. The final section examines how social and interactive gaming models could be applied to training simulations within a virtual world in relation to its benefits and expected learning outcomes for assisting students construct knowledge.

Self-Improvement for People Who Stutter: The Case for MUVEs

Stuttering is a debilitating condition that affects an estimated 1% of the human population (Andrews & Harris, 1964; Ginsberg & Wexler, 2000). As of the 23rd of March 2011 close to 69 million people (U.S. Census Bureau, 2011) the world over could be estimated to afflicted by this condition. Stuttering is an unpredictable and highly variable disorder which can significantly vary without warning during daily life activities (Yaruss & Quesal, 2001).

Stuttering is physically characterized by an individual's speech patterns being involuntarily interrupted by blocks, pauses, repetitions, prolongations of sounds and facial grimacing (La Trobe University, 2005; Teeson, Packman, & Onslow, 2003). At a psychological level stuttering can cause an individual to perceive less of their self-worth in society, influence feelings of depression and recent research has revealed that it could leave an individual 34 times more likely to meet the criteria for a diagnosis of suffering from a social phobia in comparison to a person who does not stutter (Blumgart, Tran, & Craig, 2010a; Iverach et al., 2009).

The act of stuttering may cause many afflicted people to shy away from socializing (Craig & Tran, 2006), accept less satisfying careers (Klein & Hood, 2004) and perhaps at times may lead to suicidal thoughts (Corcoran & Stewart, 1998). The prevalence and occurrence of stuttering is not bound by socioeconomic or geographical borders. Unfortunately the ability to locate and access support during and after an intensive courses of therapy or fluency shaping can be restricted by both of these factors. Factors of economics and geographical distances also may hamper a person who stutters (PWS) who wishes to simply socialize and get support from other PWS regardless of choosing to practicing any speech shaping technique or not

The World Wide Web (WWW) is the great Internet-based service that has enabled the barriers to communication, services and goods to be broken down across borders around the world. People who stutter (PWS) have followed the general populace and have embraced the WWW. The WWW has enabled PWS to communicate freely and to access services and all forms of e-commerce. People who stutter have created their own online communities of forums and blogs to talk with other like people, share experiences and to offer peer lead support (Packman & Meredith, 2011). Some examples of such are the online PWS led forum called stutteringcommunity.com and the online informational podcast site called stutteringiscool.com .

Currently most stuttering support groups are generally organized by speech pathologists and or by people who stutter themselves. Many of these support groups are aligned to a particular fluency shaping technique or support organization. These support groups are normally located in largely populated areas and may vary in their frequency of occurrence. Unfortunately many people are unable to attend these meetings frequently or even at all due to the constraints imposed by distance, cost and time scheduling. The information technology (IT) boom has seen great advancements into the provision of online stuttering therapy techniques and tentative looks into the possibilities that could result from virtual reality (VR) but there has been very little if any forward movement at all in the advancement of online self-support for a PWS. Multiuser virtual environments (MUVEs) with their three dimensional graphics and rich interactive opportunities seem to be the natural leap forward for stuttering self-support structures beyond online user text fed forums, blogs and the well meaning but conventional face-to-face models that by their very nature may hinder participation.

Online Learning: A New Platform for Teaching and Learning

The Internet has provided a unique e-learning opportunity for educators and students. Increasingly, universities are incorporating online technologies into teaching and learning strategies with the aim of creating more flexible learning activities (Taylor & Eustis, 2002). In relation to nursing, researchers suggest that the benefits of an online learning platform in tertiary nursing programs include enhancement of marketability, maximizing students' choice of learning styles, location, time and place of learning, reduce instruction time; enhancing effectiveness and mastery of learning, improvement in retention; and an increase in student motivation, satisfaction and enjoyment in learning (Farrell & McGrath, 2001; Kenny, 2002; Miller, Lee, Rogers, Meredith, & Peck, 2010). Carbonaro et al. (2008) found that students reported enjoying the novelty and flexibility afforded by the use of online technologies. Farrell and McGrath's (2001) work with nursing students produced both positive and negative responses towards online computer-based learning. Investigation revealed themes in student's responses around the development of independent learning, student engagement, and negative aspects such as feelings of social isolation and lack of applied practice.

The anecdotal and statistical evidence from studies involving nursing students suggest that computer-based simulations and online learning can expose students to meaningful experiences and assist students to develop a range of skills and knowledge. However, leading healthcare researchers argue that online learning has a limited range of training functions due to its unrealistic settings, and therefore cannot provide problem-based learning supported by cognitive science (Alinier, 2007; Jeffries, 2006).

Traditionally, online learning and computer assisted learning in healthcare has rarely been anything more than the student passively viewing information or playing a game by themselves (Jeffries, 2006). However, research has shown that effective learning does not occur in isolation, but rather, in teams working together to solve problems (Jonassen, 1998). Petty (2004) claims that on average students retain 10% of what they read and 30% of what they see, whereas students retain 50% of group interaction and 90% of what they act on. There is a need for online learning activities that are supported by cognitive science and that allow students to be actively involved in trying to solve a problem presented to them by interacting and communicating with their peers, environment, equipment and patient (Fanning & Gaba, 2008).

In the past decade online computer-based technology has become an important element in teaching the necessary skills and knowledge to develop competent healthcare students (Carbonaro et al., 2008; Farrell & McGrath, 2001; Miller et al., 2010). Studies have established that computer simulations can assist in developing both technical and nontechnical skills that can be effectively transferred into the real world (Curran, Aziz, O-Young, & Bessell, 2004; Engum & Jeffries, 2003; Jeffries, Woolf & Linde, 2003; Mili, Barr, Harris, & Pittiglio, 2008). Studies also suggest that students appreciate the flexibility and enjoyment of learning in an online environment (Farrell & McGrath, 2001; Kenny, 2002).

Virtual Worlds in Education: Creating New Opportunities for Online Simulation

The Internet in education has been largely viewed as an instructional tool for conveying information and providing more flexible learning environments (Duffy & Cunningham, 1996; Jonassen & Reeves, 1996). Recent studies have established that access to teaching materials alone is unlikely to engage learners who respond best to multimodal forms of delivery and are not effective in development of students' deeper knowledge and skills (Prensky, 2006; Rouvrais & Gilliot, 2004). Furthermore, current practices in tertiary online learning methods are limited in their ability to capture the social dimension that exemplifies learning in the real world (Jonassen, 1998; Lombardi & McCahill, 2004). Three-dimensional virtual worlds have generated significant interest with secondary and tertiary educators. Virtual worlds, such as Second Life, enable learners to interact with and manipulate information and representations of an environment and synchronously communicate with other people from a first person perspective (Dickey, 2005).

The idea of synchronous interactive virtual spaces is not new but the definition of virtual worlds has changed over the years. Early forms of virtual spaces such as text-based multiuser domains (MUDs) and MUD object orientated (MOOs) have been used in education for decades (Nykvist, 2005). Second Life can be described as an online computer-based entity that can simulate a real-world environment by representing objects to the user, giving the user the impression, as realistically as possible, of being in another place. Through the use of an avatar, a digital representation of the user, people can create, interact with and manipulate elements of the modelled world and communicate with other users (Haycock & Kemp, 2008). Communication between users ranges from text, graphical icons, visual gesture, sound and voice. Second Life is a virtual world that has gained much attention with tertiary educators as it enables its users, known as residents, to explore, socialize and participate in individual and group role play activities.

How Second Life Facilitates Self-Help and Improvement

Second Life can aid the facilitation of self-help and improvement groups in a number of ways which are clearly shown through the design of the Virtual Stuttering Support Centre (VSSC) outlined later in this chapter. Second Life gives the user the ability to create closed and private areas which can be accessed by invited members only. In the case of the VSSC this enables the creation of safe areas for members to interact with each other and run through simulations without the fear of being judged or being ridiculed by unassociated users. Second Life gives the user the ability to easily set up access lists and to create user groups in order to manage support group interactions and permissions.

The avenues of personal and social communication that Second Life offers can also lead to rich interactions within a virtual support group environment. Second Life enables both text and voice chat facilities, both

Figure 8.1. The VSSC foyer.

of which can be privatized to areas, groups or individuals. Within a self-help group there may be at times the need to discuss issues publicly and at other times perhaps privately.

Having multiple options in which to communicate leads to a more comfortable and equitable environment with users being able to communicate in an understandable and agreed manner which suits the group and at other times through ways that suit the individual's needs. Second Life also gives the user the ability to use animated gestures when interacting with others. These gestures help to give the avatar and their interactions with others more life and appeal. Each avatar comes with a standard set of gestures which include waving and pointing. These standard gestures can be incorporated into virtual communication and can help add dynamics and order to a virtual meeting. For example when holding a meeting with voice chat the user often hear multiple members trying to talk over each other. By insisting that a user raises his or her hand via a designated gesture first the user helps to restore some order to a meeting. Another example is using appropriate gesture animations during a slideshow presentation to avoid a stagnant looking delivery. The rich 3D environments that can be constructed within Second Life also enables the user to create comfortable user environments which can be interacted with and navigated through with relative ease. This leads to the creation and configuration of a very dynamic and flexible environment.

Self-help and improvement can also be gained through the use of simulations. Both the VSSC and the Critical Life project rely heavily on simulations run within Second Life to aid their associated aims and goals for the user(s). Second Life enables developers to create highly detailed and interactive simulations. These simulations once established can run without constant attention or the costs involved with constant maintenance or moderation. Simulations can also enable tasks that in the physical world can be awkward, dangerous and perhaps unethical for a user to undertake. In the case of the Critical Life simulations it is the setting up of a patient in medical distress over and over again for a group of students to diagnose and treat. In the case of the VSSC simulations an example would be the simulated shop environment which if in the physical world could lead to a very embarrassing circumstance and perhaps negative reactions from a shop assistant.

Second Life in Higher Education

Although research into Second Life as a learning environment is in its early stages; significant research exists from similar technologies such as immersive VR and multi-user domains which support constructivist learning, online communities and educational games (Bruckman, 2002; Fanderclai, 1995; Gee, 2003; Wankel & Kingsley, 2009). Many universities have a presence in Second Life, generally using these spaces to hold lectures, meet with students, host gatherings and build virtual campuses (Baker, Wentz, & Woods, 2009; Miller et al., 2010).

Research into the use of virtual worlds has been conducted in various disciplines in the higher education context. Psychology educators Baker, Wentz, and Woods (2009) found that a virtual world can allow for valuable social interactions to occur. A study by Ritzema and Harris (2008) explored the ability of Second Life to convey simplistic programming techniques. Results showed that students found the overall experience to be enjoyable, and that the virtual learning activity added to their understanding. A similar study by Haycock and Kemp (2008) investigated the use of Second Life for training off-campus librarian students, reporting that students found the Second Life learning environment highly engaging and that the course content was most effective when the constructivist activities simulated in Second Life are blended with reflective writing exercises.

Specific to healthcare, a team of medical staff at Stanford University (Heinrichs, Youngblood, Harter, & Dev, 2008) examined a series of case studies that compared learning outcomes and usability of a human patient simulator (HPS) with a virtual world similar to Second Life. The results advocate that team training and assessment are feasible using virtual world technologies and that it can be as effective as the traditional method of using human patient simulators.

Second Life also has its share of critics with some suggesting that many activities can be accomplished using a regular website or video conferencing technologies, have little value other than novelty and can have a negative effect on the learning process as a result of the steep learning curve associated with its use (Berge, 2008).

The value of Second Life in tertiary education and training is its ability to provide unique opportunities for students to engage in online simulated learning experiences. The Second Life platform allows educators and developers to create simulations in an environment incorporating rich information (Good, 2004), high immersion (Dede, 2005), development of a skill and understanding (Alinier, 2007), multiuser interactivity (Miller et al., 2010; Sontag, 2009), and gaming elements (Gee, 2003).

Current research in the field of nursing education suggests that clinical experience is an essential part of the development of competent nurses (Benson, 2004; Higgs & Titchen, 2000). Traditionally, undergraduate nursing students have engaged in experiential learning activities by participating in clinical field placements. Evidence is mounting however, which suggests these field placements are not ideal for providing exclusive learning experiences and, in the current ethical and professional climate, clinical student training in the field has been diminished due to patient safety and for ethical reasons (Ziv, Small, & Wolpe, 2000). To tackle this problem, healthcare educators have implemented activities that attempt to reproduce the desired experience through simulation. Online computer-based technologies have become an important element in teaching the necessary skills and knowledge to develop competent nursing students (Benson, 2004). Clinical simulation provides active experience by involving the student in solving a problem presented to them through interacting and communicating with their peers, clinical environment, equipment and the patient. By integrating technical skills, human factors, team management, and situational awareness concepts into a simulation, participants concurrently learn and develop clinical skills and concepts related to patient safety, reducing the risk of potential errors in the work place (Dickey, 2005).

Incorporating Constructive Gaming Theories Into Second Life

The constructivist learning theory is an important technique to consider when designing effective educational simulations. The constructivist learning theory, derived from the work of Piaget (1972), Vygotsky (1978) and more recently Lave and Wenger (1991) advocates the fundamental belief that knowledge is constructed, not transmitted, and that learners play an active role in the learning process (Jonassen, Peck, & Wilson, 1999). A clinical simulation incorporates strategies that provide students with practice in *problem-based* decision making, based on physical world situations. Central to the constructive learning theory is problem-based learning and improvement (PBLI), where the student becomes actively engaged in the learning process. Hmelo-Silver, Duncan, and Chinn (2007) suggest constructive PBLI activities allow students to learn content, strategies, and self-directed learning skills through collaboratively solving problems, reflecting on their experiences and engaging in self-directed inquiry.

Kolb (1984) described the learning process involved in problem based simulation as experiential learning. Experiential learning is particularly suited to learning a profession, in which the integration of theory and practice are performed in a cyclic process where the student develops an understanding of a concept from direct experience (Grant & Marsden, 1992). The model is based on the idea that concrete experience provides a basis for observation and reflection, suggesting that learning through observation and interaction with a simulated environment will allow the student to make illuminating discoveries and develop a superior cognitive understanding, as opposed to reading about a problem from a book (Kolb). Furthermore, Kolb explained that these observations and reflections are refined and assimilated into abstract concepts which can be applied to new experiences, suggesting the learning obtained from a simulation can be actively tested and refined and in turn result in a new experience and development of understanding.

For a simulation in Second Life to be effective it involves more than just students practicing protocols and skills; it also requires a human dimension where nontechnical skills such as teamwork, communication and leadership can be applied. Although virtual worlds can provide students with a highly interactive environment, currently in Second Life, students are rarely co-constructing knowledge in a collaborative context. One way of achieving this may be through the application of gaming theories in the design and development process of simulated environments. One such gaming theory is psychological moratorium (Smith, 2004; Taber, 2008).

Psychological moratorium is a gaming theory which essentially means a person can perform a task in a realistic environment without real consequences. It allows the student's attention to be maintained within the scenarios and keeps the learner invested in the outcome, but they are also free to take risks and try new approaches. Psychological moratorium is key for active learning in emergency response organizations that deal with dangerous, life-and-death scenarios; allowing the student to learn from their own, or others, mistakes without ill-fated consequences (Smith, 2004; Taber, 2008).

Virtual worlds provide healthcare educators with unique teaching and learning strategies that can assist in meeting the needs of today's healthcare students. Students can learn and understand new concepts and material, make meaningful social interactions and develop teamwork skills in virtual worlds such as Second Life (Baker et al., 2009; Haycock & Kemp, 2008; Ritzema & Harris, 2008). Second Life is an ideal setting for proactively engaging nursing students in constructing knowledge which relates to realistic problems as well as assisting in the development of problem solving skills in a collaborative environment without inflicting harm to patients (Heinrichs et al., 2008).

CASE STUDY ONE: THE VIRTUAL STUTTERING SUPPORT CENTRE

The Virtual Stuttering Support Centre (VSSC) is a Second Life based simulation environment designed to act as an online virtual self-help and support hub for people who stutter (PWS). The VSSC is not aligned to any particular fluency technique or self-help faction and is designed to show

PWS, speech pathologists and other interested parties how MUVEs can be designed and used to offer a richer online self-support service in comparison to what is already available via the World Wide Web.

The VSSC houses two primary modes of user interaction. These two modes are the simulation and the self-support structures. The simulations are designed to desensitize a PWS to commonly feared physical world situations and to also encourage the practice of any learnt fluency shaping technique. The self-support structures include meeting and conference rooms designed to harbor and stimulate support group meetings and informative presentations.

Development of the VSSC began in 2009 due to a funding opportunity afforded to the researchers through an internal University of Ballarat grant. Initial development proceeded by employing a small team of final year information technology students. This enabled the VSSC to be built up to a strong first prototype stage in a very short amount of time. This stage enabled a simple set of simulations and basic support group facilities to be developed. These simulations still operate and are explained in their current form later in this chapter. This first stage enabled initial user testing to be conducted on the simulations' controls and basic functionality. This prototype enabled the entire concept to be presented to the target users (people who stutter), academics and speech and language professionals at the Australian Speak Easy convention early in 2010. The concept and ideas being developed were very well received.

Currently the VSSC is in its second stage of building. This has been achieved via the help of another team of University of Ballarat students. This team of two students has continued to develop the VSSC through their yearlong capstone project courses. This second stage concentrated on refining the simulations and their associated controls. This has enabled the simulation functionalities to run even more efficiently and intuitively compared to the first build. This second stage also features enriched multimedia interactions involved with meeting rooms and support groups through the addition of interactive whiteboard facilities and WWW capabilities. Screens have been added to some virtual rooms which enable a user to browse the Internet for websites to show other members or to access multimedia content to stream. The second build has also enabled the developers to give the hosted BOTs more gesture-based interaction than originally planned for. This has been achieved by adding animated gestures to their verbal interactions with the simulations.

The next step in the life of the VSSC is to test the simulations and support structures with the intended target audience (PWS). This phase will cover researching perceptions of the worth of using this virtual world for self-help, support, and simulations. It is also planned to conduct extensive user testing to guide even further refined design and interactions.

Figure 8.2. A VSSC support group meeting room.

VSSC SIMULATIONS

The following are descriptions of the simulations that are currently operating within Virtual Stuttering Support Centre (VSSC). The simulations in practice at this stage in the VSSC's development are job interviews, a cafe and phone calls. These simulations are designed to target some key identified feared situations that people who stutter (PWS) may face and/or avoid during their daily lives. These simulations are designed to give a PWS the opportunity to practice speech and/or learnt fluency shaping techniques in a safe and stable environment with the aim of taking it out into the real world when feeling personally comfortable and confident.

The described simulations could also aid in helping to desensitize a PWS, regardless of practicing any fluency technique or not, to the simulation being presented, again with the hopeful aim of transference into the real world. All the simulations in practice are hosted by Bots. Bots in terms of the VSSC are script controlled Second Life avatars that look the same in design and nature as any "real" user (avatar) within the world would. Keeping the Bots as Second Life avatars and not simply two dimensional cut-outs or action-limited 3D models gives them more life and believability in terms of the look and behavior within the simulations. The Bots use simple standard Second Life avatar gestures during their actions. Unlike the real world the Bots do not give any form of feedback either positive or negative. At this stage in the development of the VSSC the Bots simply run through their scripts in a neutral fashion.

With all the simulations outlined, the users are able to practice their speech in a private place of their choosing and without others in the VSSC hearing them. Users may choose to use a microphone to talk within the virtual world or they may simply talk to the screen and simulation at hand without a microphone and anybody else having the chance to hear them. Users through the simulations may interact privately if they wish to do so. Entry to the VSSC is also only allowed to members who have requested to take part and who have been accepted by the moderators. This helps to ensure privacy and to regulate a like-minded community. Any users found violating the rights of other users or are found out not to be PWS, unless invited in, will have their access to the VSSC revoked. Actions and interactions within the VSSC will be constantly moderated for the sake of the health of the community.

JOB INTERVIEWS

For anyone, let alone a PWS, job interviews can be highly stressful situations (Hussain, 2008). Job interviews are situations that can be prepared for, but can still they can be variable and volatile situations with perhaps major life changes hinging on their successful outcome. Some PWS may have avoided job interviews due to their speech problems and have admitted at times that their speech problems have impacted their decisions about taking on and being accepted for promotions (Blumgart, Tran, & Craig, 2010b) and/or have impacted negatively on them being satisfied with their overall career performance (Yaruss, 1998).

The VSSC houses three job interviews all of which work relatively in the same fashion. The three interview types are:

1. Single male interviewer;
2. Single female interviewer;
3. A pairing of both a male and female interviewer.

The mixture of genders is designed to allow the PWS a level of comfort communicating with both genders under these simulated circumstances. This brings the simulated scenario back closer to that of a physical life encounter where there is always an air of uncertainty about who will host the interview and how they will act. The job interview simulations starts once the user has entered the room after being ported from the interactive navigational signboard in the VSSC's foyer. These interviews only allow one user in them at time to help ensure privacy and efficient practice. Upon entry to the interview room the interviewee is asked to start the simulation. From the start of the simulation the interviewer verbally asks the participant 10 questions, each respondent randomly drawn from individual theme-based databanks. This creates a unique interview experience every time the simulation is run and the participant will never be at total ease or able to predict the flow of the questions from interview to interview. The databanks cover 10 major themes of questions commonly asked in job interviews, which were compiled by the primary author from discussions with University of Ballarat Human Resources staff and online guides. These sets of questions will be continually refined in the future through the process of evaluation. All in all the simulation can draw from 100 plus questions.

The simulated job interview is staged in a very clear and linear fashion. The interviewee enters the room and is asked by the Bot (interviewer) to take a seat. The Bot then asks the interviewee one random question drawn from each databank. After the question has been asked the Bot waits for the respondent to give the order to proceed via a pop-up menu system. This enables the respondent to think about their answer and to verbally answer it in their own time without the pressure or stress of a real face to face encounter. The respondents are able to proceed at their own pace. Some respondents may choose to have the question repeated until they proceed to the next. This gives the user more control over their individual simulation. The Bot interviewer will only proceed upon the command of the user and while waiting will cycle through some basic animated gestures giving the impression to the user that they are still interested and active within the scenario.

The interviews have a privacy system scripted into them to enable only one real user at a time to be present within them. Other users are locked out of the simulation if it is currently running. This helps to ensure a private and safer practice environment for a user.

CAFE

The design of the simulated cafe is similar in design to that of the simulated job interviews. Some PWS find the process of ordering food in public to be a highly stressful situation and may at times avoid the process totally or simply opt for a menu option that they can usually successfully say but maybe are not totally satisfied with. Some PWS may rely on others to make their orders for them. The café is designed to both empower and desensitize the PWS to the situation. The simulated café gives the PWS the opportunity to practice ordering foods. The café is hosted by both a male and a female Bot. The user may approach whichever the Bot with whom they wish to engage. The simulation only enables one user at a time to run through the scenario but unlike the job interviews it does allow spectators.

The simulation runs through a simple three step process:

1. Service person Bot asks you for your order;
2. Service person Bot asks you if you require anything additional;
3. Service person Bot asks you for your name.

This process has been designed to address not only the issue of ordering food but also the issue of a PWS having to speak their name. Many PWS find saying their name to be a very hard process and may often either shy away totally from the ordeal and at times have reported substituting their real name for a pseudonym that is easier to say (Kamhi, 2003). The second step in the simulation adds a bit of variety to the scenario and may put some PWS' on the spot. They may have formulated a pre-planned script in their minds about what they will order and the occasional deviation from their expectations will cause them to think more spontaneously in circumstances similar to a physical world situation.

Unlike the job interviews the café enables multiple users to be present at any time. This helps to shape the café not only into a simulation environment but also into a minisocial area. Enabling other users to be present opens the simulation somewhat to a real world experience where one could expect other customers to be in the same café that you are. This could enable a PWS to get more comfortable when ordering within an active and somewhat unpredictable environment such as a busy shop. Having other like people in the simulation may also aid the peer support and encouragement process.

Telephone Calls

Another series of simulations that the VSSC holds are simulated telephone calls. Located in an annexed area of the roof top social area are two telephones. One phone simulates making a call and the other simulates taking a call. Quite a lot of people who stutter find both making and taking phone calls to be a very difficult and fearful process. Again these anxieties and speech problems may lead to many people who stutter avoiding using the telephone all together or be forced to rely on others to make essential phone calls for them. These simulations aim to continue the self-empowering and desensitizing techniques upon which the VSSC is grounded.

Figure 8.3. VSSC café.

Taking a Phone Call

The phone rings for a random number of times to eliminate the predictability of the simulation. When the PWS chooses to answer the phone the caller randomly asks them something. For example "Could I please talk to Mr. Smith?" Random responses from both telephone modes draw from a databank of different questions with different genders and nationalities acting out the parts. This again this adds to the unpredictable nature of the simulation. In the physical world when a PWS answers a phone call they cannot always predict what the nature of the call. The simulated calls do not have any form of caller id to ensure the random nature of the calls.

Making a Phone Call

The PWS chooses to make a phone call, the phone then rings for a random amount of time. A random person answers the phone enabling the PWS to practice a lead into a telephone conversation. An example response may be "Hello Ashanti's restaurant. How may I help you?" The PWS would then respond in their own time.

VSSC Self-Support Structures

The VSSC has the ability to host support group and online conference interactions.

Support Group

The VSSC enables members to host their own support groups. The main support group room is located on the second floor of the VSSC and has whiteboard and slide-show capabilities built in to enable instruction or spontaneous guided conversation. The support group room is highly configurable allowing a moderator to add and configure as many chairs as needed. The VSSC enables a support group to be run as in the physical world but without the constraints of time, distance and money.

CASE STUDY TWO: THE "CRITICAL LIFE" SIMULATION

The Critical Life simulation development began in 2008. The objective was to develop an emergency room simulation in Second Life that could be used to provide real time skills to nursing students. The initiation of the project was preceded by an extensive literature review. The simulation was developed at the University of Ballarat by a small group of students in the Graduate School of Information Technology and Mathematical Sciences (GSITMS) in conjunction with supervisors in both the GSITMS and the School of Nursing. The aim was to replace the hard copy case studies first year nursing students complete in year 1 and year 3 as a form of assessment in the two courses.

Simulations in Nursing Education

Research into nursing education indicates that clinical experience is an essential part of learning for the development of competent nurses (Benson, 2004; Fanning & Gaba, 2008). Traditionally undergraduate nursing students have gained this experience through participation in clinical field placements. However, there are inherent issues involved with clinical field placements. Current evidence indicates that traditional field placements are not adequately preparing undergraduate nursing student for the field. Opportunities for nursing students to participate in clinical practice are restricted and there is an imperative for learning activities that can provide the applied experience by other means (Alinier, 2007).

Role-play has become a popular form of providing simulated clinical experience in nursing education as it allows learners to assume and act on the behaviors of the particular character role with minimal risk (McLaughlan & Kirkpatrick, 2004). A simulation that requires students to play a role for a specific purpose encourages the student to practice sets of interests, values, and knowledge encompassed in the profession allowing them to experience and learn the adopted role in a safe, non-threatening environment (McLaughlan & Kirkpatrick).

A nurse's ability to solve problems and make accurate clinical judgments is crucial to the patient's safety and well-being. Kwan, So, Lai, and Tiwari (2006) found the use of case studies could significantly improve students' truth seeking, analytical and critical thinking and self-confidence. A similar study (Alinier, Harwood, Gordon, & Hunt, 2006) determined that simulation is a useful training technique as it enables students to practice skills and obtain hands on experience in a safe and controlled environment.

Simulations can provide nursing students with the opportunity to improve communication and thus enhance patient safety. Communication and teamwork failures have been found to be a major cause of errors and accidents within the nursing profession (Lingard et al., 2004; Sexton, Thomas, & Helmreich, 2000). There is need for simulation to assist nursing students learn both technical and nontechnical skills required in their profession (Beyea & Kobokovich, 2004).

Technology Assisted Learning and Communication in Nursing Education

Virtual reality (VR) simulations are a popular form of simulating clinical experience. VR is a technology that allows users to interact with a computer-simulated environment. VR clinical simulations are used in nursing training for a variety of teaching and learning purposes, such as teaching facts, principles and concepts; assessing the student's progress or competency with a certain skill, integrating the use of technology in the learning experience and developing problem solving and diagnostic reasoning skills in a safe, nonthreatening environment (Jeffries, 2006; Lee et al., 2007; Miller et al., 2010; Taber, 2008; Waltz, Strickland, & Lenz, 2004).

VR simulators allow nursing students to practice a range of skills and improve techniques without consequences. VR simulations also reduce the potential risks involved in training nurses and can help develop standards and optimize procedures. Engum and Jeffries (2003) found engagement and skill development through the use of VR simulation. In a similar study (Jeffries, Woolf, & Linde, 2003) the researchers concluded that a skill learnt from a computer simulated experience can be similar to a skill learnt via traditional methods, and that simulated experiences can be effectively transferred into the real world.

Building a Simulation for Nursing Education

The small project team, comprised of final year IT students developed a computer-based clinical simulation (CBCS) in Second Life with the intention of extending a text-based simulation into a virtual learning activity (Rogers, 2008). The skills that were required within the team to allow for the effective development of the simulation included a project manager, a graphic/multimedia designer, and a programmer who was familiar with the Linden Scripting Language (LSL), similar to C++. Funding was provided to purchase some virtual furniture and other aesthetic parts of the simulation, but in most cases, the simulation equipment and other aspects were built by the team. A location within Second Life had to be secured for the Emergency Room to be built, and was later moved to the Island that the university purchased.

The development of the Critical Life simulation encouraged students to virtually interact with apparatus and patients to identify, solve and perform various tasks; allowing them to construct their own understanding of the problems at hand, rather than simply reading about them. These learning tasks were used at two separate campuses at the university and were expected to be completed outside of class; requiring the student to explain why the steps taken in the scenario were necessary and to discuss the acute care of the patient from their own perspective. Using the text-based scenarios as a guide; a simulation was created within Second Life, called Critical Life.

The Critical Life simulation takes place in a virtual clinical room, where a team of students are faced with one of six clinical problems relating to a patient. Rather than outlining the status and steps required to assist the patient, as presented in low fidelity or text-based simulation, the students are required to perform several tasks: identify the problem, develop a solution, perform and observe the procedures and reflect on the outcome as a team through interacting with the patient and apparatus, and communicating, collaborating and cooperating with their peers. The simulation includes six different scenarios containing various clinical diagnostic problems involving clinical documentation, respiratory, pharmacological, electrocardiogram, and family issues.

Briefing Stage

The briefing stage involves the nursing students becoming familiar with the Critical Life simulation. Students spend time creating their own avatar allowing them to familiarise themselves with the Second Life interface and developing an identity with their avatar. The participants then view videos either in-world or on YouTube introducing them to the Critical Life simulation. The videos demonstrate how to interact, copy and view objects, navigate the simulation, and communicate with peers. This process can be repeated until the student is ready to enter the clinic, allowing them the opportunity to develop a sense of trust and association with the simulation and their peers.

Presenting a Clinical Problem Which Requires Action

The students enter Critical Life by teleporting into the clinical room. A simple detection script is used to count the number of people in the room, and once the maximum number of four is reached no more students can enter the room. The students are presented with an in-world video which mimics the process of a verbal handover for the particular scenario. The student is also provided with accompanying documentation. This process allows the student to adopt the role of a nurse and presents them with background information detailing the patient's ailment. The problems are organized around relevant, authentic problems or questions from the original case studies, where the student is required to develop a hypothesis of the required patient care by using the information provided and interacting with patient and apparatus. For example the student can take the patient's blood pressure, examine oxygen saturation, listen to breathing patterns and ask the patient questions via the instant message interface. Students can leave and enter the simulation at

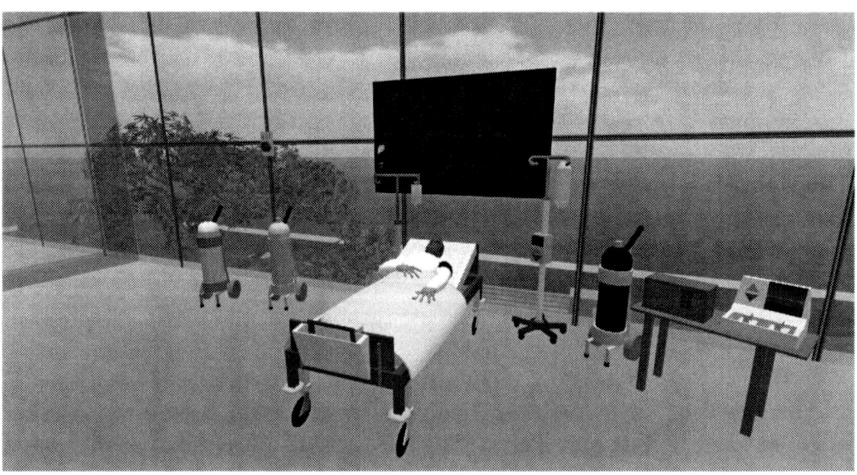

Figure 8.4. The critical life ER.

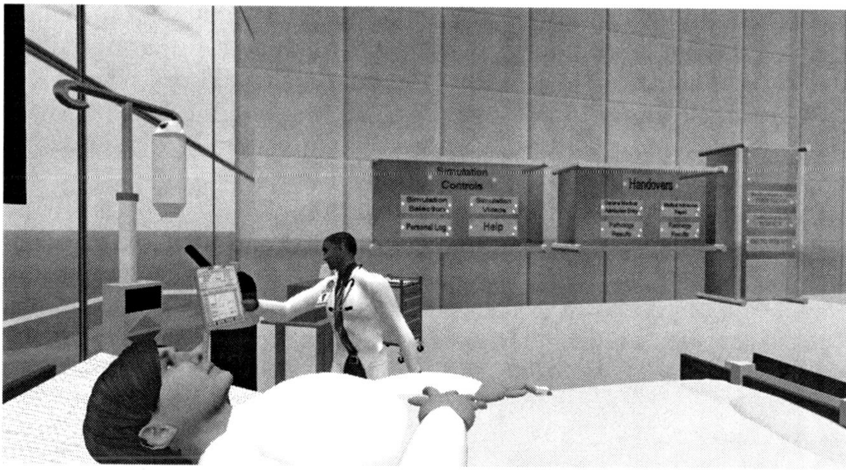

Figure 8.5. A student completing the simulation.

any time (if the maximum number of avatars has not already been reached) which simulates the process of debriefing as students who enter late after the simulation have to be briefed by the other participants.

From a design perspective, the problem then becomes how to encourage the students to articulate a hypothesis and co-develop a solution as a team in a collaborative environment. Previous studies have shown that although there are not any particular differences in the amount of communication between students in virtual environments and a face-to-face environment; there is a significant reduction in the amount of collaborative reasoning and as a result new concepts may not be elaborated and understood (Gao, Noh, & Koehler, 2008; McLaughlan & Kirkpatrick, 2004; Whitton & Hynes, 2006). Therefore the Critical Life design required a model in which team work and collaboration were encouraged and no single student could control the simulation themselves; as it was believed that this affected the learning process negatively.

Controlled by the simulation, no action is taken in a scenario without a unanimous decision. This process engages the students to develop evidence based explanations, articulating their ideas and reflecting on discussions made. Many of the scenarios in Critical Life require multiple steps, some of which need to be performed in a particular order. It is therefore crucial that group discussion occurs before any action is taken to ensure the team develops an understanding of the reasoning behind any decisions made. For example an action may include turning the oxygen tank on to increase the patient's oxygen saturation, or giving drugs to the patient to reduce pain.

The Critical Life simulation will not perform an action until the entire group agrees on a decision by detecting the unique identification of an avatar (the avatar name) and storing this information in a list. When a student uses their avatar to perform a particular action the name is cross-checked with an action list and checked off. Once all names in the action list are checked off; the action is performed, thus ensuring that a group decision is reached before performing an action.

Unlike traditional CBCS, this model supports effective clinical judgement as it not only encourages students to develop a prediction by applying the concept of a problem to their own knowledge; it also encourages students to develop a prediction by applying the concept of a problem to their peer's knowledge. With traditional simulations, if a student is unable to form a hypothesis they can simply obtain the desired result with a few clicks of the mouse.

However, the student does not develop a cognitive understanding of the underlying concept that can be transferred into a clinical practice by learning which combination of buttons to press to achieve the desired results. In the Critical Life simulation if a student is unsure they understand the problem and are not able to derive their own prediction, the student can collaborate with their peers to acquire abstract information and then relate it to their experience obtained from the simulation. This process assists the student to grasp the concept and develop a co-constructed mental model of the problem.

Once a concept has been agreed upon and an action has been carried out by all team members; information from the particular apparatus object is passed to the patient object—which, in most cases, will change the state of the patient. For example if the oxygen tank is turned on the patient's visual and auditory attributes such as facial color and breathing patterns may change. When the student observes the patient's vitals after turning the oxygen on they see changes in the patient's heart rate, blood pressure, oxygen saturation, and so forth. This monitoring allows the group to observe their actions and determine whether their predictions are valid. The entire process can be repeated many times allowing the students to experience clinical practice and the required skills for effective teamwork. After each scenario is completed in Critical Life, the simulation prompts the students to answer questions in their personal log relating to their experiences of clinical judgment, team work and specific skills carried out in the simulation to encourage reflection and self-directed inquiry.

EVALUATING THE SIMULATION

An evaluation of this simulation was conducted in 2009 that included both quantitative and qualitative data collection to investigate the learning habits and perceptions of healthcare students with regard to simulation, computer games, online learning and virtual worlds (Rogers, Miller, & Firmin, 2011). Couched in a constructivist framework, this information is used to provide a descriptive profile of the participant sample and to determine the complexity level of the simulation for the students. In addition, the participants reported on which experiences they perceived could assist them to effectively develop as a healthcare professional. The research further focuses on what technical and nontechnical skills can be developed in a virtual simulation and how the participants perceived they could transfer these skills into the real world.

CRITICAL LIFE, LESSONS LEARNED

The outcomes of this simulation have been promising. Further use of the simulation for nursing practice and assessment will enable developers and researchers to fine tune the simulation and develop further learning and teaching initiatives using this framework. Several lessons were learned during this project. The foremost lesson was the need for strong integration between the subject matter experts and the developers. Another lesson was the need for the appropriate personnel to work on the project. Finally, there was a need to establish and maintain an ongoing location for the simulation is critical. Currently, the university has its own Island, and this simulation is located in a maintainable and secure environment for future use (http://tiny.cc/9mrev).

CONCLUSION

Future work in the area of MUVE simulations for healthcare students and self-supportive structures for people who stutter will need to include discussions about best practices and simulation designs in these areas. There are thousands of sites sponsored by or built by universities all over the world. Many simply reproduce the physical world university in structure, many offer some lecture capability. Equally, there are hundreds of attempts at simulations for education in MUVE's such as Second Life. The development of best practises in this area will assist in relevant, useful and educative simulations and environments that can provide impact and advancement.

Studies are currently being conducted to establish many of these best practices. The University of Ballarat is currently investigating the technology used to develop these simulations and surveying students to determine the best way to develop in the future that will encourage easy and effective use.

The use of simulation in clinical experience through a virtual world is a relatively new and complicated area of research. With regards to stuttering research it is believed that is the first such use of this technology. However an extensive investigation into various fields of healthcare literature revealed that although simulation and online learning are frequently employed in Australian Bachelor of Nursing Programs, little is understood about the phenomenon of simulating clinical experience in an online virtual world. Studies have shown that clinical simulations can equip learners with skills and experiences that can be directly transferred into the clinical setting (Clague et al., 1997; Engum & Jeffries, 2003). Studies have also established that in the past two decades computer technologies have played a large part in developing competent nurses (Jeffries, 2006; Lee et al., 2007; Taber, 2008; Waltz et al., 2004) and the Internet is changing the preferences healthcare students have towards accessing information and socialising with their peers (Carbonaro et al., 2008; Farrell & McGrath, 2001; Kenny, 2002).

Tertiary healthcare programs have seen a movement to e-learning in online environments. In an educational context, a strong relationship seems to exist between constructive learning and online virtual worlds. Research suggests that virtual worlds have the ability to assist people to learn and understand new concepts and material, make meaningful social interactions and develop team work skills (Baker et al., 2009; Haycock & Kemp, 2008; Ritzema & Harris, 2008); however there are no studies which indicate whether virtual worlds can provide meaningful simulated clinical learning experiences. Further research is needed to investigate the experiences and attitudes of healthcare students in higher education in the use of virtual worlds as a simulation tool to determine what educational potential and benefits it has for developing competent nurses.

This chapter has presented a clear case for the use of simulation in MUVE's in healthcare education and the support of people who stutter. Two different applications of MUVEs were presented; these applications whilst at different levels of maturity and being conceived for different reasons each take advantage of a single critical aspect of MUVEs in that they provide a safe environment for people to practice skills which would otherwise be difficult in the physical world. A review of the relevant literature on constructivist learning methods was presented. The developing Virtual Stuttering Support Centre with its focus on simulation, skills transference and support was first discussed in detail, outlining simulation design and best practice techniques for holding virtual support meetings. Discussion of research showing the usefulness of simulation based education for healthcare students was also given. Best practices for ongoing construction and relevancy were reviewed, and suggestions for development were made based on the current research and the authors findings.

This is an exciting and important area of education and support that both provides for real-world simulation and transcends the boundaries of time and distance for today's fast moving and globally minded society.

APPENDIX: CRITICAL LIFE LESSON PLANS

Background

The Critical Life Simulation was developed through the conversion of an existing set of paper-based case studies that were used for assessment within a Second year Nursing Course. This was done to enhance the learning potential of these case studies by adding an element of reality and hand-on learning to this segment of the course. The lessons below were taken directly from the existing paper case studies. A subject matter expert in Nursing guided the development of the simulation. Testing was conducted after development to ensure accuracy and engagement. The eight lesson plans presented below are provided to students as a teaching aid along with a series of 29 videos (located in the Critical Life YouTube Channel—http://www.youtube.com/user/CriticalLife) that instruct in the use of the simulation. Videos are also accessible from within the Second Life environment. The eight lessons below take students from the point of learning the simulation environment and meeting their team through six scenarios within the emergency room. The Simulation completes with a personal reflection opportunity for students.

Lesson 1: Briefing and Familiarization

Videos to view

1. Introduction (2 Videos)
2. Starting Simulation (4 Videos)

Actions

1. Create your Avatar and familiarise yourself with the environment
2. Receive an introduction to the operation of the simulation
3. Get briefed on the patient through the video previews
4. Locate the pathology, radiology, medical and general admission notes stored in inventory
5. Fill out your personal log to track your activity in the simulation

Instructions

Enter the simulation room, introduce yourself to the rest of the student team and familiarise yourself with the patient (Lionel) and general environment. Read and develop an understanding of the emergency room, including admission notes. (This may take some time, so do not rush this and use as many resources as you can). Adjust your avatar to suit your personal preferences, get familiar with the required communication tools. Find and get used to working with the viewing tool. Learn how to use the equipment in the room through the operational flow chart.

Outline discussions, decisions made and steps taken by your team as well as personal lessons learned in your Personal Log. When you complete Lesson 1, please move on to Lesson 2.

Lesson 2—Simulation One

Videos to view

1. Simulation 1 (3 Videos)

Actions

1. Scan your environment and meet your team members
2. Review pathology, radiology, medical, and general admission handover reports for this simulation
3. Work through Simulation 1 with your team
4. Fill out your personal log to track your activity in the simulation

Instructions

Your team should aim to assess and solve the situation presented to you. In your team, discuss this patient's presentation and the way in which you will have to cope with the process of getting this patient stable. Discuss the challenges of working in a team quickly to assist a patient. Outline the key aspects to be included in a handover to the night staff.

Communication with family is an important part of working in a medical environment. In this simulation, you meet Lionel's wife Sue. Think about the best way to communicate with Sue throughout these lessons. Outline discussions, decisions made and steps taken by your team as well as personal lessons learned in your Personal Log. When you complete Lesson 2, please move on to Lesson 3.

Lesson 3—Simulation Two

Videos to view

1. Simulation 2 (1 Video)

Actions

1. Scan your environment and meet your team members
2. Review pathology, radiology, medical and general admission handover reports for this simulation
3. Work through Simulation 2 with your team
4. Fill out your personal log to track your activity in the simulation

Instructions

Your team should aim to assess and solve the situation presented to you. In your group come up with responses to the following challenges: Develop a series of questions you would ask, a physical assessment of Lionel and what you might expect to find given his diagnoses and past history. Discuss the pharmacological aspects of care and the significance of your observations.

Think about the questions you would ask Lionel and his wife during your admission interview.

Outline discussions, decisions made and steps taken by your team as well as personal lessons learned in your Personal Log. When you complete Lesson 3, please move on to Lesson 4.

Lesson 4—Simulation 3

Videos to view

1. Simulation 3 (1 Video)

Actions

1. Scan your environment and meet your team members
2. Review pathology, radiology, medical and general admission handover reports for this simulation
3. Work through Simulation 3 with your team
4. Fill out your personal log to track your activity in the simulation

Instructions

Your team should aim to assess and solve the situation presented to you. Confrontation and demands of other medical staff can put you in a difficult position. Outline the way in which you would respond to the requests made of you in this lesson. Outline discussions, decisions made and steps taken by your team as well as personal lessons learned in your Personal Log. When you complete Lesson 4, please move on to Lesson 5.

Lesson 5—Simulation 4

Videos to view

1. There is no video for this simulation

Actions

1. Scan your environment and meet your team members
2. Review pathology, radiology, medical and general admission handover reports for this simulation
3. Work through Simulation 4 with your team
4. Fill out your personal log to track your activity in the simulation

Instructions

Your team should aim to assess and solve the situation presented to you. Reflect what has happened to Lionel from a physiological and psychological point of view and the clinical reasoning behind his current management from an Oxygen, IVT, and Pharmacological perspective.

Outline discussions, decisions made and steps taken by your team as well as personal lessons learned in your Personal Log. When you complete Lesson 5, please move on to Lesson 6.

Lesson 6—Simulation 5

Videos to view

1. Simulation 5 (1 Video)

Actions

1. Scan your environment and meet your team members

2. Review pathology, radiology, medical and general admission handover reports for this simulation
3. Work through Simulation 5 with your team
4. Fill out your personal log to track your activity in the simulation

Instructions

Your team should aim to assess and solve the situation presented to you. Describe your immediate response to finding Lionel in this state. What is the likely cause of Lionel's current condition? Working with patients that are as sick as Lionel can be challenging emotionally—have a discussion with other team members about the ways in which you could manage your emotions both now and later. How can you best support Sue during this difficult time?

Outline discussions, decisions made and steps taken by your team as well as personal lessons learned in your Personal Log. When you complete Lesson 6, please move on to Lesson 7.

Lesson 7—Simulation 6

Videos to view

1. Simulation 6 (2 Videos)

Actions

1. Scan your environment and meet your team members
2. Review pathology, radiology, medical and general admission handover reports for this simulation
3. Work through Simulation 6 with your team
4. Fill out your personal log to track your activity in the simulation

Instructions

Your team should aim to assess and solve the situation presented to you. Take a breath … think about what has happened to Lionel and describe the reasoning behind his management. Look for some research material to support this treatment and Lionel's future management. Discuss the notion of delegation. How would you delegate the care of your other patients so that you are able to focus on the care Lionel needs and how do you feel about that?

Outline discussions, decisions made and steps taken by your team as well as personal lessons learned in your Personal Log. When you complete Lesson 7, please move on to Lesson 8.

Lesson 8—End of Simulation Reflection

Video to watch: Ending Video

Actions

1. Reflect on your time with Lionel, your team and your personal experience
2. Fill out your personal log to track your activity in the simulation

Instructions

Think about what has happened to Lionel and describe the reasoning behind his management. You will see a defibrillator next to Lionel's bed—at what stage do you think it would have been appropriate to defibrillate Lionel? What response would be appropriate for Lionel when he asked—"Am I going to die?" Unfortunately this is where we leave Lionel. I have always found closure one of the most difficult aspects of the job. Write about how you would deal with this situation ... there is no right or wrong answer. Reflect on using these simulations and how you personally felt while using them. Outline discussions, decisions made and steps taken by your team as well as personal lessons learned in your Personal Log.

REFERENCES

Aldrich, C. (2004). *Simulations and the future of learning*. San Francisco, CA: Wiley.

Aldrich, C. (2005). *Learning by doing: A comprehensive guide to simulations, computer games, and pedagogy in e-learning and other educational experiences*. San Francisco, CA: Pfeiffer.

Alinier, G. (2007). A typology of educationally focused medical simulation tools. *Medical Teacher, 29*(8), 243-250.

Alinier, G., Harwood, C., Gordon, R., & Hunt, B. (2006). Effectiveness of intermediate-fidelity simulation training technology in undergraduate nursing education. *Journal of Advanced Nursing*, *54*(3), 359- 369.

Andrews, D. H., & Bell, H. H. (2000). Simulation-based training. In S. Tobias & J. D. Fletcher (Eds.), *Training and retraining: A handbook for business, industry, government, and the military* (pp. 357–384). New York, NY: Palgrave Macmillan.

Andrews, G., & Harris, M. (1964). *The syndrome of stuttering: Clinics in developmental Medicine* (Vol. 17). London: Heinemann.

Baker, S. C., Wentz, R. K., & Woods, M. M. (2009). Using virtual worlds in education: Second Life (R) as an educational tool. *Teaching of Psychology, 36*(1), 59. Retrieved from http://pdfserve.informaworld.com/62692_751313631_907770663.pdf

Benson, P. (2004). Online learning: A means to enhance professional development. *Critical Care Nurse, 24*(1), 60-63.

Berge, Z. (2008). Multi-user virtual environments for education and training? A critical review of Second Life. *Educational Technology, 48*(3), 27-31.

Beyea, S. C., & Kobokovich, L. J. (2004). Human patient simulation: A teaching strategy. *AORN Journal, 80*(4), 738-742.

Blumgart, E., Tran, Y., & Craig, A. (2010a). Social disorder in adults who stutter. *Depression & Anxiety, 27*(7), 687-692.

Blumgart, E., Tran, Y., & Craig, A. (2010b). An investigation into the personal financial costs associated with stuttering. *Journal of Fluency Disorders, 35*(3), 203-215.

Bronack, S.C., Riedl, R., & Tashner, J. (2005). Teaching in 3D: Developing learning communities in a multi-user virtual environment. In C. Crawford, D. A. Willis, R. Carlsen, I. Gibson, K. McFerrin, J. Price, & R. Weber (Eds.), *Proceedings of the Society for Information Technology and Teacher Education International Conference 2005* (pp. 2166–2170). Chesapeake, VA: Association for the Advancement of Computing in Education.

Bruckman, A. (2002). The future of E-learning communities. Supporting community and building social capital: *Communications of the Association for Computing Machinery, 45*(4), 60-63. Retrieved from http://portal.acm.org/citation.cfm?id=505274

Carbonaro, M., King, S., Taylor, E., Satzinger, F., Snart, F., & Drummond, J. (2008). Integration of e-learning technologies in an interprofessional health science course. *Medical Teacher, 30*(1), 25.

Clague, J. E., Reed, P. G, Barlow, J., Rada, R., Clarke, M., & Edwards, R. H. (1997). Improving outpatient clinic efficiency using computer simulation. *International Journal of Healthcare Quality Assurance, 190*(4-5), 197–201.

Corcoran, J., & Stewart, M. (1998). A qualitative analysis of interview narratives. *Journal of Fluency Disorders, 23*(4), 247-264.

Craig, A., & Tran, Y. (2006). Fear of speaking: Chronic anxiety and stammering. *Advances in Psychiatric Treatment, 12*, 63-68.

Curran, V. R., Aziz, K., O'Young, S., & Bessell, C. (2004). Evaluation of the effect of a computerized training simulator (ANAKIN) on the retention of neonatal resuscitation skills. *Teaching and Learning in Medicine, 16*(2), 157-164. Retrieved from http://pdfserve.informaworld.com/718526_751313631_785833716.pdf

de Freitas, S. (2008). *Serious virtual worlds: A scoping study.* Bristol, England: Joint Information Systems Committee. Retrieved from http:// www.jisc.ac.uk/media/documents/publications/seriousvirtualworldsv1.pdf

Dede, C. (2005). *Planning for neomillennial learning styles: Implications for investments in technology and faculty.* Cambridge, MA: Harvard University Press. Retrieved from: http://www.gse.harvard.edu/~dedech/DedeNeoMillennial.pdf

Dickey, M. D. (2005). Three dimensional virtual worlds and distance learning: Two case studies of active worlds as a medium for distance education. *British Journal of Educational Technology, 36*(3), 439-541.

Duffy, T. M., & Cunningham, D. J. (1996). *Constructivism: Implications for the design and delivery of instruction.* In D. H. Jonassen (Ed.), Educational communications and technology (170-199). New York, NY: Simon & Schuster Macmillan.

Engum, S., & Jeffries, P. R. (2003). Intravenous catheter training system: Computer-based education vs traditional learning methods. *American Journal of Surgery, 186*(1), 67–74.

Fadde, P. J. (2004, October). *How real does virtual reality training really need to be? Variables of display and input in media-based simulation.* Paper presented at the 2004 Association for Educational Communications and Technology International Convention, Chicago, IL.

Fanderclai, T. L. (1995). MUDs in Education: New Environments, New Pedagogies. *Computer- Mediated Communication Magazine, 2*(1). Retrieved from: http://www.ibiblio.org/cmc/mag/1995/jan/fanderclai.html

Fanning, R., & Gaba, D. (2008). Simulation-based learning as an educational tool. In J. Stonemetz & K. Ruskin (Eds.), *Anesthesia informatics* (pp. 459–479). New York, NY: Springer.

Farrell, M., & McGrath, I. (2001, December). *Online Learning in an undergraduate nursing program: The perceptions of the learning technology mentors.* Proceedings of the Annual Conference of the Australasian Society for Computers in Learning in Tertiary Education, Melbourne (pp. 197-205). Melbourne, Australia: ASCILITE

Gao, F., Noh, J., & Koehler, M. (2008). *Comparing student interactions in Second Life and face-to-face role-playing activities.* Proceedings of Society for Information Technology and Teacher Education International Conference 2008 (pp. 2033-2035). Chesapeake, VA: AACE. Retrieved from: https://www.msu.edu/~gaofei1/research/08slpaper.pdf

Gee, J. P. (2003). *What video games have to teach us about learning and literacy.* New York, NY: Palgrave Macmillan.

Ginsberg, A. P., & Wexler, K. B. (2000). Understanding stuttering and counselling clients who stutter. *Journal of Mental Health Counseling, 22*(3), 228-239.

Good, R. (2004, September 27). 3D virtual spaces for learning and collaboration [Web log post]. Retrieved from http://www.masternewmedia.org/2004/09/27/3d_virtual_spaces_fo r_learning.htm

Grant, J., & Marsden, P. (1992). *Training senior house officers by service-based training.* London: Joint Centre of Medicine Education.

Haycock, K., & Kemp, J. W. (2008). Immersive learning environments in parallel universes: Learning through Second Life. *School Libraries Worldwide, 14*(2), 89-97. Retrieved from http://asselindoiron.pbwiki.com/SLW+14:2+Kemp+and+Haycock

Hays, R., & Singer, M. (1989). *Simulation fidelity in training system design.* New York, NY: Springer-Verlag.

Heinrichs, W. L., Youngblood, P., Harter, M., & Dev, P. (2008). Simulation for team training and assessment: Case studies of online training with virtual worlds.

World Journal of Surgery, 32(2). Retrieved from http://www.springerlink.com/content/82211030u48h0115/fulltext.pdf

Higgins, G., & Champion, H. (2000). The military simulation experience: charting the vision for simulation training in combat trauma. TATRC Report No. 01-03x. U.S. Army Medical Research and Material Command, Fort Detrick, MD. Retrieved from http://www.tatrc.org

Higgs, J., & Titchen, A. (2000). Knowledge and reasoning. In J. Higgs & M. Jones (Eds.), *Clinical reasoning in the health professions* (23-32). Oxford, England: Butterworth-Heinemann

Hmelo-Silver, C., Duncan, R. G., & Chinn, C. A. (2007). Scaffolding and achievement in problem-based and inquiry learning: A Response to Kirschner, Sweller, and Clark. *Educational Psychologist, 42*(2), 99–107.

Hussain, A. (2008). *Job interviews and stuttering* [Web log post]. Retrieved from http://ezinearticles.com/?Job-Interviews-And-Stuttering&id=1208930

Iverach, L., O'Brian, S., Jones, M., Block, S., Lincoln, M., Harrison, E., ... Hewat, S. (2009). Prevalence of anxiety disorders among adults seeking speech therapy for stuttering. *Journal of Anxiety Disorders, 23*(7), 928-934.

Jeffries, P. (2006). Designing simulations for nursing education. *Annual Review of Nursing Education, 4,* 161–177.

Jeffries, P., Woolf, S., & Linde, B. (2003). Technology-based vs. traditional instruction: A comparison of two methods for teaching the skill of performing a 12-lead ECG. *Nursing Education Perspectives, 24*(2), 70-74.

Johnson, J., Zerwic, J., & Theis, S. (1999). Clinical simulation laboratory and adjunct to clinical teaching. *Nurse Educator, 24*(5), 37–41.

Jonassen, D. (1998). Designing constructivist learning environments. In C.M. Reigeluth (Ed.), *Instructional theories and models*. Mahwah, NJ: Lawrence Erlbaum Associates.

Jonassen, D. H., Peck, K. L., & Wilson, B. G. (1999). *Learning with technology: A constructivist perspective*. Upper Saddle River, NJ: Merrill/Prentice-Hall.

Jonassen, D. H., & Reeves, T. C. (1996). Learning with technology: Using computers as cognitive tools. In D. H. Jonassen (Ed.), Handbook of research for educational communications and technology (pp. 693-719). New York, NY: Macmillan.

Kamhi, A. G. (2003). Two paradoxes in stuttering treatment. *Journal of Fluency Disorders*, 28, 187-196.

Kenny, A. (2002). Online learning: Enhancing nurse education? *Journal of Advanced Nursing, 38*(2), 127.

Klein, J. F., & Hood, S. B. (2004). The impact of stuttering on employment opportunities and job performance. *Journal of Fluency Disorders, 29*(4), 255-273.

Kolb, D. A. (1984). *Experiential learning: Experience as the source of learning and development*. Upper Saddle River, NJ: Prentice-Hall.

Koutantji, M., McCulloch, P., Undre, S., Gautama, S., Cunniffe, S., Sevdalis, N., ... Davis, R. (2008). Is team training in briefings for surgical teams feasible in simulation? *Cognition Technology and Work, 10*(4), 275-285. Retrieved from http://www.springerlink.com/content/ 335k4272p1x1w526/fulltext.pdf

Kwan, Y., So, M., Lai, P., & Tiwari, A. (2006). A comparison of the effects of problem-based learning and lecturing on the development of students' critical thinking. *Medical Education, 40*(6), 547-554.

Lave, J., & Wenger, E. (1991). Situated learning: Legitimate peripheral participation. Cambridge, England: Cambridge University Press.

La Trobe University. (2005). What is stuttering? [Web log post]. Retrieved from http://www.latrobe.edu.au/istutter/?q=node/67

Lee, C. H., Liu, A., Del Castillo, S., Bowyer, M., Alverson, D., Muniz, G., & Caudell, T. P. (2007). Towards an immersive virtual environment for medical team training. *Studies in health technology and informatics, 125,* 274-279. Retrieved from http://simcen.org/pdf/lee_mmvr07%20ver%202.pdf

Lingard, L., Espin, S., Whyte, S., Regehr, G., Baker G. R., Reznick, R., ... Grober, E. (2004). Communication failures in the operating room: an observational classification of recurrent types and effects. *Quality and Safety Health Care, 13*(5), 330-334. Retrieved from http://qshc.bmj.com/cgi/ reprint/13/5/330.pdf

Lombardi, J., & McCahill, M. P. (2004, January). *Enabling social dimensions of learning through a persistent, unified, massively multi-user, and self-organizing virtual environment.* Proceedings of Creating, Connecting and Collaborating through Computing (pp. 166-172), Kyoto, Japan.

McLaughlan, R. G., & Kirkpatrick, D. (2004). Online roleplay: Design for active learning. *European Journal of Engineering Education, 29*(4), 477-490.

Mili, F., Barr, J., Harris, M., & Pittiglio, L. (2008). Nursing training: 3D game with learning objectives. Proceedings of First International Conference on Advances in Computer-Human Interaction (pp. 236-242). Sainte Luce, Martinique: IEEE Computer Society.

Miller, C., Lee, M., Rogers, L., Meredith, G., & Peck, B. (2010). Enhancing tertiary healthcare education through 3D MUVE-based simulations. In Vincenti, G., & Braman, J. (Eds.), *Teaching through Multi-User Virtual Environments* (pp. 341-364). Hershey, PA: IGI Global.

Nykvist, S. (2005). *Establishing a sense of community within an online learning environment using open source software.* In Proceedings WCCE 2005— The eighth IFIP World Conference on Computers in Education, Cape Town, South Africa.

Orlansky, J., Dahlman, C. J., Hammon, C. P., Metzo, J., Taylor, H. L., & Youngblut, C. (1994). The value of simulation for training. (IDA P-2982). Alexandria, VA: Institute for Defense Analyses.

Packman, A., & Merideth, G. (2011). Technology and the evolution of clinical methods for stuttering. *Journal of Fluency Disorders, 36*(2), 75-85.

Petty, G. (2004). *Teaching today: A practical guide* (3rd ed.). Cheltenham, England: Nelson Thornes.

Piaget, J. (1972). *Psychology and epistemology: Towards a theory of knowledge.* Harmondsworth, England: Penguin.

Prensky, M. (2001). Digital natives, digital immigrants. *On the Horizon, 9*(5). Retrieved from http://www.marcprensky.com/writing/ Prensky%20-%20Digital%20Natives,%20Digital%20Immigrants%20-%20Part1.pdf

Prensky, M. (2006). *Don't bother me Mom—I'm learning: How computer and video games are preparing your kids for 21st century success and how you can help!* St. Paul, MN: Paragon.

Ritzema, T., & Harris, B. (2008). The use of second life for distance education. *Journal of Computing Sciences in Colleges, 23*(6), 110–116.

Rogers, L. (2008). *Virtual worlds: a new window to healthcare education*. Proceedings of Australian Society for Computers in Learning in Tertiary Education, Melbourne, Australia. Retrieved from http://www.ascilite.org.au/conferences/melbourne08/procs/rogers-poster.pdf

Rogers, L., Miller, C., & Firmin, S. (2011). Evaluating the impact of a virtual emergency room simulation for learning. In D. Holt, S. Segrave, & J. Cybulski (Eds.), *Professional education using e-simulations: Benefits of blended learning design*. Hershey, PA: IGI Global.

Rouvrais, S., & Gilliot, J. (2004). *A pedagogical canvas for on-line simulation-based lessons*. In Proceedings of the World Conference on Educational Multimedia, Hypermedia and Telecommunications (EDMEDIA) 2004, Lugano, Switzerland.

Saunders, R. L. (2007). The genesis of a virtual world revisited. *International Journal of Web-Based Communities, 3*(3), 271–282.

Sexton, J. B., Thomas E. J., & Helmreich R. L. (2000). Error, stress, and teamwork in medicine and aviation: Cross sectional surveys. *British Medical Journal, 320*(7237), 745-749.

Smith, D. (2004). For whom the bell tolls: Imagining accidents and the development of crisis simulation on organizations. *Simulation & Gaming, 35*(3), 347-362.

Sontag, M. (2009). A learning theory for 21st-century students. *Innovate Journal of Online Education, 5*(4). Retrieved from http://innovateonline.info/index.php?view=article&id=524

Taber, N. (2008). Emergency response: E-learning for paramedics and firefighters. *Simulation & Gaming, 39*(4), 515-527. Retrieved from http://sag.sagepub.com/cgi/reprint/39/4/515

Taylor, C., & Eustis, J. (2002). Assessing the changing impact of technology on teaching and learning at Virginia Tech: A case study. *New directions for Institutionalised Research, 1999*(102), 55-70.

Teeson, K., Packman, A., & Onslow, M. (2003). The Lidcombe behavioural data language of stuttering. *Journal of Symbolic Interaction, 46*, 1009-1015.

U.S. Census Bureau. (2011). World POPClock projection. Retrieved from http://www.census.gov/ipc/www/popclockworld.html

Vygotsky, L. S. (1978). *Mind and society: The development of higher mental processes*. Cambridge, MA: Harvard University Press.

Waltz, C. F., Strickland, O., & Lenz, E. R. (2004). *Measurement in nursing and health research* (3rd ed.). Berlin: Springer.

Wankel, C., & Kingsley, J. (Ed.). (2009). *Higher education in virtual worlds: Teaching and learning in Second Life*. Bingley, England: Emerald Group.

Whitton, N., & Hynes, N. (2006). Evaluating the effectiveness of an online simulation to teach business skills. *e-Journal of Instructional Science and Technology,*

9(1). Retrieved from http://www.usq.edu.au/electpub/e-jist/docs/vol9_no1/papers/current_practice/whitton_hynes.pdf

Wolburg, J., & Pokrywczynski, J. (2001). A psychographic analysis of Generation Y college students. *Journal of Advertising Research*, *41*(5), 20–52.

Wright, W. (1989). *SimCity* [computer software]. Moraga, CA: Maxis.

Yaruss, J. S. (1998). Describing the consequences of disorders: Stuttering and the international classification of impairments, disabilities and handicaps. *Journal of Speech, Language and Hearing Research*, *41*, 249-257.

Yaruss, J. S., & Quesal, R. W. (2001). The many faces of stuttering. *ASHA Online*, *6*(14).

Ziv, A., Small, S. D., & Wolpe, P. R. (2000). Patient safety and simulation-based medical education. *Medical Teacher*, *22*(5), 489-495

CHAPTER 9

TRAINING FOR TECHNOLOGICAL DEMOCRACY

A Civic Engagement Class Example

Michelle K. Gardner and Thomas A. Bryer

Implementing a civic engagement course in one of the lowest youth-engaged states was going to be a challenge. Adding a layer of technology use to such a program presented the necessary hook to get the students involved. Overall, the two civic engagement courses taught online at the University of Central Florida (UCF) in Spring and Summer 2010 were successful at introducing students to the potential for utilizing Internet based collaborative technologies for Civic Engagement. In this chapter we describe a special course offered at a 4-year college in conjunction with the Orange County Supervisor of Elections Office. With an emphasis on civic engagement through technology, each student was trained in a virtual environment to serve Orange County as poll workers for the November 2010 general election.

Engaging the Avatar: New Frontiers in Immersive Education, pp. 255–275

Introduction

College-aged youth are engaging communities and democratic institutions in new ways. Dalton (2009) finds that this age demographic tends to be more "engaged" as opposed to "dutiful" citizens. They are more culturally and globally aware than older generations, and they see their responsibility as reaching beyond the ballot box and toward pursuit of social ends such as justice and equity. Younger citizens are more likely than older peers to seek opportunities for engagement with the government through the Internet, social media, and networking tools (Smith, 2010). The new generation of citizens is more inclined to trust and be influenced by Internet based collaborative technologies than by traditional methods of civic communication (Raphael, Bachen, Lynn, Baldwin-Philippi, & McKee, 2009, p. 205). Among the tools made available through the Internet are those that enable and facilitate individual and group-level empowerment. For example, wiki technology allows for co-creation and production of content, thus democratizing information sharing and dissemination. Overall, these tools make the costs of participation (Cooper, 1979) exceedingly low.

In Florida, location of this study both young and established citizens, are some of the least engaged in the United States (National Conference on Citizenship, 2009). Indicators of engagement captured by the National Conference on Citizenship include voting, attendance at public meetings, volunteering, donating to charity, and helping a neighbor in need. The potential for engagement provided by social media and networking technologies is not being realized even though research has demonstrated a clear connection between students highly engaged in online political groups and nongovernment agencies and a high level of offline participation (Calenda & Meijer, 2009, p. 890). Joseph Kahne, Ellen Middaugh, and Chris Evans (2008) demonstrated that 70% of students who were actively involved in game based communities (offering advice, posting to game discussion boards, etc.) go online to research political and current event information.

The purpose of the civic engagement class was to leverage Internet based collaborative technologies to engage college-aged youth in Florida. This call to action formed during a statement from former Governor Bob Graham (2010), who suggested students use technologies to design and implement a civic initiative aimed at a perceived problem in their community (e.g., illegal dumping, neighborhood beautification, street lighting). Our response to this challenge was to introduce an Internet based virtual world to train poll workers for the November 2010 elections in Orange County, Florida. Social media, networking tools, and the virtual world known as Second Life (SL), were selected to teach the class. These tools form part of the technology choices that local government agencies are evaluating for educational and engagement purposes. One of goals was to ensure students demonstrated competency in the use of these tools as they prepare to be effective active citizens in a technology rich world.

Our civics engagement course was designed entirely as a web-based course, with the core instruction set embedded into the Blackboard Learning Systems. In Blackboard, course content is delivered through modules. Personal communication is conducted using the mail system. Other class activity occurs within publicly accessible spaces: a shared Internet-based wiki, Facebook, and Second Life. The course was designed to teach students how to use social media and networking tools for civic engagement through the instructor's use of these during the instructional process. Once completed with the course, students would be expected to engage effectively as active citizens through various social media and networking tools.

The use of virtual space, such as Second Life, can be valuable as an alternative to, not a replacement for, face-to-face space for training and teaching. Teaching using publicly open and accessible spaces presents unique challenges for faculty and students (Bryer & Chen, 2010). Faculty cite concerns regarding privacy, technology complexity, and limited time for learning as reasons to not use social media and networking tools for teaching. At the same time, faculty recognize virtual world research results suggest that student learning outcomes are comparable to those achievable in face-to-face environments; although they understand the research is fairly limited at this writing.

Virtual spaces, as unique spaces within the wider array of social media and networking tools available, may be most beneficial in terms of achievement of learning outcomes. Learning, according to Vygotsky, Vygotsky, and Cole (1978) is fundamentally a social process. Virtual spaces with simulated environments allow users to interact both through spoken and/or written word as well as through visual imagery, such as avatar movement or gesture. Other tools, such as Facebook or wiki technologies allow for co-creation of content and efficient sharing of information; however, they lack the intensity of interaction that exists in virtual spaces, including opportunity for role play simulations. Virtual environments have the added benefit of engaging the spoken and written word as potential simulations to reinforce the learned skills (Gee, 2007, p. 218). This course facilitated student exposure to unique environments, diverse cultures, institutions, and, for the purposes of the class, civic spaces.

Government Uses of Virtual Environments

The vast landscape of SL lends itself to exploration. Locating, visiting and documenting all governments builds within SL involves a tremendous amount of time. We are not so bold as to claim an infinite number of builds but a large enough number that exploration of them all is not feasible. Of the four builds described below, there are two federal agencies and two local agencies. None of these builds are involved in supporting or funding this written work and are only suggested here due to the authors' knowledge of SL.

The two most elaborate government builds in SL are the National Aeronautics and Space Administration (NASA) and the National Oceanic and Atmospheric Administration (NOAA). Both of these builds focus on education, historical content, and engagement. The NASA CoLab features a museum of the space program, immersive, 3D models of various spacecraft and satellites, interactive learning tools based on the space program, and a live-stream video of the astronauts on the International Space Station. Weather is a focus at the NOAA build featuring an interactive, 3D live weather map of the United States. Beyond weather the NOAA sim features a tsunami experience, a flight into the eye of a hurricane and an immersive underwater exhibit complete with animated sea-creatures that are clickable links to Wikipedia pages of that particular creature. Both organizations feature many more interactive spaces in SL than space here allows.

Smaller municipal organizations located in SL as of November 2010 are the Virtual Town Munigov2.0 and the Central Florida Memory supported by the Orange County Florida Library System. Virtual Town planned by the members of MuniGov, represents a typical town with interactive municipal

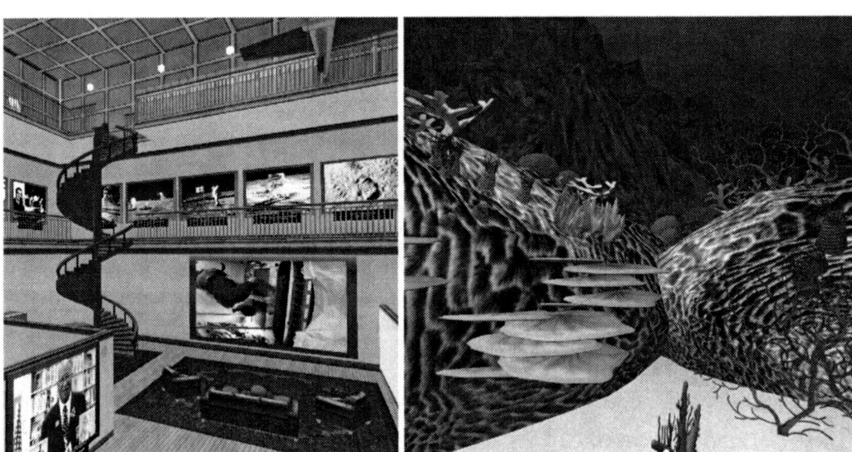

Figure 9.1. Image on left is NASA SL build and image on right is the NOAA Ocean build.

buildings such as paramedic training simulator and a library with Web 2.0 resources for governments, all surrounding a grassy town square. Central Florida Memory is a collaborative project designed to tell the story of Central Florida through the digitization of historical documents searchable through the SL space. Similar to the NOAA build, this space is full of clickable linked items with full explanations of the history of that particular item as well as the historical value or purpose of the clicked item.

Description of Virtual Space

Located at the online@UCF space in SL, the UCF Virtual Voting Center is patterned after a typical one-story community center found in Central Florida. Outside the building, there are several native palm trees and low ground shrubs. An asphalt-parking surface is clearly marked with disabled and nondisabled parking spaces. Entering the building via the ramp or the steps the SL visitor encounters his/her first real documentation from the Orange County Supervisor of Elections Office (OCSoE). Prominently displayed on easels outside the main entrance are two boards. One board displays a map of Orange County with that particular precinct highlighted and the other board in both English and Spanish are the Voter's Bill of Rights and Responsibilities, a notice about voting fraud and a notice referring to the Florida statute banning photography within 100 feet of the voting space. All of these documents are Orange County requirements.

Figure 9.2. Image on the left is Munigov2.0 Virtual Town. Image on right is Central Florida Memory.

Figure 9.3. UCF Virtual Voting Center.

Figure 9.4. Virtual Voting Center.

Inside the front door is one of two existing, fully immersive, interactive voting spaces constructed in SL. (The other virtual voting space was established by Alameda County California several months before UCF). Set up in the exact formation of a perfect voting space, the UCF Virtual Voting Center is a state of the art system. To the left of the door is a near perfect replication of a DS200 Optical Scanner, which is the device that a voter inserts their marked ballot in once it is completed. Beyond the scanner, along the left and opposite wall are voting booths. Typical in design to most precincts, these booths are programmed in SL to animate an avatar to vote while s/he stands in front of it. The voting avatar will mimic writing on the surface of the voting booth. On the wall opposite the front door is a touch screen tabulator for disabled voters. Both the DS200 Optical Scanner and the touch screen tabulator can be unplugged and plugged into wall sockets by the virtual machines. In the far corner diagonal to the door is a table set up with a laptop computer, several pads of ballots, pens, pencils and various documents that are exclusive to the OCSoE voting process. This is the Internet Oath Person's (IOP) table.

On the avatar's right, the door is the focal point of our training sessions in SL. Two laptop computers are set on two folding tables and are surrounded by a sea of papers. These two computers are the ePoll Book Inspector's work location. UCF and VCC students were trained as ePoll Book Inspectors for the November 2, 2010 general election. Each table in this ideal voting center is outfitted with all the proper documents from the OCSoE office.

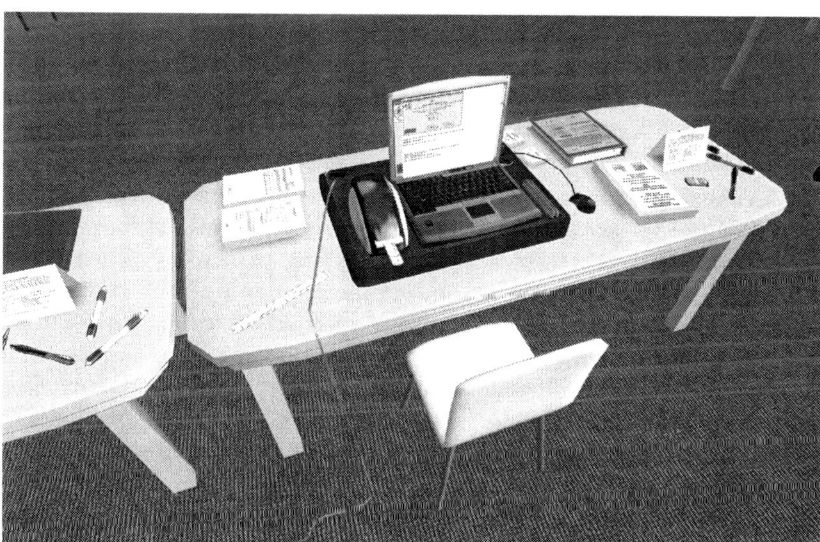

Figure 9.5. Image is the ePoll Book Inspectors work area.

As with other objects in this space, the computers have interactive screens. This particular feature was not available with our first class in Spring of 2010. Fortunate for this project, Linden Lab updated the SL viewer to a version that allowed access to and interaction with websites external to SL. This update enhanced our users' hands-on experience and the summer and fall 2010 students were able to have a completely immersive experience with their training. Using one name in a fictitious database, the avatar is able to check in voters, confirm IDs, addresses, and even handle voters with mismatched information. (A video tour of the UCF Virtual Voting Center is on YouTube at this link: http://www.youtube.com/watch?v=igTH3B9Bfqc)

Instructional Use of Virtual Space

The in world sessions last for a total of six hours. Students are sent a SLurl through the Blackboard Learning System e-mail and on a specific date and time on where to meet at the Online@UCF SL space (http://slurl.com/secondlife/arts%20and%20letters/200/185/22/). The first part of the training includes a brief SL orientation session to introduce students to the environment and impart the basic skills necessary to fully engage the training. Then, students gather together in the UCF Treetop Classroom to participate in the Poll Workers Sensitivity Training. The first session in world lasts approximately two hours. The second part of the training takes approximately four hours. The second session includes two ½-3 hours of lecture covering how to be a poll worker on election day. The length of part 2 varies based on technical difficulties and questions from participants. Part 2 concludes with an interactive, role-play simulation in the UCF Virtual Voting Center, which involves students engaging the ePoll Books by checking in each other, as potential voters.

In a virtual environment we are able to add layers to an online course that would not be possible on a two-dimensional social media platform. The course design attempts to incorporate four principles of instruction for virtual worlds: immersion, integration, identity, and interaction (Atkinson, 2009, p. 31). Unlike latitude granted in typical instructional design, this class had to meet certain Florida Statutes and as a result could not clearly integrate all four of these principles. A secondary goal required a safe introduction to virtual environments with an emphasis on learning the platform and the potential of virtual environments for future civic participation.

Immersion encourages focus in the learning environment (Atkinson, 2009, p. 31). Students in these three courses are required to create accounts and download the SL platform. In our experience, we discovered that too much class time is devoted to more traditional slide and lecture style of training. Since our end result was aimed at serving in the November 2, 2010 elections, the instructor had to conform to the instructional requirements of the OCSoE office. Beyond the basic instructions, the students were encouraged via a class assignment to travel to and interact with at least five different government and education based sims in SL —essentially known as the scavenger hunt learning archetype. The scavenger hunt reinforced the basic skills learned in the orientation session. Research into civic learning based game design indicates games that incorporate civic experiences do increase the likelihood of civically engaged youth (Raphael et al., 2009, p.201). After analyzing student reflection papers, it was clear that the students appreciated the potential depth for creativity and engagement afforded by a virtual environment. A large number of students were surprised by the way government was using the SL space and reported that the scavenger hunt gave them a better appreciation for the environment. Touring SL allowed for student immersion into the virtual environment that is not possible through a traditional classroom setting.

Interactivity is a key tool necessary in a three-dimensional educational setting. Kapp and O'Driscoll (2010) explain utilization of both interactivity and immersion in virtual environments promotes student engagement on a new level of learning. The UCF Virtual Voting Center replicated an OCSoE polling location, emphasizing interactivity while remaining within our budget requirements, a top design priority. Our designed included such detail as ensuring that the electrical equipment could be plugged and unplugged into wall sockets, the lights turned on and off at the wall switch, the binders on the ePoll Book Inspector's table opened and closed with an avatar's touch, and the screens on the ePoll Books could access the Internet directly. With the introduction of the new SL viewer 2.0 for the second and third semester, we were able to simulate the experience by responding to student input in the ePoll Books directly. Instead of a static image on the two computers, the buttons on the screens reacted to a mouse click and enabled the students to advance the screen as the live system does on election day. Similar to results found by the Situated Avatar-Based Immersive Learning program (SAIL) based at North Carolina State, we discovered students were highly engaged in the interactive scenarios (Bares, Zettlemoyer, & Lester, 1998, p. 84). The simulation experience, as explained by Gee (2007) and through student self-reports, solidified much of the content covered in the previous 5 hours of training. Having to manipulate the SL browser in conjunction with an Internet browser screen caused the first class tremendous frustration as seen in many of the self-reported technological issues. The advancement of the SL viewer capabilities enabled us to immerse the students without interruption into the voting simulation.

Embodying student personality into an avatar encourages meeting synchronously supporting a sense of community not possible in an asynchronous, two-dimensional, web-based course. By observing avatar behaviors, we have developed a hypothesis that students able to identify with their avatars might become more involved in the virtual environment and which could improve their learning. Gee (2007) discusses the intersections of three different identities to create the learning experience: a virtual identity, a real identity, and a projective identity (pp. 49-50). Assimilating these three identities, according to Gee is powerful "because it is both active and reflexive" (p. 54) in the learning process. The students in the civic engagement course were not involved in SL long enough to create a projective identity or feel a sense of ownership of their avatars, however, which gives rise to the need for deeply studies to investigate the effectiveness of synchronous interactivity. Several students did take time between the first and second sessions to modify and upgrade their appearance. Few students inquired about the potential appearance modifications but through conversation established a desire to find items to upgrade their avatars. Creating a sense of assimilated identity was not critical for the poll worker training. Given the limitations of a 6 hour course, students were not required to discuss the effect of avatar appearance and did not mention an attachment to their avatar. For this class, students did not have to identify with their in world selves in order to synthesize the content.

Several technologies were integrated into the structure of this course to enhance the learning experience as well as introduce new experiences to the students. Housed on the well-known learning management system, Blackboard Learning Systems, online discussions were used to enhance group interaction. E-mail was also used to support communication. In an effort to address the larger community, students posted their community projects on a public wiki. This exposure opened the content up to a broader community to encourage public participation. We do not have any evidence whether participation increased or decreased as a result of using the wiki. We believe this is also an area to research regarding student engagement in virtual worlds.

Why Here? Why Now?

A team of forward thinking academics at the University of Central Florida collaborated to create an innovative way to promote civic engagement on multiple technological platforms. As previously discussed, the use of several Internet based collaborative technologies intersected in this program to educate students on the potential for engagement in many different electronic platforms. This program was developed as a grant proposal to the U.S. Elections Assistance Commission in the summer of 2009 and was awarded the highest dollar amount of the thirteen distributed that year.

Working very closely with the OCSoE office, we discovered a need for an online poll worker training program. They had been approached by several different independent agencies to create an online program. The UCF Virtual Voting Center proved to be the most intriguing idea. Once the Virtual Voting Center was built inworld, a presentation was made to a majority of the staff at the OCSoE office. Overwhelmingly, the space elicited responses regarding the advancement of 3D technology for learning and the ability to create replicated space.

Virtual environments are not considered mainstream technologies. Second Life has not enjoyed the global acceptance of other social media platforms such as Facebook. It is, however, a technology with merit and shows potential promise. One of its key features is the ability to provide learning in a safe classroom environment. Students are able to practice using the concepts of civic engagement in a physical world setting. They can participate without the fear of asking questions or overtalking another speaker, or even looking at something with their avatar. This provides feedback to the instructor in a meaningful way. Students reported they felt confident in their training and felt their ability to serve as a poll worker improved for the November 2010 elections because they were able to "see" the environment in which they would be working in. They felt that moving around in a physical space and getting to understand the placement of the machines, watching the flow of people traffic coming in and out, and visually placing the materials was a benefit in creating a mental version of their workplace. With the exception of the technical issues associated in the simulation, most students reported this part of the training as the most beneficial because they were able to role-play the voter check in process.

Recent research into online engagement points to the idea that students who are engaged online carry this engagement into the offline community. "The new participatory culture offers many opportunities for youth to engage in civic debates, to participate in community life, to become political leaders, even if sometimes only through the "second lives" offered by massively multiplayer games or online fan communities." (Jenkins, 2006, p. 10). Our virtual world civics engagement class gave the students a rare opportunity to engage a virtual environment, practice how to interact with other avatars, as well as become poll workers for the November 2010 election. Most students reported no desire to continue using Second Life for personal interaction but almost all of them realized the potential and can see the value in a virtual environment for educational and business collaboration.

Student Feedback

Several reports of classes taught in SL have published in recent years (e.g., Wankel & Kingsley 2009). Student surveys and self-reports all have near consistent findings: students can see the value of SL for collaboration and e-learning, but the technological hurdles of individual computer issues and lack of intuitive features in SL prevent a large percentage of the students from wanting to participate beyond classroom requirements in SL. Student sentiments expressed in reflection journals from the civic engagement class echo these findings.

Students in the spring and summer 2010 completed reflection journals following their training in SL. Specifically, they responded to the following questions:

(a) Is this a program you would consider using for personal socializing outside of class activities? Why or why not?
(b) Is this a program you would consider using for professional networking or meeting? Why or why not?
(c) Assess the poll worker training (the second class meeting in Second Life). Did the technology permit you to receive information in a clear and coherent manner?
(d) Did the technology permit you to ask any question you had?
(e) Given what you know about Second Life, what recommendations do you have for improving the training?
(f) Last, what would you say are the benefits and drawbacks of using Second Life on a larger scale to train poll workers, instead of the traditional face-to-face training?
(g) Was what you learned in the training well tested in the simulation? Give some examples.
(h) Did you feel prepared for the simulation, and do you feel prepared to work an actual face-to-face election? Why or why not?
(i) Do you have any other comments regarding the use of Second Life in government or for civic engagement?

The authors content coded student responses to these questions to identify themes, including the perceived benefits and drawbacks of SL for teaching and learning. Following a coding process used by Bryer (2009), each coder sorted student statements for each question into preliminary categories based on similarity of idea expressed. The coders subsequently met to discuss the emergent categories, explore differences, and create a set of agreed upon category labels. Each coder then re-coded all statements into the final categories. Across questions, the intercoder agreement rate was 93% at this point, allowing for a high degree of confidence in interpretation of the data.

Usability Outside of Class

The majority of students expressed a willingness to continue using SL outside of class for professional purposes. In stating this willingness, students cited four specific reasons: (1) the technology is interactive, (2) SL allows for global connectivity, (3) the training provided through SL is convenient, and (4) it is efficient, in that individuals can save money that otherwise would be spent on travel. One student brought several of these perspectives together, noting:

> I would definitely like to see more people use the program for professional and business purposes. I feel that training a large group of people that are not all in the same location would be a perfect application to use Second Life. It could save the business money and time by training everyone at one time and not having to travel. It reminds me a lot of three-way-calling, but now you can put a picture with the words.

The fourth justification was also exemplified by one of the tasks students had to complete during a SL scavenger hunt: find and wear a t-shirt from the Munigov2.0 build that reads "I saved my taxpayers $ by meeting in Second Life."

There were students who felt the technology would not be appropriate for professional development and training. Some noted a simple preference for face-to-face meeting; a handful found widespread application unrealistic or too complicated; one student considered the use of avatars to be unprofessional. One student integrated a couple of perspectives, noting:

> Since I do not plan to use Second Life in my social life I do not care to use it in the professional world either. It seems very impersonal and people do not know the real individual they are communicating with. I believe I make a much better impression on an individual in the real world than in the virtual world. Second Life seems to be slightly dated in the business world.

Whereas the majority of students expressed that they would likely use SL for professional development and training, the vast majority expressed an unwillingness to use it for personal socializing. Among those who were not willing to use SL personally, the majority cited two reasons: (1) preference for face-to-face interaction (2) the tool is too complicated. Other reasons cited include a felt lack of control in the environment, insufficient free time to explore the vastness of the environment, unrealistic engagements, and a perception that the environment is boring. One student bridged some of these perspectives, noting:

> Personally I would not use this program. It requires numerous amounts of computer space and it's an altered reality. I'm trying to live the best life I can in the real world. I know it provides ways to socialize, however there are many other ways that have already been put in place. There is so much out there on the internet and you never know what or who you can trust. If I really want to socialize I can use the common networks such as face-book, my-space, and twitter.

The small handful of students who thought SL is desirable for use as a socializing tool identified three reasons: (1) it enabled cultural learning experiences, (2) it is easy to use, and (3) it allows the user to be creative. However, most students did not express these perceptions.

Assessment of the Training

Students perceived communication to flow efficiently and effectively in SL, both in receiving information and in asking questions of others. One student observed,

> One component that I really liked about Second Life was that the technology did permit me to ask any questions I had. One could choose to interact and ask questions using text or through voice chat. While participating in the simulated elections, I experienced difficulties with text, because my internet connection was not fast enough. However, I was able to use voice chat, which allowed me to actively participate and communicate with the citizens wanting to vote.

Expressing initial doubts but a realization that face-to-face training can be simulated, a student noted,

> Prior to the Second Life training, I presumed that Second Life was not the proper way for the poll worker training to be conducted. The reason that I made such a statement is because I perceive myself and many other students as hands on learners. However, the technology did not create any limitation in experiencing the full intended experience of the training. The environment was like any classroom environment where student was given the opportunity to ask questions. In addition, students were given the opportunity to play the role of a poll worker, which provided the hands on learning that I had hoped for. This hands on learning experience is no different than the one that students would have received in the classroom.

Students who experienced technological and connection speed challenges were more likely to assess the training negatively. For instance, a student noted,

> The technology itself caused several issues that affected my experience in using Second Life. I am always hesitant about what new programs I add onto my computer so that it does not become infected with a virus or cause the computer to slow. Despite this usual precautionary measure, I had to download Second Life. After doing so my computer became very slow. I managed to complete the first training session even though it was difficult to navigate the avatar since my slow computer made for a delayed response. Before downloading I made sure that my computer met the system requirements. Even though it did, I still experienced problems with the software.... The morning of the second poll worker training session, my computer crashed when I tried to open the software. The window opened up, but it never completely loaded my avatar or her surroundings. After several attempts the computer gave up. This was an extremely frustrating situation since I did not have enough time to move to another location to try and access the program.

For those students who experienced few if any technology barriers, they assessed the training positively.

Recommendations for Improved Training

Students offered recommendations for how the training might be improved. They focused on five areas for possible improvement: (1) alter the training protocols in some way, (2) increase participation/interaction within the environment, (3) allow for better or more control within the environment, (4) schedule different times for training, and (5) alter the technology used.

With regard to altering technology, one student observed that the simulation might have been enhanced with more interactive features. The student noted:

> If I were to make any adjustments to the poll worker training in Second Life I would place particular emphasis on the interaction of the e-poll machine. The primary deficiency in the training was the lack of interaction with the actual machine. Though we were given slides on what we would be seeing, those really did not give a lot of insight to the actual use of the machine. At the very least the training needs to find a way to have fully interactive e-poll machine software of some kind. When I say fully interactive I would like to emphasize that rather than looking at simple slides, the program would allow the user to type in voter names and scan through voter data bases rather than

> just load the same screen shots over and over again. Also I think this would permit the trainee to simulate several different types of voter situations and give them more of a feel for the voter's who need to visit the IOP. If there could be a way to integrate all of the slides and training manuals within Second Life without having to load them in outside pages that would be incredibly beneficial to the user simply because their Second Life screen would have to be the only window loaded on their computer. Another aspect I would improve would be how the mach election is run. If actual elections could be carried out and Second Life had its own version of an e-poll machine loaded into the virtual simulation, I think students would find it much easier to learn the ins and outs of the machine as well as have more of a purpose for being there rather than just to earn a grade. Our mach election was marginal at best and could only use one certain voter name <s> to correctly complete the series of slides. As I stated before if this situation could be rectified the election process would seem a little more legitimate.

Other students proposed alterations to the training itself, such as options for mixed-mode delivery and longer or shorter training sessions.

Perceived Benefits and Drawbacks Versus F2F

Students observed several benefits and drawbacks of SL for training purposes. Drawbacks were primarily two: technology needs/challenges of users and no face-to-face interaction. Students accessing SL on a computer more than three years old noticed lag issues with the software. Bandwidth proved to be the other challenge with the sessions. We encouraged the class to use a high-speed connection, not public Wi-Fi. Once students engaged the SL browser, many felt the software was not intuitive. SL presented a steep learning curve even with our orientation session. One final drawback mentioned by students was verifiability of a student completing training; not an issue with face-to-face engagement.

Benefits included three primary areas: convenience, efficiency of training, and an ability to reach new audiences. We offered sessions three different nights during the week as well as Saturday mornings. All students were able to meet within the scheduled times. Beyond the time convenience, many students felt SL a well-situated meeting platform accessible from anywhere. Students concluded the training was efficient in both use of time and also in the ability to educate new and diverse audiences. Overall, students appreciated the benefits of SL for the training but struggled as first time users of the technology. Once technology concerns were overcome, students seemed more appreciative of the potential.

Conclusion

Virtual worlds have been touted as the next big thing for several years. Experiences of the authors teaching in world, and analysis of student feedback, suggests both significant potential for these environments as well as potential barriers to effective, widespread use. The SL platform proved to be effective, allowing the vast majority of students to engage in training and interaction in a socially adaptable environment. Some students suffered technology capacity issues, thus limiting the benefits of the virtual environment. Overall, though, students were able to receive information, engage socially, demonstrate their competence through role play simulation, and ultimately apply their virtual training in a real world election.

Transition from SL viewer 1 to 2 presented both new opportunities for interaction and simulation as well as new technology capacity issues. As teachers using this technology, it is important to remain adaptable to changing platforms and to have institutional information technology support to enable necessary changes to computer systems used by faculty and/or students.

Unlike other VW courses, this class did not engage the environment – the students did not have to create anything and had minimal need to manipulate the environment. Thus, this course lacked a constructivist emphasis preferable in a three-dimensional environment. This class mostly recreated the classroom environment in a three-dimensional platform, which was one of the most enjoyable components of the class for students. Future work in using virtual spaces for civic engagement needs to consider the efficacy of constructivism as pedagogical technique.

ACKNOWLEDGMENTS

We are grateful to our project partners, Dr. Terri S. Fine, UCF Political Science Department, Adrienne Matthews, Valencia Community College, and the Orange County Supervisor of Elections Office for their work on this project. We are also grateful to our students at the University of Central Florida who willingly joined us in testing Second Life as a tool for civics training.

This material is based on work supported by a grant from the U.S. Election Assistance Commission (EAC). Any opinion, findings, conclusions, or recommendations expressed in this material are those of the author and do not necessarily reflect the views of the EAC.

APPENDIX: LESSON PLAN

Course Title: Civic Engagement

Course Description

Formal and informal civic engagement methods the United States across levels of government, sectors of society, and modes of interaction.

Course Objectives

This course examines civic engagement in the United States across levels of government, sectors of society, and modes of interaction. Students are encouraged to critically reflect on citizens' responsibilities in a democracy in their various roles across societal spheres, as students, voters, family members, renters, homeowners, drivers, consumers, employees, and employers. Students will also be required as part of the course to enroll in a poll worker training program in Orange County, Florida, and to serve as a poll worker in Orange County on November 2, 2010. The training will occur on Second Life, a simulated environment modeled after the training facility found in downtown Orlando, Florida.

Course Texts

Dalton, R. J. 2009. *The Good Citizen: How A Younger Generation is Reshaping American Politics,* Washington, DC: CQ Press.

Graham, B. 2009. *America, The Owner's Manual: Making Government Work for You,* Washington, DC: CQ Press.

Course Modules

Module 1—Reflections on Citizenship

Students read "The Ones Who Walk Away from Omelas" a short story by Ursula K LeGuin. Highlights from the 2009 Florida Civic Health Index are presented to the class. Students are to reflect on the LeGuin reading and the statistics on Florida in a public post on the course wiki.

Module 2—Meanings of Citizenship

Students read chapters from *The Good Citizen*. Students are asked to reflect with family and friends on the question, "What does it mean to be a citizen in the United States?" All responses are posted on the public wiki. Students are asked to download and install Second Life on their computers.

Module 3—Cases on Meanings

More detailed information from the 2009 Florida Civic Health Index read by the class. Facebook is discussed as a source for civic engagement. Students are asked to engage the Facebook page for the course as an additional online sharing resource. Students are asked to enhance the course Facebook page through contributions such as links to other civic resources.

Module 4—Challenges of Citizenship

Students are assigned to explore blogs attached to news articles. Address questions based on the content of the blogs. Specifically focus on tolerance, civility, and bias.

Module 5—The Role of Government With Citizens

Students are to reflect on (1) trust in government and perception of societal problems, (2) citizen desired functions of government, outside civic engagement, (3) images of government, and (4) preview of next week on forms of engagement and taking action. We will examine some different data that suggests a gulf between our government institutions and the citizenry.

Module 6—Second Life Poll Worker Training

Students participated in two Second Life sessions for a total of six hours to learn all the skills to be a poll worker for the November 2010 General Election.

Module 7—Taking Actions and Defining Problems

Students begin the task of identifying a civic project. Identify a problem on campus, in their apartment building, neighborhood, community, or some other location. Post the problem and discussion of the issue on the public wiki.

Module 8—Informed Citizenship

Students update their problem with data collected about their specific civic project. Further identification of which government agency their will address to assist with the problem. All information is posted on the public wiki.

Module 9—Building Support for your Civic Initiative

Students are assigned readings from *America, The Owner's Manual*. Students are to identify specific populations of citizens and their concerns related to the problem. The project wiki is updated to include letters written or a summary of telephone calls made by the student to a decision maker.

Module 9—Future of Citizenship and Democracy

Students are to reflect on the value of tools such as social media for future citizen engagement.

Course Assessment

Assignment	*Percentage of Grade*
Reflection Essay on Poll Worker Training	30%
Participation on Wikispace, Facebook, or YouTube	20%
Midterm Exam	10%
Class Exercises: Civic Project	30%
Final Exam	10%
Total	100%

REFERENCES

Atkinson, T. (2009). Teaching and learning. *TechTrends: Linking Research & Practice to Improve Learning, 53*(3), 30-32.

Bares, W. H., Zettlemoyer, L. S., & Lester, J. S. (1998, August). Habitable 3D learning environments for situated learning. In B. P. Goetti, H. M., Halff, C. L. Redfield, & V. J. Shute (Eds.), *Proceedings of the 4th International Conference on Intelligent Tutoring Systems*. San Antonio, TX. Berlin: Springer-Verlag.

Bryer, T. A. (2009). Explaining responsiveness in collaboration: Administrator and citizen role perceptions. *Public Administration Review, 69*(2), 271-283.

Bryer, T. A., & Chen, B. (2010). Social networking in teaching public administration. In C. Wankel (Ed.), *Cutting-edge social media approaches to business education: Teaching with LinkedIn, Facebook, Twitter, Second Life, and blogs* (241-267). Charlotte, NC: Information Age.

Calenda, D., & Meijer, A. (2009). Young people, the Internet and political participation. *Information, Communication & Society, 12*(6), 879-898.

Cooper, T, L. (1979). The hidden price tag: Participation costs and health planning. *American Journal of Public Health, 69*(4), 368-374.

Dalton, R. J. (2009). *The good citizen: How a younger generation is reshaping American politics*. Washington, DC: CQ Press.

Gee, J. P. (2007). *What video games have to teach us about learning and literacy*. New York, NY: Palgrave MacMillan.

Graham, B. (2010). *America, the owner's manual: Making government work for you*. Washington, DC: CQ Press.

Jenkins, H. (2006). *Confronting the challenges of participatory culture: Media education for the 21st century*. Chicago, IL: MacArthur Foundation. Retrieved from http://digitallearning.macfound.org/atf/cf/%7B7E45C7E0-A3E0-4B89-AC9C-E807E1B0AE4E%7D/JENKINS_WHITE_PAPER.PDF.

Kahne, J., Middaugh, E., & Evans, C. (2008). *The civic potential of video games*. Chicago: MacArthur Foundation. Retrieved from http://www.civicsurvey.org/Civic_Potential_of_Games.pdf

Kapp, K., & O'Driscoll, T. (2010). *Learning in 3D: Adding a new dimension to enterprise learning and collaboration.* San Francisco, CA: Pfeiffer.

National Conference on Citizenship. (2009). *Florida Civic Health Index 2009: Communities and the State's civic destiny.* Retrieved from http://www.ncoc.net/index.php?tray=content&tid=top57&cid=2kc29

Raphael, C., Bachen, C., Lynn, K. M., Baldwin-Philippi, J., & McKee, K. (2009). Games for civic learning: A conceptual framework and agenda for research and design. *Games and Culture, 5*(2), 199-235.

Smith, A. (2010). *Government online: The Internet gives citizens new paths to government services and information.* Washington, D.C: Pew Internet & American Life Project. Retrieved from http://pewinternet.org/Reports/2010/Government-Online.aspx

Vygotsky, L. S., Vygotsky, L. S., & Cole, M. (1978). *Mind in society: The development of higher psychological processes.* Cambridge, MA: Harvard University Press.

Wankel, C., & Kingsley, J. (Ed.). (2009). *Higher education in virtual worlds: Teaching and learning in Second Life.* Bingley, England: Emerald Group.

CHAPTER 10

GENOME ISLAND

A Virtual Laboratory Environment in Second Life

Mary Anne Clark

Genome Island was created as a virtual laboratory environment in the Second Life metaverse. The island contains over 60 interactive experiments and exhibits, from Mendelian genetics to bioinformatics. Although it was first developed to provide a laboratory experience for online students, it is also used as a supplemental learning space for face-to-face students. It is used most intensively in a non-majors class that meets exclusively in Second Life. The in-world course includes 8 content modules, including a course orientation. Students get activity guidelines and receive and return assignments via a Blackboard interface, and meet weekly on Genome Island for questions and trouble shooting. Data collected from the fall 2009 class includes visitor records for the island, object interaction records for the class, performance on learning objectives, and student comments.

THE GENESIS OF GENOME ISLAND

Genome Island is a learning environment that resides in the virtual world of Second Life developed by Linden Lab. The learning activities are virtual experiments, models and exhibits related to genetics (Clark, 2008,

Engaging the Avatar: New Frontiers in Immersive Education, pp. 277–308

2009). Much of science exists in virtual space—in the theoretical framework that connects, explains and predicts observed phenomena. When Gregor Mendel discovered the laws of inheritance in the nineteenth century, he did not provide any kind of physical description for the inherited elements whose transmission he followed from generation to generation. Genome Island is a mental landscape populated by objects representing the ideas of Mendel and many of the scientists who followed him in articulating the principles of genetics. Some of the organisms represented reflect real experiments with familiar organisms like garden peas, fruit flies, or bacteria. Others are imaginary organisms that would not be found in anybody's laboratory. The laboratory space itself is not a reproduction of any real laboratory. Mendel's Abbey on Genome Island is represented only by a conference and reading room, a greenhouse and a few gardens. Unlike the real abbey (now the Mendel Museum in the Czech Republic), the Abbey on Genome Island enjoys an ocean view. The laboratories in the Tower have carpeted floors and no walls. The cats live in a trellised area that would never meet the requirements of an institutional animal care and use committee. The focus of Genome Island is less on training students to perform specific experimental manipulations than it is on encouraging students to practice scientific thinking and data analysis.

The activities on Genome Island were first created as a laboratory experience to support an online science class for general education students at Texas Wesleyan University. The online class had previously been taught with a laboratory component consisting of a combination of online simulations and databases and home laboratories conducted using a packet of materials provided by the university's Department of Biology. In 2004, the author was introduced to Second Life metaverse, which offered the opportunity to build virtual laboratories in which online students, in the person of their avatars, could actively perform experiments. Genome Island was begun late in 2004 as a personal project on a small section of the Second Life mainland. In 2006, after a number of activities had been built and tested with a few adventurous genetics students, the project was moved to a private island, which became part of the SciLands, a continent of science-related islands in Second Life. The project was initially supported by the author's tax refund, a grant from the Sam Taylor Fellowship (awarded by the General Board of Higher Education and Ministry of the United Methodist Church) and the generous educational discount offered by Linden Lab up to 2011. After a proposal to relocate the non-majors genetics course to Second Life was approved, the island server fees were incorporated into the budget of the Biology Department, and the activities on Genome Island are now used by several classes.

CLASSES ON GENOME ISLAND

At Texas Wesleyan University, Genome Island is used primarily for a non-majors genetics class that is held entirely in-world, and also for occasional "field trips" for students in the freshman and junior genetics classes for majors. Other regular users of the island from other institutions include students from general biology classes, students from health professions programs, and graduate students in science education. The island is also open to the general public. This chapter will focus on the in-world genetics class (NSC 1414 at Texas Wesleyan University), a natural science elective for students not majoring in one of the sciences. Student access to hardware capable of running the Second Life software can be a limiting factor in using Second Life as a learning environment. Because NSC 1414 is only one of several courses available to meet the university's science requirement, enrollment can be restricted to students able to download and run the Second Life viewer. Most class assessment data was collected from the class enrolled in 2009.

THE GEOGRAPHY OF GENOME ISLAND

Genome Island is divided into several activity areas, the Abbey, the Tower, the Terrace and the Gene Pool and Atelier, which can be seen in aerial view in Figure 10.1. The Abbey (Figure 10.2) and an adjacent greenhouse and gardens hold basic experiments in Mendelian genetics. Historical objects related to Mendel are found in the Abbey, which also serves as a conference area for meeting students in groups. The author is grateful to the Mendel Museum in Brno, Czech Republic, for granting permission to use images from their website (http://mendel-museum.com/) in constructing the historical objects in the Abbey. Short slideshows in the Abbey also summarize the Mendelian principles of inheritance: paired genes, dominance, and the random sorting of genes into the gametes and their random recombination in fertilization. In the Abbey greenhouse are simulations of some of the experiments (Figure 10.3) that led to the formulation of these principles.

The Tower (Figure 10.4) is a multistory structure housing experiments related to gene structure and function. This section includes experiments leading to the identification of DNA as the genetic material (Figure 10.5), the development of the double helix model of DNA, genetic coding, and the organization of genes and genomes. A human chromosome gallery (Figure 10.6), experiments in human inheritance (Figure 10.7), fruit fly genetics and bacterial genetics are also located in the Tower. A link from one of the Tower exhibits goes to an adjacent Bioinformatics Platform,

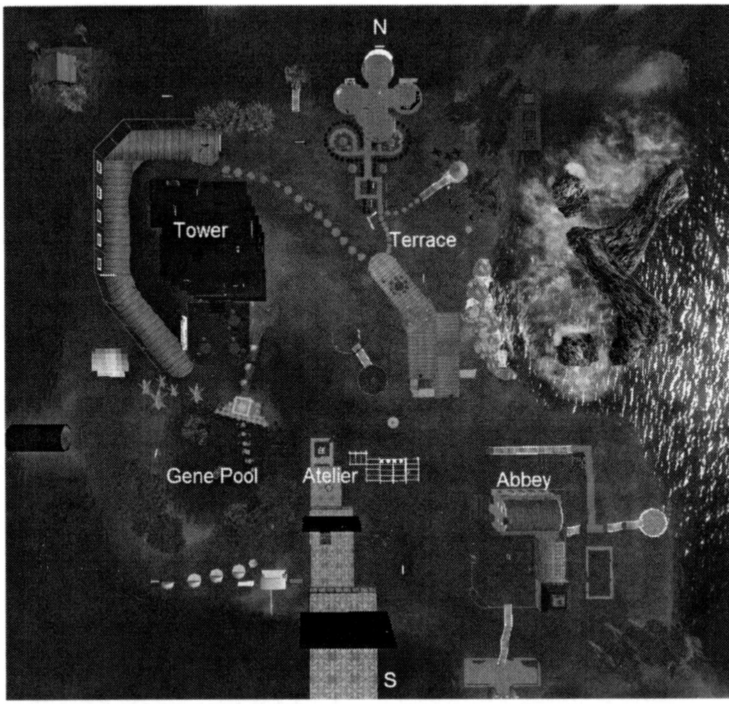

Figure 10.1. Aerial view of Genome Island, showing the major activity areas.

Figure 10.2. The Abbey, gardens and greenhouse on Genome Island.

Figure 10.3. A monohybrid cross between green and yellow peas.

which includes several activities that use information from various publically accessible genetics databases (Figure 10.8).

Between the Abbey and the Tower is a fairly broad area called the Terrace. The Terrace includes the Cattery (Figure 10.9), which illustrates various types of inheritance in cat coat colors and patterns, modeled after Judith Kinnear's CatLab (http://www.emescience.com/bio-software-catlab.html) simulations. There are two genome galleries, the Garden of Prokaryotic Genomes (Figure 10.10) and the Eukaryotic Genome Boardwalk. Each includes a sampling of sequenced genomes from representative prokaryotes and eukaryotes. Upper and lower terraces include models of eukaryotic and prokaryotic cells, the Bunny Farm, and the Mixollama Herd (Figure 10.11), which illustrate more complex inheritance patterns. The Mixollamas are mythical creatures, but their imagined genome is similar to that of other camelids.

Below the Tower are the Gene Pool (Figure 10.12) and Atelier. The Gene Pool contains experiments in population genetics. The Atelier is the student sandbox or building area, where students in the Genetics Atelier group can build objects or examine objects from their inventories. The Atelier includes 5 levels about 40 meters apart.

On Genome Island, social areas are interspersed with the activity areas. An expandable conference table in the Abbey (with virtual champagne, tea and nibbles), holds up to 40 students, and several more casual seating

Figure 10.4. The Tower, which houses molecular and human genetics, on Genome Island.

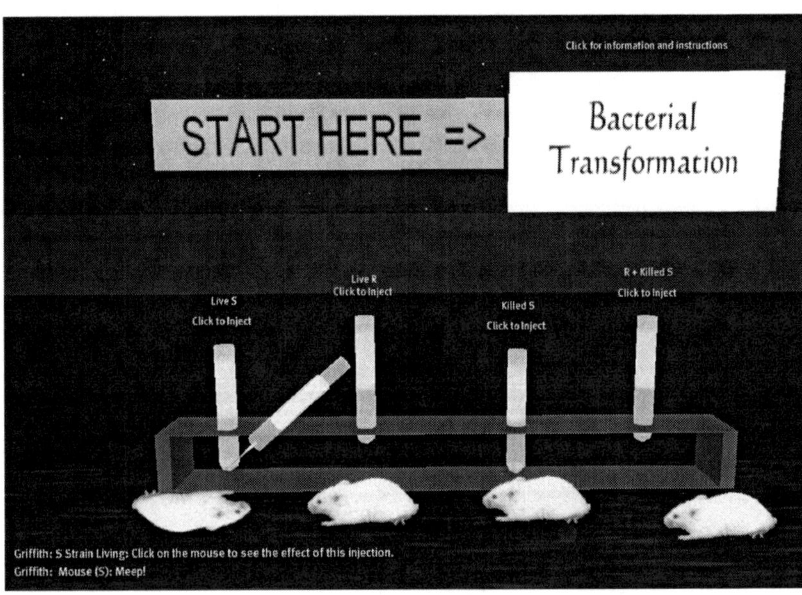

Figure 10.5. The Griffith experiment on bacterial transformation.

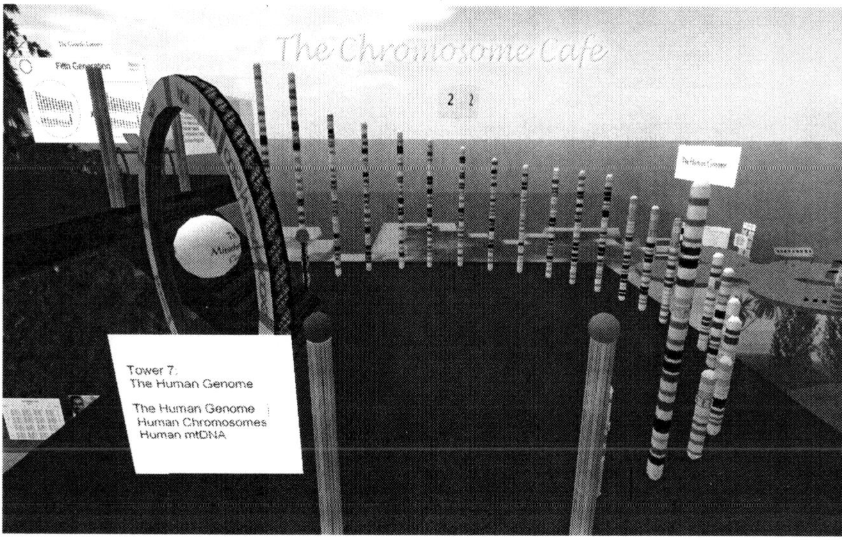

Figure 10.6. The Gallery of Human Chromosomes in the Tower on Genome Island.

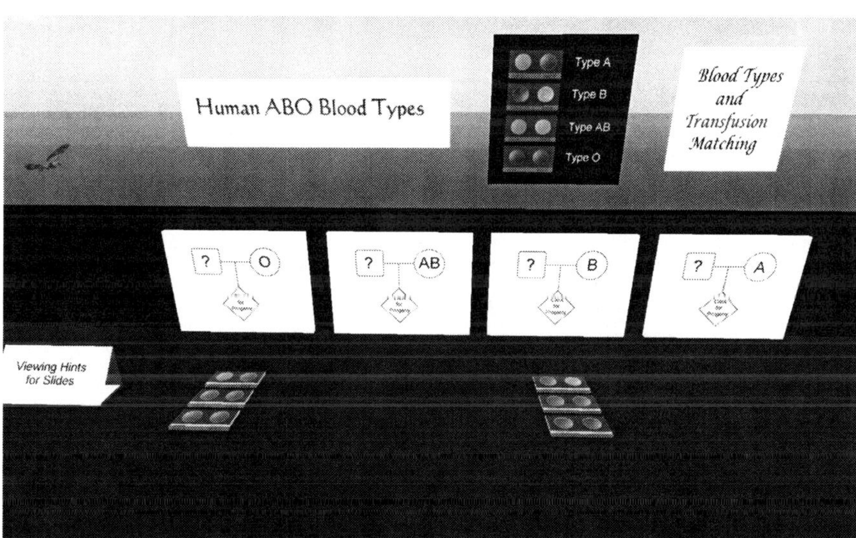

Figure 10.7. Inheritance of human blood types in the Tower on Genome Island.

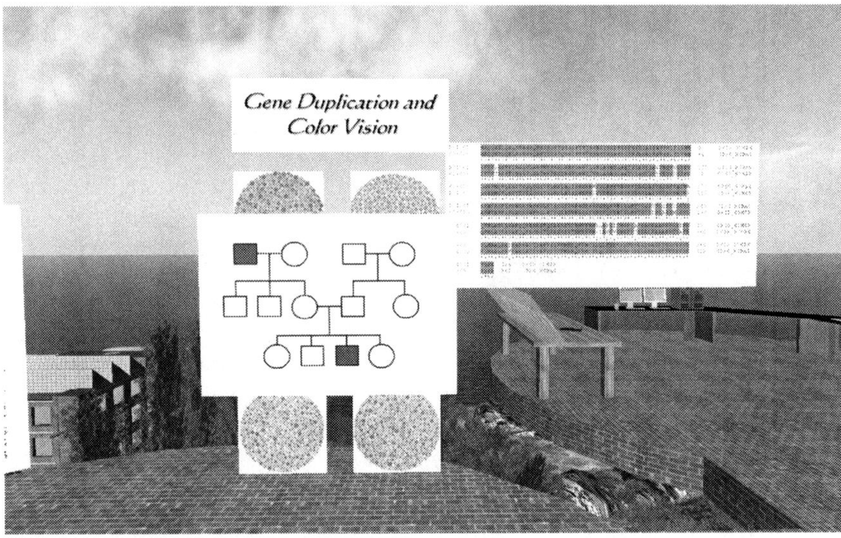

Figure 10.8. Gene duplication and the inheritance of color vision.

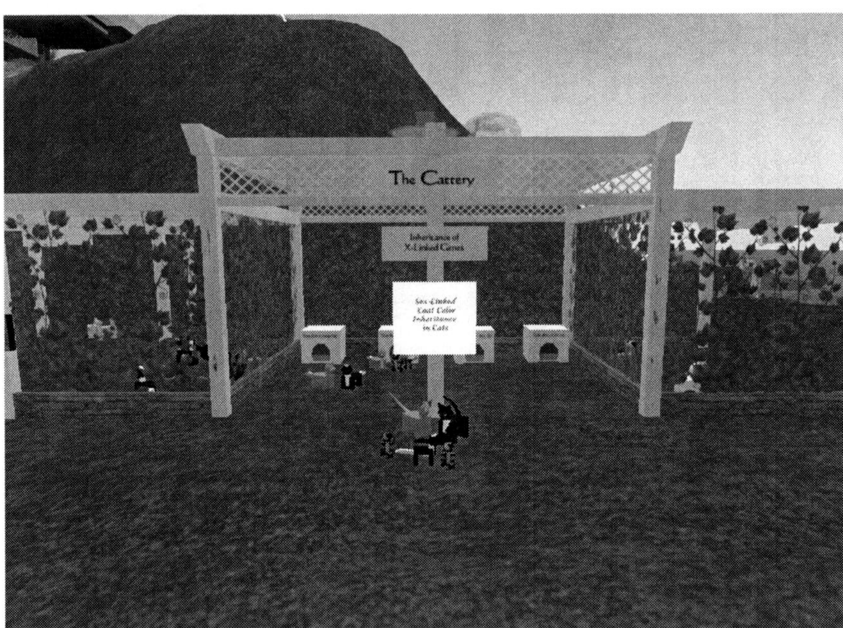

Figure 10.9. The Cattery on Genome Island.

Figure 10.10. The Garden of Prokaryotic Genomes on Genome Island.

Figure 10.11. The Mixollama Herd on Genome Island.

Figure 10.12. The Gene Pool on Genome Island.

areas for 4-10 students are scattered around the island. The melding of laboratory space with social space on Genome Island may encourage a freer student-instructor interaction. Mazar and Nolan (2008) have noted that virtual worlds tend to obliterate the typical power structure of a classroom with the instructor as a privileged dispenser of information. Boon and Sinclair (2009) have pointed out that in Second Life the identities of both student and instructor are submerged in their avatars, which may also contribute to a more democratic learning environment. On Genome Island, the author has observed that students in her in-world class converse with her more frequently and for longer than students in her on-campus classes, and that group discussions often continue well after the official end of the scheduled discussion period.

COURSE ORGANIZATION

The NSC 1414 course described in this chapter is interfaced with a Blackboard (Campus Edition 8) website. Students can connect directly with the course Blackboard site via the computer in the Genome Island Tower office (Figure 10.13). As Burgess, Slate, Rojas-LeBouef, and LaPrairie (2010) have pointed out, Second Life has limited facilities for document storage apart from notecards. The Blackboard site is used for student grading records, for off-world communication with students, and for organization of the course modules. The course content is divided into eight modules: Introductions and Orientation, Genes and DNA, Genes and Proteins, Genes and Genomes, Principles of Inheritance, Patterns and Pedigrees, Human Variation, and Historical Genetics. Each of the 8 modules includes a brief introduction, guidelines for activities on Genome Island, and an assignment. Although each of the 60 or so activity stations on Genome Island includes background information and instructions, there is no fixed path for students to follow through the activities. About 40 of the activities on Genome Island have been selected for the NSC 1414 class. The guidelines for activities for each course module direct the student to a sequence of activities, and indicate what types of information must be collected from each station. The guidelines also include landmarks for the map coordinates at which each activity is located on the island. Students keep a laboratory folder for each module in their Second Life inventories, in which documentation for each activity

Figure 10.13. The Blackboard portal in Max Chatnoir's office on Genome Island.

is filed. Types of documentation include the informational notecards offered at each station, data generated at experimental simulations, data collected from databases, and answers to questions posed during the activity. The assignment for each module is based on information collected during the activities. Examples of activity guidelines and their related assignment are given in sample Lesson Plan for this chapter. Students can do the activities at their convenience, either alone or with another member of the class. The guidelines for several activities are available simultaneously, so that students can proceed at their own pace. However, there are deadlines throughout the course term by which each of the course modules must be completed.

In addition to the Genome Island activities, the students meet with their instructor in small discussion groups for 1½-2 hours each week (Figure 10.14). Students select a regular group discussion time but may attend any of the discussion groups each week. Discussions are held in-world at one of several conference locations around the island. The most frequent discussion sites are the Abbey conference table, the Tower terrace, the Cell terrace and the student project gallery. Group discussions are for reflection on the Genome Island activities, for answering questions about course content or the operation of Second Life, and for introducing new modules. All group discussions are held using text chat. Using text allows all students to contribute, sometimes simultaneously, to the discussion. The chat logs are posted in the Discussion area of the Blackboard site. Students can therefore review discussions from their own or other groups. Students who may have to miss an in-world discussion can view and comment on information from the chat records. In addition to meeting with the discussion groups, the instructor frequently drops in and out of Genome Island to observe student activity and to answer questions. These drop-in visits may range from a few minutes to several hours. Conversations during a drop-in may be conducted either in text or in voice. Chat records from drop-in visits are not saved and posted by the instructor, although students are welcome to do so.

Two online exams are given in Blackboard: a midterm and a comprehensive final exam. Each exam is available for a period of 4-5 days, but must be completed by each student within a fixed time period once the exam has been opened. In addition, each of the students selects a human or other gene in which he or she is interested and prepares a project poster in which information collected about the gene is presented. A presentation session is held near the end of the semester, and the posters remain on display in the student project terrace (Figure 10.15) on Genome Island for the remainder of the academic year.

Figure 10.14. A discussion group on Genome Island.

Figure 10.15. Student projects on Genome Island.

STUDENT INTRODUCTION TO GENOME ISLAND

The course is given in the fall term, which begins in mid-August and continues through mid-December. At the beginning of the previous summer, information about the course is e-mailed to the students in the class who have enrolled during preregistration. These e-mails direct the students to online information about the course and to websites where they can obtain their free Second Life accounts and download the free Second Life viewing software. The students are also directed to do a brief self-orientation at an orientation walkway located in the SciLands—the group of islands in which Genome Island is located. They are then invited to visit Genome Island. As other students enroll in the course during later registration periods, additional e-mail messages are sent, containing the information described above. If the students reveal their avatar names, they are added to the instructor's Second Life friends list and invited to the Genetics Atelier group, which gives them building privileges in some areas of the island. About a week before the class begins, the students get access to the Blackboard site, and are sent their Blackboard login information. A course information section on Blackboard also includes the information they need to obtain their Second Life accounts and viewer and the location of the SciLands orientation site. The days, times and location of the first class meetings are both e-mailed to the students and posted on the Blackboard site. Although the procedure above is designed to give students the opportunity to acquire some advance experience in Second Life, many of them do not venture in-world until the first class session. During the initial class meeting, most of the students report having spent little previous time in-world.

The first course module is devoted entirely to skills practice and an introduction to Genome Island. The first class meeting is held in the conference area of the Abbey. Bell (2009) has pointed out the value of familiar settings for easing the transition of students into the Second Life environment. During the first class meeting, the students are invited to add one another to their Second Life friends lists and to set their "home" to Genome Island. Having the island selected as their home location allows them to return to it quickly from any other Second Life location and gives the students more confidence to explore. Because there are several sections of the first class meeting, I post the full list of avatars to the Blackboard site so that students can recognize and "friend" all of their classmates. The first class meeting is devoted to personal introductions, questions about Second Life, a brief description of course procedures and the beginning of the Scavenger Hunt. The Scavenger Hunt is not, strictly speaking, a Scavenger Hunt, but an orienteering activity. Each station of the Hunt gives the students instructions about how to find the next station. The Hunt begins at the island Landing Point. Other stations are in various locations around the island. At each station, the student is given instructions and practice in a Second Life skill: moving around (walking, flying, and teleporting), using the minimap and location coordinates, changing clothes, taking and saving a snapshot, sending postcards, sending an instant message in-world, using external links, creating new folders and notecards, copying and saving data from the chat record, and browsing their inventory. Several free t-shirts are given during the Scavenger Hunt, including their university t-shirt. Students are asked to wear their university t-shirts through the first few weeks of the course to allow them to recognize one another easily when they are working on the island. The students are followed through the first two stations of the Scavenger Hunt to be sure they understand the procedure. Most of the students return later to finish the Scavenger Hunt on their own. At the final station, the students are given a landmark to a "freebie" Island, where they can shop for free clothing, hair and other items. Most of the students report completing the Scavenger Hunt in less than an hour and a half.

Up to the end of 2010, the restriction of Second Life residency to individuals 18 and over limited direct access to the educational islands of the main Second Life grid mostly to university students. In 2010, Linden Lab announced that, after the closing of the Second Life Teen Grid at the end of the year, students 16-17 would also be admitted to the main grid, where they can benefit from its extensive educational resources. The presence of this younger group of students in Second Life may revive discussion of concerns raised by Bugeja (2007, 2008a), who described potential legal complications for universities exposing students to (virtual) assault, harassment, racism or homophobia in Second Life. Shortly after publication of Bugeja's first article, the issues were discussed at length by the SLED (Second Life Education) group (see the archive at https://lists.secondlife.com/pipermail/educators). Although external scripts and damage are blocked on Genome Island, it is a public access island and strangers can drop in at any time, including class time. However, only once in the last three years has it been necessary to eject a visitor for disrupting class activities. The students are advised to be cautious, and to report any untoward incidents on or off the island.

STUDENT ACTIVITY ON GENOME ISLAND

Burgess et al. (2010) have pointed out the importance of involving all students in class activities. I have used two different software applications to help monitor activity on the island. General visitor activity is recorded using commercial software available from Maya Realities. This software samples visitor presence on the island at 1 minute intervals and covers the entire island. Reports are sent weekly, although reports covering any specific time interval can be collected from the Maya Realities website. Reports include a map that locates the position of visitors on the island, so that concentrations of activity can be identified. They also include numbers of unique visitors during the monitored period, total number of visits, total and average time spent by visitors to the island, numbers of return visits, and visitor country of origin. Although the Maya Realities software allows the collection of useful data, it does not distinguish students enrolled in the class from other visitors to the island. Selected objects on the island are also programmed to monitor visitor interactions. Objects must be touched to activate their data production. Visitor touches of the programmed objects are reported to me by e-mail, which does identify specific users as well as the date and time of the interaction. I collect the object touch data several times each week, and record relevant object interactions by members of the class.

An interesting feature of teaching in Second Life is that each student has a dual identity, and class records are associated both with their real names and with their avatar names. When students first introduce themselves in the orientation session, they may or may not have previously revealed their real world identities. Some students seem to enjoy a period of anonymity as they begin to inhabit their avatars. Until avatars are firmly connected to the real world student roll, separate records are kept for activities associated with avatar names (object and chat records, lab notebooks) and activities associated with the real world class roll (assignments and exams). Since the class is relatively small, the bookkeeping is not onerous, and the two types of records can usually be melded well before midterm, when midcourse grades must be reported to the university. As Reuters (2008) has reported, Second Life residents often choose avatars physically similar to themselves, and many students even choose their avatar names as some variant of their real names.

Although the class cohort cannot be distinguished from other visitors to the island, the overall contribution of the class to island activity can be assessed by comparing visitor data from the fall (August-December) term, when the class is in session, to the spring term (January-May), when it is not. As Figure 10.16 illustrates, even though there were more total visitors in the spring of 2010, the number of visits, average time per visitor, and the total time on the island were greater in the previous fall term, suggesting that the class contributes disproportionately to visitor activity.

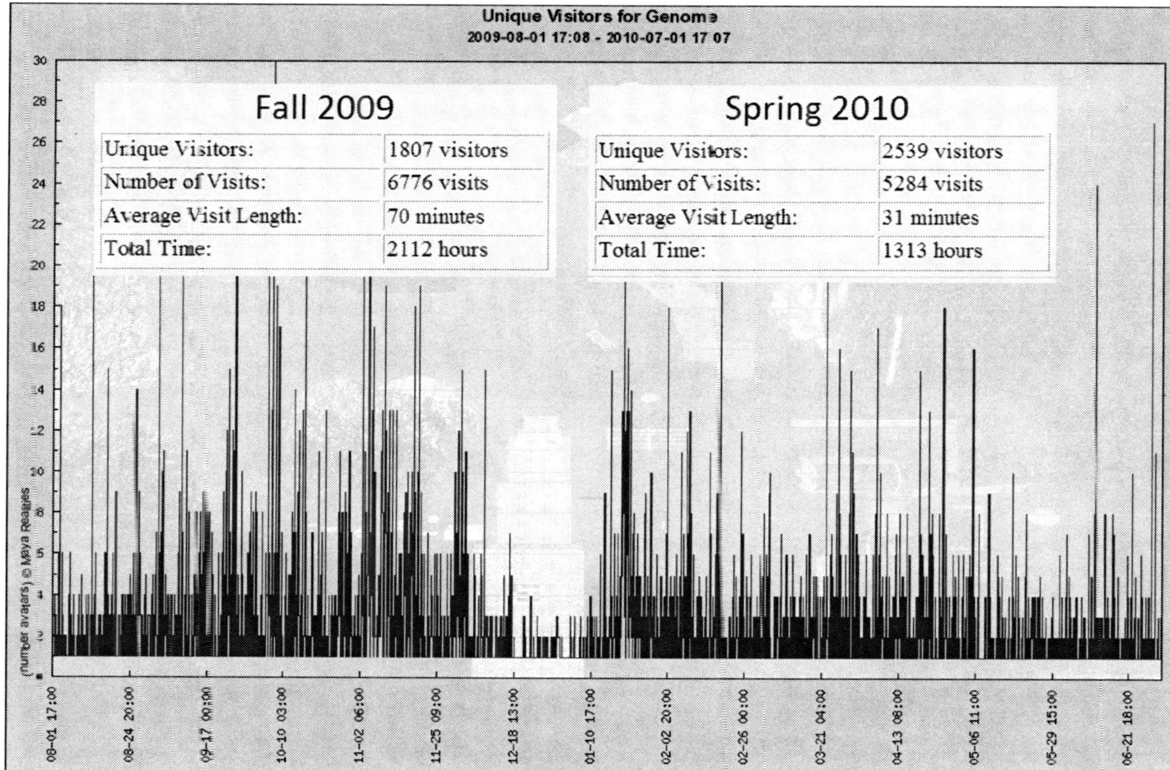

Figure 10.16. Genome Island visitor data for the 2009-2010 academic year. Background shows unique visitors over the year. Insets show data from the first to last days of the fall and spring semesters.

STUDENT PERFORMANCE ON GENOME ISLAND

Because of its superficial similarity to multiuser games like World of Warcraft, the popular conception of Second Life is that it is a game. Educators interested in the exploration of Second Life and other virtual worlds as educational venues are quick to point out that they are not games. However, in the case of Second Life the confusion is understandable. Second Life is a kind of frontier town, and yes, it does have brothels, saloons and gunfights as well as churches, professional organizations and schoolhouses. Aldrich (2009) has classified virtual worlds, games, and simulations as points along a continuum of interactive virtual environments. In virtual worlds, the focus is on social interactions. In games the focus is to pile up points (or wolf pelts) toward a defined goal. In simulations, the focus is on the acquisition of skills and information with real world applicability. In spite of these differences in focus, any virtual environment will have elements of all three. Where do students place Genome Island along this continuum? Because their expectations, and perhaps their performance, will be influenced by the kind of environment they think they are working in, it is useful to have some agreement on the point. Cheal (2007) has noted that discordance between instructor and student views of the learning environment can contribute to student dissatisfaction. Because Genome Island was conceived as a laboratory (and therefore a simulation by Aldrich's classification), the students are asked to do experiments, collect data, keep a notebook, and apply what they have learned to addressing related questions. About midterm in the 2009 class, the NSC 1,414 students were asked whether they experienced Genome Island as a game, a laboratory, a classroom or something else. A sampling of their responses can be seen in Table 10.1. Although the students recognize that the Second Life is complex, their view of the working environment concords well with that of the instructor.

An important learning outcome for simulations is the acquisition of skills applicable in other situations. What skills do students learn on Genome Island? Course skills can be divided into two groups: Second Life operational skills needed to engage with the learning objects, report data and assemble projects; and content related skills specific to the course. Most students acquire the basic operational skills during the 10-day orientation module. Some skills, like basic communication and navigation, are quickly acquired, and students develop enough proficiency even during the hour or so they spend on the Scavenger Hunt that no additional training is needed to engage with the course materials. Additional skills, like controlling the lighting, changing clothes or turning off the typing animation, are acquired during "trick of the week" sections in the weekly

Table 10.1. Sampling of Student Perceptions of the Nature of Genome Island

Genome is more of a laboratory for me.
Genome Island is more of a classroom in my opinion because when I go there I feel like I am stepping into a classroom, plus the fact that the teacher is there makes it seem more structured like a classroom environment.
I think that Genome Island is like a lab and a game rolled into one. The game aspect makes it more fun, but the work is just as difficult.
Genome Island is an environment that is a combination of classroom, laboratory and great game. It definitely is a classroom. However, the labs are surrounding the learning materials that helps enforce learning. The whole island is a big game to me that I very much enjoy.
Genome island is a new and different type of environment. It is WAY more challenging mentally than any game (any game I have played!). Not quite as formal as a classroom and a very controlled laboratory.
Well, it's all of the above. It functions like a video game though, which is not so great for people like myself who are video game challenged.
This application is really all three rolled into one which helps by keeping things fun. It is cartoony like a game, but with relevant information provided to help learning.

class discussions. Advanced skills, like building or script writing, are not generally required for the course.

Some object manipulation skills are needed for preparation of the student gene project posters and for a family pedigree construction project. For the gene project posters, the students are supplied with a modifiable poster object for which they prepare images, using PowerPoint software. Students send their images to the instructor, who uploads them to avoid upload fees for the students, and then returns the images to them in world for adding to the posters. For the pedigree project, the students are given a Pedigree Kit, which contains objects representing male and female parents, male and female progeny, and the necessary connecting lines. The students are given a training session in moving objects in the three-dimensional environment, and in linking them together. Students then assemble their family pedigrees using the objects and adjust the color of the symbols to represent the inheritance of a family trait of their choice (Figure 10.17).

Content-related skills include the intellectual skills defined in the Revised Bloom's Taxonomy (Anderson & Krathwohl, 2001): remembering, understanding, applying, analyzing, evaluating and creating (Figure 10.18). Examples of Genome Island activities that foster the skills of Bloom's taxonomy are given in Table 10.2. To these I would add academic skills like communicating their ideas in both private and public formats, and basic quantitative reasoning. From its beginnings with Mendel to bioinformatics, genetics has been a quantitative science, which offers many

Figure 10.17. Student pedigree building.

opportunities for students to practice quantitative thinking. Because the NSC 1414 course is one of several that can be used to fulfill the university science requirement, it emphasizes both scientific thinking and a limited domain of scientific knowledge. Student apply their skills to answering questions about the phenomena they encounter as they proceed through the assigned activities, to answering the questions posed by the related assignments, to discussing questions raised during the in-world class meetings, to the two semester exams, and in selecting and processing the information for their gene projects. Several of the students have used their gene projects to learn more about genetic disorders found in their own families see (Figure 10.18).

All of these features are commonly found in any university science class. What is unusual about Genome Island is the pervasiveness of the data generated by different activities. Any experiment on the island is only a mouse-click away. As Robbins-Bell (2008) has observed, the advantage of virtual worlds is that even in the absence of other students or the instructor, the laboratory activities are always there. Genome Island is never closed. The ability to immerse students in an experimental environment is one of the chief affordances of Second Life for teaching science. Thomas (2010) has argued that Second Life allows the creation of active learning environments in which student engagement is an emergent property. While students in the class are given guidelines about which set of activities to engage, clicking almost any object produces some result

Table 10.2. Genome Island Activities and Bloom's Taxonomy

Cognitive Skill	*Genome Island Activity*	*Question*
Remembering	Base Ball (Tower)	What four bases are found in DNA?
	Mendel (Abbey)	What organism did Mendel use for his experiments?
Comprehension	Human Karyotypes (Tower)	Which of these karyotypes represents trisomy 21?
	Monohybrid Cross (Greenhouse)	What two genotypes are possible in a yellow pea?
Application	Chromosome Gallery (Tower)	What is the gene density of human chromosome 7?
	Genetic Code (Tower)	Write a message to a friend using genetic code.
Analysis	Pedigrees (Tower)	What type of inheritance is suggested by this pedigree?
	Cattery (Terrace)	A black and gray cat were mated and produced both black and gray kittens. What were the parents' genotypes?
Evaluation	Color Vision (Bioinformatics)	From what ancestral gene was the human red opsin most probably derived?
	Chimp Karyotype (Tower)	What chromosome rearrangements might have produced the differences seen between human and chimp chromosomes?
Creation	Create a presentation about the human gene of your choice, using information from five different genetics databases.	

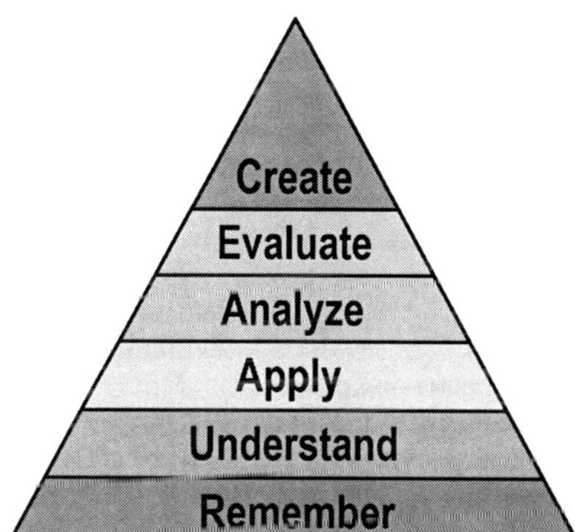

Figure 10.18. Bloom's Taxonomy as revised by Anderson and Krathwohl (2001).

that the student is asked to think about. Active engagement with the information is almost inescapable. Even though it would be possible for one student to follow another around the island and simply observe what happens when you click on a cat, a garden plot or a chromosome, most students collect their own data. Table 10.3 presents information about student interactions with selected objects assigned in each module. The data shows that most of the students interact with most of the objects most of the time. Records on individual students (not shown) indicate that some students interact with all of the objects all of the time.

Student retention and learning are summarized in Table 10.4. Although there is no appropriate control group for the class at my university, the comparison between pretest and posttest performance demonstrates learning. Most of the students leaving the course did so in the first week of class, citing difficulties with downloading or running the Second Life software on their home computers. A few of these students tried proceeding with the course using the computers in the university laboratories. Two of the university computer laboratories have Second Life as resident software on the computers. The program can also be run on most campus computers from Second Life software downloaded onto a flash drive. However, being restricted to the campus computers during the usual laboratory hours removed most of the advantage of taking an online course—flexible scheduling—and all of the students in this group dropped the course before taking the pretest.

The experience of these students underlines the most serious barrier to the routine use of Second Life in university classes. Kirriemuir (2010) has discussed some of the technical issues arising from the use of Second Life for teaching. The program makes heavy demands on its computer for the Second Life system requirements), and requires a game quality video card. At any university there may be students who do not have the necessary computer access from home. At Texas Wesleyan University, the activities on Genome Island have been used in other biology classes, with students in a common computer classroom. The program has been run satisfactorily in this environment, although lag and program freezing occasionally interrupt the experience. While the NSC 1414 class meets in small synchronous groups for the weekly discussions, activity records suggest that most of the activities are done individually. This pattern may be due to a relatively small class composed of students with different working patterns. With only one copy of most activities, even half a dozen students trying to do the same experiment at the same time would interfere with one another. However, the activity guidelines could be modified to facilitate group activity.

With most students working through the activities individually, the weekly group meetings offer opportunities for the students to connect

Table 10.3. Student Interactions With Learning Objects on Genome Island

Course Module	*Number of Index Objects*	*Number of Students Interacting With Any Object*	*Student Interactions Per Object.*
Module 1	13	20	19
Module 2	12	18	16.6
Module 3	12	17	13.2
Module 4	10	16	10.3
Module 5	15	14	11.6
Module 6	12	14	12
Module 7	14	14	12.4
Module 8	12	12	10.1

Table 10.4. Student Retention and Achievement in NSC 1414.

	Number of Students	*Mean Score on Assessments*
Enrolled	23	
Pretest	18	35 +/- 11
Midterm Exam	17	72 +/- 32
Final Exam	16	73 +/- 12

with the instructor and with one another. In the opinion of the author, regular interaction with others is an important element in the success of Second Life educational activities. Even with frequent instructor visits at various times of day, evenings and weekends, the chance of encountering any specific student at random is relatively small. Chance student-instructor encounters may result in chat or question sessions ranging from a few minutes to several hours, but asking the students to attend one of the group meetings each week greatly increases the probability of the instructor seeing all of the students in any week, gives the students a sense of community and provides fertile ground for the emergence of teaching moments. In the fall class of 2009, there were 13 weekly meetings. Students were given credit for participating in these meetings, which encouraged attendance. Average weekly attendance was 14.7 +/- 1.9. Discussion attendance tended to fall off in each of the weeks before the two major exams, possibly because the last assignment preceding each of these exams would already have been completed.

SCIENCE IN THE NEIGHBORHOOD

In September, 2010, Linden Lab reported a total of over 30,000 active regions and a total virtual land area of over 2000 square kilometers (http://gridsurvey.com/index.php). Some islands are isolated; others are gathered into archipelagos or continents. Genome Island is one of 68 islands currently located in the Second Life SciLands continent. There are many other science islands located elsewhere in Second Life. Troy McConaghy (Troy McLuhan in Second Life) lists 120 island or mainland regions in his "Science Related Places" (https://sl-science-places.dabbledb.com/page/sl-science-places/vbxOGovy#). Most of the science regions have an educational function, and represent universities (like the University of Cincinnati's Galapagos Tour), museums (like the Exploratorium), governmental agencies (like the many islands sponsored by the National Oceanographic and Atmospheric Administration), scientific journals (like the Nature Publication Group's Elucian Islands), professional organizations (like the American Chemical Society Islands), professional schools (like the Imperial College of London's medical complex) libraries (like the National Library of Medicine's ToxTown) and individual science promoters (like the International Spaceflight Museum). Some of the science places are used primarily for public information, while others are also used for university classes or professional training. AstroSim (Nakasone et al., 2009), a collaborative project of American and Japanese research facilities, uses real astronomical data for generating interactive models of star systems, whose behavior investigators can modify by changing different parameters. Chemistry and biochemistry classes can benefit from molecular modeling tools developed by Andrew Lang, Erich Bremer and Peter Miller (Lang & Bradley, 2009). In the Virtual Biosciences Laboratory of the University of East London (Cobb, Heaney, Corcoran, & Henderson-Begg, 2009), students can learn to perform PCR (a technique for amplifying specific DNA sequences), and apply their learning in a real world setting. At the Galapagos project of Cincinnati University (Collins & Millard, 2008) visitors can tour the islands and observe their occupants aboard a replica of the Darwin's Beagle, which also contains documents from Darwin's own explorations. Several science regions are devoted to health education. The University of Plymouth's Sexual Health SIM (Boulos & Toth-Cohen, 2009) focuses on sexual health education. The National Library of Medicine's HealthInfo Island and the University of Plymouth's Virtual Neurological Education Center (Boulos, Hetherington, & Wheeler, 2007) are both places where physicians, patients and families can meet to discuss health issues. At the island managed by Coventry University (Stott, 2007) or the Imperial Hospital London's large medical complex in the SciLands (Bradley, 2009), students can practice clinical management, diagnostics, triage or patient care. Students visiting the Ohio State University Medical Center can take a tour of a giant testis. In the nursing program at Kansas State University (Digman, 2008), students can visit an adult living center constructed in Second Life, and study simulated medical histories to better understand the needs of the center's residents. Other nursing programs using Second Life simulations include the University of Wisconsin Oshkosh and Tacoma Community College (Skiba, 2009), whose students can work with nursing faculty on activities related to patient care or public health. Cyberpsychiatrists are even venturing into psychotherapy in Second Life for treating phobias and post-traumatic stress disorder (Stott, 2007; Murphy, 2010).

Second Life also seems broadly applicable across the rest of the academic spectrum. Courses or projects in archeology (Edirisingha, Nie, Pluciennik, & Young, 2009), communication (Jarmon, Traphagan, & Mayrath, 2008), computer science (Crellin, Duke-Williams, Chandler, & Collinson, 2009), English composition (Davis & Smith, 2009), intercultural awareness (Diehl & Prins, 2008), geography (Dittmer, 2010), literature (Ritter-Guth, Nicosia, & Pasteur, 2008), English for Chinese students (Wang, Song, Stone, & Yan, 2009) and film production (Foss, 2009) have all been held in Second Life. It is also possible for students to experience historical or cultural events at Second Life, like a visit to the Basilica of St. Francis, the temple of Isis, the Globe Theatre or ancient Rome (Lamb & Johnson, 2009).

TEACHING IN VIRTUAL WORLDS

Warburton (2009) comparing Second Life with other virtual worlds has identified Second Life as the current "platform of choice for education" (p. 416). At the 2010 Virtual Worlds Best Practices in Education Conference, Terrence Cummings (as Terrence Linden, 2010) reported that over 600 colleges and universities, including 60% of top ranked universities in 40 countries, are actively operating educational regions in Second Life. Linden also noted that education is the largest and fastest growing community in Second Life. The Second Life ISTE (International Society for Technology in Education) group currently lists over 6,000 members and maintains an active schedule of professional presentations and social events (Trotter, 2008).

Although Young (2010) has expressed the concern that many educators may find the Second Life learning curve too steep for routine use, the educational community at Second Life seems to be alive and well. Kelton (2008) points out that there are two learning curves to negotiate in Second Life: one short, but steep, in which new users become comfortable with basic operational skills, and a later learning curve for building and scripting. For the second curve, the advantage of SL community emerges. Educational tools, many of which can be freely obtained or purchased for under $2, have already been developed by more experienced builders, and are adaptable for use in classes. Both the ISTE headquarters and the orientation center of the New Media Consortium (http://virtualworlds.nmc.org/) are good sources for free educator tools.

Donald Welch (interviewed by Waters, 2009) has commented on the advantages of Second Life for professional collaboration and for content sharing. On Genome Island, several of the objects and activities were the results of collaborative efforts with other residents, notably Elizabeth Gloucester's notecards in the prokaryotic and eukaryotic genome galleries, Ambrosia Lytton's Bunny Farm, and Simone Gateaux's Caminalcules. Quizzers and games created by Eloise Pasteur and Hiro Sheridan's molecule rezzer Orac are also used on Genome. Many educational regions are open for use by students other than those of the hosting institution. On Genome Island, the Protein Primer directs students to the Amino World built on ACS Island by Joan Slonczewski of Kenyon College.

Ferry, Gelfand, Peterman, and Tomren (2008) have described Second Life as a type of grey literature—information that is not commercially published, whose authorship is sometimes uncertain, and that, while it is generally open source, may not be easy to search out. Second Life is (so far) persistent but not fixed. Genome Island is a small library of information to which anybody who visits has access, and which is under continuous revision. The "text" of Genome Island is dispersed in its objects. Objects and their related documents can change as new activities are introduced or even in response to a student puzzled by a question. Genome Island is itself reconstructed as students construct their own understanding of genetic concepts. Although a short Guide to the Island is delivered to visitors at the Landing Point, and a more comprehensive User's Guide is available at http://faculty.txwes.edu/mclark/Genome/Guide.htm, there is no textbook or laboratory manual associated with the island. A selected set and order of activities is used for the NSC 1414 class, but other instructors are free to select and order the activities to meet their own objectives.

Robbins-Bell (2008) has characterized education as virtual conversation and says that "The more conversations that an instructor can encourage, the more involved students may become" (p. 26). As mentioned above, NSC 1414 students on Genome Island seem to initiate conversations more frequently than on-campus students do. It is possible that, in interacting with the learning objects on Genome Island, the students have begun a dialogue that they continue later in class or private conversations. The NSC 1414 class meets weekly for discussion, previews of the next module or questions about the current one, but the class includes no lectures or formal instructor presentations. The class sometimes wanders quite far afield chasing a particular line of inquiry. A student in the current class captured this sense of exploration when, during a conversation that began with another student's question about whether dogs are colorblind, he said "this is insane ... that you can learn so much about something that would otherwise seem simple."

Boettcher (2007) has identified four elements of an effective learning environment: the learner, the content, the mentor and the environment. The central actors are the students, with the instructor acting as director, and in the case of Second Life, perhaps writer and set designer as well. Each student on Genome Island participates in an ongoing narrative, adding to it as he or she puts the results of experiments together with information from notecards or slideshows and conversations with the instructor to solve little mysteries. Why did that mouse die? What does this genetic message mean? Who were the parents of those cats? Second Life seems to blend the four elements Boettcher describes in a way that engages both students and instructors.

As Kirriemuir (2010) has noted, many of the barriers to the academic use of Second Life are gradually being overcome, both by the efforts of persistent educators, by increased technological support on university campuses, and by changes within Second Life itself. It is regrettable that, because of differences in computer access for the students, the adventure of Genome Island cannot be shared more widely with the biology majors. The advanced majors have been brought into Second Life during their laboratory period to review basic genetics or to do selected experiments, like the gene mapping experiment in the Fly Room of the Tower. Dr. Alan Hudson, Professor of Immunology and Microbiology at Wayne State University School of Medicine, was able to do a research presentation to the student research class, live from Genome Island. Recently Genome Island has been used in demonstration mode to present genetics principles for the freshman majors. However, none of these groups is in-world long enough or frequently enough to become, not just competent, but comfortable in the Second Life environment. Ironically, the non-majors (and their instructor) are having all the fun. Perhaps, in future, it will be possible to spread the joy.

ACKNOWLEDGMENTS

This research was supported by a grant from a Sam Taylor Fellowship awarded by the General Board of Higher Education and Ministry of the United Methodist Church to Mary Anne Clark.

APPENDIX: LESSON PLAN

Sample Instructional Materials for Genome Island

Sample Guidelines for Activities (From Course Module 3)

Protein Primer: http://slurl.com/secondlife/Genome/83/169/65 . If this slideshow is not showing its title slide "Proteins: A Primer," click on the slideshow until the title slide is showing, and then go through the slides in order. The slideshow only contains basic information; more details are given in the "Genes, Proteins and Information" link you saw at the beginning of Module 3. For your lab notebook, enter the common features of amino acids, the number of different amino acids found in proteins, and how amino acids are connected to make proteins. Describe the picture of the beta globin molecule in some way that makes sense to you.

Normal and Sickle Cell Hemoglobin: http://slurl.com/secondlife/Genome/78/170/65 . Read the informational notecard from the "Genes and Proteins" sign and add a copy to your notebook. The poster with the purple spots displays data from an experiment that showed the difference between normal hemoglobin and hemoglobin from people with sickle cell disease. Examine the poster and be sure that you understand (from the information in the notecard) how the spot pattern was produced. Click on the poster to see the difference between the spots produced by normal hemoglobin and the spots produced by sickle-cell hemoglobin. Then click on the two small flasks to see the difference in amino acid composition between normal and sickle-cell hemoglobin. Describe this difference in a notecard and add it to your notebook.

The Genetic Code: http://slurl.com/secondlife/Genome/79/172/65 . Click on the Genetic Code chart in the northwest corner of the lab to get the informational notecard. Read the notecard and add it to your lab notebook. Examine the genetic coding table to be sure that you understand how to read it. Then read the informational notecard from the "The Genetic Code" sign to see how the genetic code was figured out. Read the information in the "Cell Free System" flask. Put both notecards into your notebook. To test your understanding of the code, look at the first three test tubes of the Code Breaking experiment: AAA, GGG, and AGG. What amino acids are found in the proteins encoded by each of these artificial messages? Click on each of the first three tubes to get this information and put it into your lab notebook.

Sample Assignment Questions (from Course Module 3)

Both DNA and proteins are informational molecules. That is, the order of the subunits is not just random. This means that both of them need a source of sequence information when a new molecule is made. New DNA molecules can use old DNA molecules as a pattern, but proteins can not get information from old proteins. Explain why this is the case. HINT: What is it about DNA structure that allows it to serve as its own pattern?

The Protein Primer slideshow at Second Life gives you the amino acid sequence for the protein beta globin. If the amino acids that compose beta globin were to be taken apart and then the exact same amino acids were "restrung," but in a different order, would the resultant molecule still be beta globin? Why or why not?

In beta globin the normal DNA codon for amino acid #6 (glutamate) is GAG. In the sickle cell mutation, the amino acid valine is substituted for glutamate. Examine the attached genetic code table and predict the most probable way to convert a glutamate codon to a valine codon. HINT: most substitution mutations are due to just a single nucleotide change.

Beta globin contains 146 amino acids. In normal beta globin, amino acid #6 is glutamate. In sickle cell beta globin, amino acid #6 is valine. This single amino acid substitution is the only difference between the two hemoglobin molecules. However, if you were to compare human beta globin with cat beta globin, you would find 23 amino acid substitutions between the two molecules. Cat and human beta globin both have 146 amino acids, with 123 matching amino acids. Why do you suppose that a single amino acid change can have such a negative effect on the protein, while 23 amino acid differences can still produce a normally functioning molecule? HINT: Think about how proteins fold into their functional form.

REFERENCES

Aldrich, C. (2009). Virtual worlds, simulations, and games for education: A unifying view. *Innovate, 5*(5), n.p. Retrieved from http://www.innovateonline.info/index.php?view=article&id=727

Anderson, L. W., & Krathwohl, D. R. (Eds.). (2001). *A taxonomy for learning, teaching, and assessing: A revision of Bloom's taxonomy of educational objectives*. New York, NY: Longman.

Bell, D. (2009). Learning from Second Life. *British Journal of Educational Technology. 40,* 515–525. doi:10.1111/j.1467-8535.2009.00943.x

Boettcher, J. V. (2007). Ten core principles for designing effective learning environments: Insights from brain research and pedagogical theory. *Innovate, 3*(3). Retrieved from www.innovateonline.info/index.php?view=article&id=54

Boon, S., & Sinclair, C. (2009). A world I don't inhabit: Disquiet and identity in Second Life and Facebook. *Educational Media International, 46*(2), 99-110. doi: 10.1080/09523980902933565

Boulos, M. N. K., Hetherington, L., & Wheeler, S. (2007). Second Life: An overview of the potential of 3-D virtual worlds in medical and health education. *Health Information & Libraries Journal, 24*(4), 233-245. doi:10.1111/j.1471-1842.2007.00733.x

Boulos, M. N. K., & Toth-Cohen, S. (2009). The University of Plymouth Sexual Health SIM experience in Second Life: Evaluation and reflections after 1 year. *Health Information & Libraries Journal, 26*(4), 279-288. doi:10.1111/j.1471-1842.2008.00831.x

Bradley, J. (2009, March 30). Can Second Life help teach doctors to treat patients? *CNN Tech*. Retrieved from http://articles.cnn.com/2009-03-30/tech/doctors.second.life_1_second-life-medical-students-virtual-hospital?_s=PM:TECH

Bugeja, M. J. (2007). Second thoughts about Second Life. *Chronicle of Higher Education, 54*(3), C2-C4.

Bugeja, M. J. (2008a). Second thoughts about Second Life. *Education Digest, 73*(5), 18-22.

Bugeja, M. J. (2008b). Second Life, revisited. *Education Digest, 73*(5), 18-22.

Burgess, M. L., Slate, J. R., Rojas-LeBouef, A., & LaPrairie, K. (2010). Teaching and learning in Second Life: Using the Community of Inquiry (CoI) model to support online instruction with graduate students in instructional technology. *Internet & Higher Education, 13(*1/2), 84-88. doi:10.1016/j.iheduc.2009.12.003

Cheal, C. (2007). Second Life: Hype or hyperlearning? *Future Survey, 29*(12), 18.

Clark, M. A. (2008). Genome Island. *Educause Review, 43*(5). Retrieved from http://www.educause.edu/EDUCAUSE+Review/EDUCAUSEReviewMagazineVolume43/GenomeIsland/163176

Clark, M. A. (2009). Genome Island: A virtual science environment in Second Life. *Innovate. Journal of Online Education, 5*(6). Retrieved from http://www.innovateonline.info/pdf/vol5_issue6/Genome_Island_A_Virtual_Science_Environment_in_Second_Life.pdf

Cobb, S., Heaney, R, Corcoran, O., & Henderson-Begg, S. (2009). The learning gains and student perceptions of a Second Life virtual lab. *Bioscience Education, 13*. Retrieved from http://www.bioscience.heacademy.ac.uk/journal/vol13/beej-13-5.aspx

Collins, C., & Millard, R. W. (2008). Galapagos Islands in Second Life. *EDUCAUSE Review 43*(5).

Crellin, J., Duke-Williams, E., Chandler, J., & Collinson, T. (2009). Virtual worlds in computing education. *Computer Science Education, 4*, 315-334. doi:10.1080/08993400903384950

Davis, M. G., & Smith, C. E. (2009). Virtually embedded: Library instruction within Second Life. *Journal of Library & Information Services in Distance Learning 3*(3/4), 120-137. doi:10.1080/15332900903375465

Diehl, W. C., & Prins, E. (2008). Unintended outcomes in Second Life: Intercultural literacy and cultural identity in a virtual world. *Language & Intercultural Communication, 8*(2), 101-118. doi:10.1080/14708470802139619

Digman, B. (2008). KUMC Isle. *EDUCAUSE Review, 43*(5).

Dittmer, J. (2010). Immersive virtual worlds in university-level human geography courses. *International Research in Geographical & Environmental Education, 19*(2), 139-154. doi:10.1080/10382046.2010.482222

Edirisingha, P., Nie, M., Pluciennik, M., & Young, R. (2009). Socialisation for learning at a distance in a 3-D multi-user virtual environment. *British Journal of Educational Technology, 40,* 458-479. doi:10.1111/j.1467-8535.2009.00962.x

Ferry, K., Gelfand, J., Peterman, D., & Tomren, H. (2008). Virtual reality and establishing a presence in Second Life: New forms of grey literature? *International Conference on Grey Literature, 9,* 113-118.

Foss, J. (2009). Lessons from learning in virtual environments. *British Journal of Educational Technology, 40,* 556-560. doi: \10.1111/j.1467-8535.2009.00955.x

Jarmon, L., Traphagan, T., & Mayrath, M. (2008). Understanding project-based learning in Second Life with a pedagogy, training, and assessment trio. *Educational Media International, 45,* 157–176. doi:10.1080/09523980802283889

Kelton, A. J. (2008). Virtual Worlds? *Educause Review, 43*(5), 15-22.

Kirriemuir, J. (2010). UK university and college technical support for Second Life developers and users. *Educational Research, 52*(2), 215-227. doi:10.1080/00131881.2010.482756

Lamb, A., & Johnson, L. (2009). The potential, the pitfalls, and the promise of multi-user virtual environments: Getting a Second Life. *Teacher Librarian, 36*(4), 68-72.

Lang, A., & Bradley, J. C. (2009). Chemistry in Second Life. *Chemistry Central Journal, 3*(14). Retrieved from http://journal.chemistrycentral.com/content/pdf/1752-153X-3-14.pdf . doi: 10.1186/1752-153X-3-14.

Linden, T. (2010, March). Education and the Second Life ecosystem. *Virtual Worlds Best Practices in Education Conference*. Video recording by Treet TV, Episode 38, Broadcast 13. Retrieved from http://business.treet.tv/shows/bpeducation

Mazar, R., & Nolan, J. (2008). Hacking say and reviving ELIZA: Lessons from virtual environments. *Innovate, 5*(2). Retrieved from http://www.innovateonline.info/pdf/vol5_issue2/Hacking_Say_and_Reviving_ELIZA-__Lessons_from_Virtual_Environments.pdf

McConaghy, T. (n.d.). *Science related places in Second Life.* Retrieved from https://sl-science-places.dabbledb.com/page/sl-science-places/vbxOGovy#

Murphy, S. (2010). Avatar therapy: From couch to cyberspace. *New Scientist, 2778,* 44-45.

Nakasone, A., Prendinger, H., Holland, S., Miura, K., Hut, P. & Makino, J. (2009). AstroSim: Collaborative visualization of an astrophysics simulation in Second Life. *IEEE Computer Graphics & Applications, 29*(5), 69-81.

Reuters. (2008, January 31). Poll: Most adults don't want fantasy avatars. [Weblog entry] *Reuters/Second Life Blog*. Retrieved from: http://secondlife.reuters.com/stories/2008/01/31/poll-most-adults-dont-want-fantasy-avatars/ .

Ritter-Guth, B., Nocosia, L., & Pasteur, E. (2008). Literature Alive! *Educause Review, 43*(5). Retrieved from http://www.educause.edu/EDUCAUSE+Review/EDUCAUSEReviewMagazineVolume43/LiteratureAlive/163184

Robbins-Bell, S. (2008). Higher education as: Virtual conversation. *Educause Review, 43*(5), 24-34.

Skiba, D. J. (2009). Nursing education 2.0: A second look at Second Life. *Nursing Education Perspectives, 30*(2), 129-131.

Stott, D. (2007). Learning the second way. *British Medical Journal, 335,* 1122-1123.

Thomas, H. (2010). Learning spaces, learning environments and the dis "placement" of learning. *British Journal of Educational Technology, 41,* 502-511. doi:10.1111/j.1467-8535.2009.00974.x

Trotter, A. (2008). Educators get a "Second Life." *Education Week, 27*(42), 1-17.

Wang, C. S., Song, H., Stone, D. E., & Yan, Q. (2009). Integrating Second Life into an EFL program in China: Research collaboration across the continents. *TechTrends: Linking Research & Practice to Improve Learning, 53*(6), 14-19. doi:10.1007/s11528-009-0337-z

Warburton, S (2009). Second Life in higher education: Assessing the potential for and the barriers to deploying virtual worlds in learning and teaching. *British Journal of Educational Technology, 40,* 414-426. doi:10.1111/j.1467-8535.2009.00952.x

Waters, J. K. (2009). A "Second Life" for educators. *T H E Journal, 36*(1), 29-34.

Young, J. R. (2010). After frustrations in Second Life, colleges look to new virtual worlds. *Chronicle of Higher Education, 56*(23), A14.

PART IV

IMMERSING THE AVATAR

CHAPTER 11

USING VIRTUAL SIMULATIONS IN SECOND LIFE FOR TEACHING AND LEARNING IN NURSING EDUCATION

Michelle Aebersold and Dana Tschannen

Educating nursing students presents a challenge in the complex healthcare environment in which students will practice. To be effective in the field, assisting students to achieve a high level of skill development in a safe environment is critical. Second Life provides a training environment in which virtual simulations can be conducted to allow students to practice integrating their knowledge and skills in the delivery of patient care. Virtual simulations are effective when developed using a comprehensive approach that incorporates elements of online teaching pedagogies and the five step process outlined in this chapter. Issues and challenges in using virtual environments should also be anticipated and addressed to ensure student learning is achieved.

Training professionals in health care effectively requires both didactic and clinical training components, including the use of case based learning and simulations. This chapter will outline the development of virtual training scenarios that can be used with nursing students at both the undergraduate (prelicensure) and graduate level. The ability to incorpo-

Engaging the Avatar: New Frontiers in Immersive Education, pp. 311–337

rate virtual simulations into a comprehensive simulation program can offer diversity in learning methods and can specifically target leadership and problem-solving skill development. The chapter will begin with background information on educating nurses, including how simulations can be incorporated, in addition to an overview of relevant teaching and learning theories. The chapter will also include a detailed discussion of the framework used at the University of Michigan School of Nursing (UMSON) for simulation scenario development, as well as an overview on evaluation methodology, which can be used with a variety of simulation techniques. The University of Michigan developed the virtual scenario highlighted here in Second Life (SL) on the university owned space called "Wolverine Island." The chapter includes a description of an eight bed virtual hospital unit, the amount of startup time and investment required, and issues and challenges in using SL in education.

BACKGROUND

Developing simulations for use in virtual environments, and in particular SL, can be complex, due to the need to integrate theories of teaching and learning in on-line technologies with theories in simulation techniques. Simulations provide a basis for experiential learning which is critical in the development of professional knowledge and behaviors. Tacit knowledge, knowledge based on experience and intuition, is more difficult to transfer from teacher to student and requires experiences to facilitate learning. In the profession of nursing much of what distinguishes an expert nurse from a novice is the amount of tacit knowledge they have acquired through their experiences. However, ensuring all nursing students have the 'right' experiences to facilitate this knowledge gain is difficult.

According to the recent publication from the Carnegie Foundation (Benner, Sutphen, Leonard, & Day, 2009), nurses are in a critical role in health care as they deliver much of the care to our patients. They perform most of the bedside monitoring, make almost all of the home care visits, and are tasked with ensuring patient safety while managing multiple intrusive technologies where the margin of error is extremely narrow in very complex and chaotic environments. To continually perform in this environment nurses need to be both reflective practitioners and expert learners and engage in lifelong learning activities. To develop these skills, students need to engage in activities in the learning environment, which enable them to bridge the gap from theory to practice. Nursing, like all practice disciplines, relies on situated cognition and action to perform (Benner et al., 2009).

Students learn well in the clinical environment where they can practice hands on skills; however, there is no guarantee that students will be able to engage in all of the clinical learning situations necessary to develop the key skills they will need to practice effectively as a health care professional. Classroom teaching is often based on content delivery to the students, with lectures and slides as a common delivery method of choice. This pedagogy does not provide situated teaching that Benner et al. (2009) states is necessary to teach for a sense of salience. Salience is the ability of the nurse to know what is more or less important in a clinical situation and how to use the knowledge they have acquired to plan and deliver care to the patient. To assist nursing students to develop a sense of salience educators need to engage students in clinical type learning situations that require them to use their knowledge and apply it to different patient care situations (Benner et al., 2009). This is where simulation scenarios can become integral in situated teaching and teaching a sense of salience to the students.

Simulation has long been used in the military and airline industry as a primary method of teaching skills such as tactical decision making and actual hands on skills (flying the plane). Simulation is a proven methodology for both skill development and knowledge transfer from the simulated environment to the actual clinical environment (Day, Iles, & Griffiths, 2009; Maithel et al., 2006; Rosenthal et al., 2006; Van Sickle, McClusky, Gallagher, & Smith, 2005; Wayne et al., 2008). Simulations are generally set up to mimic real life situations and need to have a high degree of fidelity or realism to be effective. Simulations in nursing and medicine have typically been developed for use with computer driven high fidelity mannequins; however, a movement towards using virtual environments as a platform for simulations is emerging, and frames the discussion here.

Virtual simulations have been developed in SL for skill development such as identifying certain heart sounds, assessing patient conditions and engaging in reflective practice. SL, developed by Linden Labs, is the most popular and mature multiuser virtual world environment (MUVE) used in education (Warburton, 2009). SL virtual simulations provide an environment for students to participate in realistic situations or scenarios with other students through the use of avatars in an environment constructed to support learning. These simulation scenarios allow for students to play roles and practice their skills in a safe environment with a facilitator present who can guide learning and give immediate feedback on performance. The use of avatars allows students a feeling of being "physically" present in the SL environment, allowing training to be in a safe, controlled setting where students may practice and enhance their skills (Burgess, Slate, Rojas-LeBouef, & LaPrairie, 2010). Students are able to practice skills and make clinical decisions based on their knowledge while avoiding any potentially harmful consequences for patients (i.e., incorrect procedure, wrong medicine administered). The use of virtual learning environments creates a more realistic environment than classroom case studies. Paramedic students using SL for problem based learning found the environment more authentic and collaborative than using paper-based problem solving scenarios. They also found the SL environment allowed for feedback from the "patients" who increased the benefit of the learning environment (Conradi, et al., 2009). In our own work in SL we have found that students were more comfortable taking risks in the simulations and trying new skills when they were in SL because they assumed the role of the avatar (whose real life identity was not known) during the scenario and thus were not inhibited by concerns of what peers might think. This risk taking was also enabled because no real patients were at risk.

The use of virtual reality simulation in an environment such as SL can be used to support the knowledge and skill development of nursing students at both the undergraduate (prelicensure) and graduate level; however, this is not a simple task. Educators need to understand the theoretical underpinnings of simulation based learning to ensure virtual simulations scenarios are well planned and executed to ensure optimal student learning. To support the journey towards achieving expertise in practice by using simulation scenarios in virtual environments, it is necessary to determine which pedagogical theories are important for teaching and learning and developing expertise, as well as using simulation in education.

LEARNING THEORIES

There are two primary theories that guide this work—the community of inquiry (Garrison, Anderson, & Archer, 2000)—which guides the design and structure of the virtual world training scenarios, and Ericsson's (2004) deliberate practice, which guides the actual content of the training. Each of these theories is briefly described here.

Ericsson's (2004) work on expertise is helpful in guiding educators in facilitating focused experiential learning. Ericsson's framework is based on the belief that to acquire expert performance one must engage in deliberate practice activities that are focused on improving some aspect of performance. All individuals tend to improve performance with experience, however, Ericsson's theory posits that experts are those individuals who continue to improve beyond the level needed to perform adequately and become recognized as experts in their domain. This is not gained through additional generalized experience but rather through deliberate efforts focused on improving selected skills or tasks. This process occurs when individuals are instructed to improve certain aspects of their performance for a well-defined task. They are then given immediate detailed feedback on their performance which they can reflect upon and continue to practice during subsequent training sessions. Individuals who do not engage in deliberate practice activities often find their skills decay after time. Key to this work is the ability to determine the types of representative tasks (knowledge and skills) that define the essence of the domain of nursing practice.

In an effort to identify the skills requiring "deliberate practice" in nursing, the American Associate of the Colleges of Nurses (AACN) began an initiative, titled Quality and Safety Education in Nursing (QSEN), to identify key knowledge, skills, and attitudes (KSAs) needed by graduate nurses to ensure high quality nursing practice. There were KSAs identified around six key areas: patient-centered care, team work and collaboration, evidence-based practice, quality improvement, patient safety and informatics. Each of these areas identified specific skills to focus on such as; communicate with team members, adapt own style of communicating to needs of the team and situation and demonstrate commitment to team goals, which are part of teamwork and collaboration (Cronewett et al., 2007). In addition, the *Essentials of Baccalaureate Education for Professional Nursing* (2009) document provides a framework for nursing curriculum which incorporates recommendations from the Institute of Medicine around core knowledge required for healthcare professionals. The work, sponsored by AACN, includes the concepts of: patient centered care, interprofessional teams, evidence-based practice, quality improvement, patient safety, informatics, and clinical reasoning/critical thinking. Both of these bodies of work provide a clear view of what nursing students most clearly understand and be able to perform upon graduation.

The second theory that guides this work is the community of inquiry (CoI) (Garrison et al., 2000). The CoI model, a framework developed to guide educators in the area of on line learning pedagogy, is useful in understanding the elements involved in online learning. The model contains three constructs which are core elements of a collaborative constructivist learning environment: cognitive presence, social presence, and teaching presence. Cognitive presence reflects the learning and inquiry process and contains four tasks (definition of a problem or task, exploration for relevant information/knowledge, making sense of and integrating ideas and testing possible solutions). Social presence is the ability of participants to identify with the community, communicate purposefully in a trusting environment, and develop interpersonal relationships. Finally teaching presence is required to establish and maintain the learning community of inquiry. This type of presence includes the design, facilitation, and direction of the cognitive and social processes for the educational experience (Garrison, Cleveland-Innes, & Fung, 2010). Specifically the faculty has four roles; pedagogical, social, managerial, and technical (Burgess et al., 2010). In particular faculty needs to highlight areas in a virtual environment that may not be necessary to do in a classroom environment, where some of these roles or norms have already been established. For instance, setting up classroom etiquette needs to be addressed in a virtual world since students may not have any context from which to draw from to guide their behavior.

Technical management is particularly challenging when using new technology that the information technology (IT) department may not be familiar with. A recent study done in the United Kingdom highlights these challenges in universities and colleges. According to Kirriemuir (2010) SL and other virtual environments require greater technical commitments than other teaching and learning online technologies. In particular some technology related issues with SL, when using university owned computers are: ensuring the correct SL viewer is downloaded, technical capability of the computers to run the program, firewall or proxy issues and lack of knowledge in IT departments about SL.

One final area of consideration is feedback or debriefing after the virtual simulations. Feedback, including debriefing, is considered the most important feature of simulation based learning (Fanning & Gaba, 2007). The debriefing approach recommended is based on Schon's (1983) work on reflective practice and is called "debriefing with good judgment" (Rudolf, Simon, Dufresne, & Raemer, 2006). This approach values both the expert opinion of the instructors and the perspective of the learners. It creates transparency in the communication and focuses on creating a context that will help learners move forward to achieve key objectives. It involves both advocacy and inquiry as a means of helping the learner reflect back on their performance during the simulation so they can both understand their actions and hear the feedback from the instructor.

As seen throughout the preceding discussion, there are many pedagogical and technical areas to consider when deciding to use virtual simulations as a teaching methodology. It is important to understand these to enable faculty to design effective simulations for student education. This is more complex that just "running a simulation on the fly" as can happen when careful planning is not part of the process. The next section will outline the process used at the UMSON in using virtual simulations in SL.

SECOND LIFE VIRTUAL NURSING UNIT DEVELOPMENT

The University of Michigan owns land (Wolverine Island) in SL where members of Wolverine-Community, which is for anyone in SL who is currently affiliated with the University of Michigan as faculty, staff, students, and alumni, can interact with one another in a social or academic manner. Features of Wolverine Island include a library with information and items to help new members of SL get started, a large sandbox area for building within the virtual environment, an assortment of both lecture and small group meeting spaces spread over the island for group learning activities, and a medical building structure that is under development. Current use of the island has included educational sessions as well as course-related learning experiences for students within the SL environment.

In 2008 the UMSON undertook a project to establish a nursing acute care unit on Wolverine Island to be used as a training environment for nursing students. An eight-bed virtual hospital unit was built on the fifth floor of the partially furnished medical building (see Figure 11.1). The virtual hospital unit was designed to provide a more realistic setting for role play type scenarios than just a classroom setting. As noted in Table 11.1, features of the virtual nursing unit were aligned with the design considerations for SL educators and designers as identified by Minocha and Reeves (2010). In addition to the nursing unit, the medical building also includes a clinic and urgent care area. The exterior setting of the hospital unit includes an ambulance and a medical helicopter. The new unit occupies the entire two wings of the fifth floor which are separated by a stairway. Both wings are subdivided into three semiseparated rooms respectively. The decision was made to not put doors on the rooms to ensure the avatars do not feel "closed in" within the space and to allow for optimal camera maneuverability. The left wing consists of a conference room and two clinical rooms for patients (two patients per room). The right wing is made up of a nursing station and two additional clinical rooms. Each clinical room has two beds and is equipped with wall-mounted blood pressure gauges, bed-side cardiac monitors, medical supply cabinets, a sink, a computer desk, and chairs. The nurses' station is equipped with a central desk and computer workspace, crash cart, medication dispensing unit, x-ray view boxes and wheelchairs (see Figure 11.2). This layout is similar to what nursing students would find in an acute care hospital setting. Special outfits for avatars were purchased to replicate appropriate attire for the clinical setting, including stethoscopes and name badges.

Figure 11.1. Wolverine Island virtual hospital.

Table 11.1. Design Principles for UMSON Virtual Nursing Unit

Design Theme	*Key Design Principle*	*UMSON Unit Example*
Interpretation of Learning spaces in SL	Consider indoor spaces to support formality and authority relationships similar to traditional learning spaces in RL	The debriefing area and layout of the unit was in alignment with RL unit structure (i.e., patient room, conference room)
	Avoid designing spaces that could potentially trigger phobic reaction or do not provide an easy exit	Ceiling height and hallway access were intentionally exaggerated to ensure space for avatars
	Consider the SL skills and backgrounds of the users while designing learning spaces	Learners were provided education on basic functionality in SL prior to simulations. In addition, all "props" for scenarios were created in alignment with new SL members
Relationship between the pedagogy and the design of learning spaces	The learning activity and any underlying pedagogical underpinning should guide the design of the learning spaces	The environment was created to accommodate several different types of simulations

	Design learning activities that exploit the affordances of SL for education	Facilitators of simulation were able to manipulate the environment quickly to simulate RL (i.e., patient fall)
	Exploit the flexibility and ease of bringing out objects from the inventory to set up learning spaces in real time in SL to match the learning activity	Notecards were developed for a variety of patients that were distributed on cue for various simulations scenarios (i.e., patient assessment information changed depending on the scenario)
Visual realism and nonrealism in the design of spaces and activities	While considering the visual and nonvisual realism in the design of objects in the learning spaces, or the design of learning spaces themselves, take into account the learning activity and the discipline	Objects were created at appropriate height with adequate functionality. Realism was enhanced through incorporation of medical supplies and equipment (i.e., BP cuff, computers, sink) in the layout.
	Consider visual realism in the designs of spaces and activities to provide familiarity and comfort to users who are new to SL	Medical supplies and equipment common to the healthcare environment (i.e., code carts, computers, etc.) were incorporated into the design.
Design of learning spaces within an island	Design spaces to match the educator's authority that needs to be represented	The conference room included a round table and chairs. Layout was created to facilitate informal discussion during debriefing
	To articulate the aim of the learning space, choose objects in the design of the space that match their intended function and purpose	Patients were placed in beds with appropriate medical equipment (BP cuff, room numbers) located appropriately in the room.

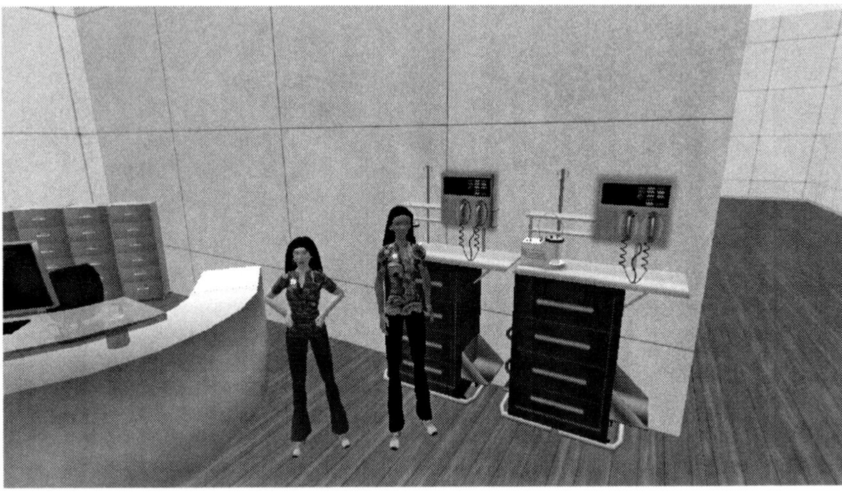

Figure 11.2. Nurses' station.

To effectively run the virtual simulation scenarios some patient interaction was needed. However, it was important for these patient interactions to work in an asynchronous manner, so each mannequin in the hospital beds within the nursing unit produces an attached note card when clicked—which highlights pertinent patient information—as well as a respective assessment box with current vital signs and assessment findings (which are modified for various simulations). This design consideration supported previous research which suggested that active engagement with learning objects increased the likelihood that the learner both retained and was able to use information and skills at a later date (Zull, 2002). Throughout the development of the nursing unit, the site was critically analyzed for functionality and fidelity in order to maximize use and value. In addition, flexibility in the learning space, a key element in the design of physical learning spaces (Oblinger, 2006), served as a critical component of the design.

An SL designer developer constructed the walls and floors in the medical building to provide an increased level of building realism. Most of the equipment used to furnish the unit was purchased through SL vendors. The cost of most of the equipment was reasonable and was purchased through Second Life Marketplace (https://marketplace.secondlife.com/) which is an online shopping venue for SL. Creating objects to interact with avatars required some additional work in scripting the objects. The more complex the interaction the more scripting was required; however, the faculty did the basic scripting using commercially available resource guides on scripting basics in SL.

SECOND LIFE SCENARIO DEVELOPMENT

Success of simulations in health care has varied, primarily due to the effectiveness of the scenario proposed and implemented. According to Jeffries (2007), design features of a well-developed simulation include clearly written objectives, fidelity or realism that mimic real life situations, building a level of complexity, providing cues for participants as the simulation progresses, and debriefing during and after the simulation is finished.

Tschannen and Aebersold (2010) designed a five step process for simulation development, which included: (1) key concept identification, (2) competency and standard mapping, (3) scenario building, (4) debriefing development, and (5) beta testing and refinement (as needed) of the scenario. In the key concept identification phase, course objectives are reviewed to determine key concepts to be explored in the simulation. The key concepts are mapped in phase 2 to clinical standards (e.g., Baccalaureate Essentials for Nursing) and competencies (e.g., Quality and Safety in Education for Nurses) to ensure that the simulation focus is in alignment with current requirements (See Table 11.2). Scenario building (phase 3) includes brainstorming clinical scenarios that will stimulate the desired response/behavior. For example, if one of the key concepts is interprofessional communication, the scenario would be developed in a manner that would require the student nurse to communicate with a physician or other member of the healthcare team. Part of the scenario building requires due diligence in ensuring the desired behaviors are in alignment with current evidence based practice. In the example given above (i.e., interprofessional communication), this may include requiring students to communicate using a structured approach, such as SBAR (Situation, Background, Assessment, Recommendation), a proven communication technique developed by the Navy and adapted widely in healthcare. In addition, scenario building should resemble a collaborative approach, with inclusion of experts in both clinical practice and education.

Once the scenario has been developed, a set of questions for debriefing (phase 4) is created. Debriefing sessions allow students to gain useful insights into behaviors and errors that contribute to the initiation and evolution of a crisis. The students also learn communication, teamwork, and leadership skills (Wong, Ng, & Chen, 2002). Feedback provided during simulation has been shown to significantly improve performance over feedback given in the actual clinical site (Day et al., 2009). In the final phase of development, the simulation is then beta tested with a group of faculty, nursing staff, and/or students (depending on the topic). Participants should be asked to provide feedback on all aspects of the simulation (i.e., scenario, implementation process, fidelity, etc.). Knowledge gained from the beta testing assists with the

Table 11.2. Virtual Simulation Scenarios in Second Life

Scenario	*Major Theme and Snapshot of Scenario*	*Key Course Objectives*	*Baccalaureate Essentials*
Nurse-Physician Communication and Conflict Management	Theme: Conflict management situation between the registered nurse and the physician	1. Analyze communication, human resource, and operational systems that promote safety and quality of care	1. Essential 2: Basic organizational and systems leadership for quality care and patient safety. Knowledge and skills in leadership, quality improvement, and patient safety are necessary to provide high quality health care
	Expectations: During this exercise each student was expected to adequately manage the physician-nurse conflict situation in a professional manner	2. Develop group facilitation skills that foster collaboration in the clinical management setting	2. Essential 6: Interprofessional communication and collaboration for improving patient heath outcomes. Communication and collaboration among healthcare professionals are critical to delivering high quality and safe patient care
	Participants: Nurse manager (faculty facilitator), medical doctor (faculty facilitator), registered nurse (student), charge nurse (student), nurse aide (student) and family member (student)		3. Essential 8: Professionalism and professional values. Professionalism and the inherent values of altruism, autonomy, human dignity, integrity, and social justice are fundamental to the discipline of nursing.

development of future implementation and evaluation of the simulation scenario. Refinements should be made based on the testing.

Several simulations (including virtual and high fidelity simulations) have been developed by faculty at UMSON using the above scenario development process. For the purpose of this chapter, an overview of one of the developed simulation scenarios will be described in relation to the five phases of simulation scenario development. This virtual simulation scenario was developed for the senior level Leadership and Management course.

Phase 1: Key Concept Identification

The *Essentials of Baccalaureate Education for Professional Nursing* (American Association of Colleges of Nursing, 2009), which addresses stakeholders' recommendations for required core knowledge of all health care providers, provides a framework for building the baccalaureate education for the twenty-first century, and thus, identifies potential opportunities for educational growth. Nine essentials are included in the document covering a range of topics, including six which describes incorporation of "effective communication techniques, including negotiation and conflict resolution to produce positive professional working relationships" (p. 22). For this reason, faculty teaching the senior level Nursing Leadership and Management course were interested in creating a simulation scenario that would allow students 'deliberate practice' of such interactions. As noted from Table 11.2, key threads in this course aligned well with the Baccalaureate essential related to interdisciplinary communication.

Phase 2: Competency and Standard Mapping

These key themes were mapped to the Quality and Safety Education for Nurses recommendations (Cronenwett et al., 2007) and The Joint Commission's National Patient Safety Goals (http://www.jointcommission.org/PatientSafety/NationalPatientSafetyGoals/). This phase is critical to ensuring the approach taken within the simulation scenario will assist in the development of the preferred competency. In addition, review of the current standards and literature related to the topic of interest can assist in beginning the brainstorming component of scenario building.

Phase 3: Scenario Building

Throughout the scenario building phase of simulation scenario development, the competencies were continually reviewed to ensure alignment with the scenario and key behaviors students were to perform. Experts in communication and patient safety participated in the brainstorming to ensure the most relevant evidence related to the themes was being highlighted in the scenario. The scenario developed included three phases: introduction, nurse physician interaction, and nursing team feedback phase.

In the introductory phase, students should be assigned roles in the simulation (staff nurse, charge nurse, family member, and nurse aide). The role of the physician should be played by one of the facilitators to ensure fidelity and standardization of the role. Once roles are assigned, the facilitator reminds students of the basic techniques and skills needed to function in the environment (i.e., how to speak, move, etc.) and provides them with any "role-specific" instructions. For example, the student playing the role of the family member is given a cue card stating:

> You are the wife of Mr. Jones who is being discharged within an hour. The nurse comes in to the room to start to give you discharge instructions and tracheostomy education, you just go along with what she says as if you understand everything. A physician will come in to the room and start 'yelling' at the nurse. Wait about 2-3 minutes and if the nurse doesn't get the physician to stop yelling or leave the room either with her or without her, please say "*You are making my husband very uncomfortable.*" If they don't stop and/or leave, state it again in 1-2 minutes.

Once assignments are made and cue cards (as applicable) are given, the facilitator "sets the stage" for the upcoming simulation. In this scenario, the facilitator states,

> We are just beginning the day shift. There are currently eight patients on the unit. You have been assigned your roles. Please listen to the following notes that should assist you with determining next steps. Staff nurse, you have just received a report on your four patients (bed 1-4). At the conclusion of report, you decided the first priority was to see Mr. Jones (bed 2). He is being discharged and still needs tracheostomy care. His wife is currently in and will be leaving very shortly. How will you proceed? Charge Nurse, you also have just received report on your patient assignment (beds 5-8) and should begin your assessments. How will you proceed? Please begin.

In the next phase of the scenario, the Nurse Physician Interaction Phase, the staff nurse—who is engaged in discharge teaching at the bedside—is confronted by a physician who is upset about the care related to a different patient. The physician (played by the facilitator) abruptly interrupts the staff nurses and begins speaking in a very nonprofessional manner. ("How come Ms. Jones ate? I wrote an order yesterday stating she was NPO for this test? I told the night RN, and yet when I went in the room Ms. Jones said she just ate a wonderful egg and cheese omelet. I can't believe how incompetent the staff is here.")

The physician continues to berate the staff nurses in the presence of the other family until the staff nurse removes herself from the environment. Throughout this interaction, the charge nurse (due to proximity in space and use of voice) would overhear the entire conversation and should facilitate further discussion. If the charge nurse does not involve herself, the expectation would be that the nurse would request help (since the physician continues to act nonprofessionally). The primary expectation of the staff nurse is to remove the discussion from the patient's room and suggest a quiet place to discuss the physician's concerns. In addition, the nurse (and charge nurse) should consider use of effective negotiation practices which include: (1) do not attack their position look behind it; (2) do not defend your ideas, rather invite criticism and advice; (3) recast an attack on you as an attack on the problem, and (4) ask questions and pause (Fisher & Ury, 1991).

In the final phase, nursing team feedback phase, the facilitator gives the following instruction:

> You were trying to figure out where the miscommunication occurred for Ms Smith (i.e., receiving tray). It has come to your attention that the nurse aide gave her the tray despite knowing she was NPO. How will you proceed?

In this phase, the staff nurse would be expected to communicate with the nursing assistant in a structured process, using the elements of successful feedback: (1) understand all potential causes of poor performance; (2) manage physical and psychological distance; (3) engage in active listening; (4) give effective nonthreatening feedback; and (5) develop a 'learning' mindset (Moss & Sanchez, 2004).

Phase 4: Debriefing Development

Once the simulation is complete, the students are asked to return to the virtual conference room. Once all students are seated, the debriefing session begins. For this scenario, a focus would be on understanding how the interaction played out. Questions such as "What went well during this scenario? What could have been done differently? How would you discuss the interaction had by the nurse and physician with Mr. Jones and his wife?" In addition, the debriefing session would highlight the critical points associated with conflict management, effective communication, and team feedback. The students who participated in roles are given the opportunity to speak first and to reflect on their performance in the scenario.

Debriefing can also include feedback to participants based on their individual or group performance in the scenario as compared to a standardized measurement scale or a list of expected behaviors. Educators can use standardized scales such as the Emergency Medicine Crisis Resource Management (EMCRM) tool, developed by Youngblood et al. (2008). This tool was developed to evaluate participant's crisis management skills in a virtual emergency department. The EMCRM assesses subject's team leadership skills, including knowledge of the environment, utilization of information and resources, and overall ability to communicate and facilitate task completion. Initial reliability testing of the tool was supported with a Crohnbach's alpha of 0.96 (Youngblood et al, 2008). Educators can also develop their own set of behaviors based on "best practice" standards or current evidence-based practice guidelines.

Phase 5: Beta Testing and Further Refinement

Prior to implementing this scenario with the senior students, this scenario was beta tested using a sample of second career students and faculty. The purpose of the beta testing is to test the process of setting up and running the scenarios and to pilot the actual scenarios. As noted in the literature there can be system issues and frustrations for students using SL in courses and any opportunity to proactively identify and minimize frustrations results in a more positive experience.

The beta testing occurred in a computer lab within the School of Nursing. For this reason, students were asked to "text chat" any and all communications (rather than use the voice chat function in SL). The scenario was implemented as described above. Some of the students were assigned roles while others participated as observers. The simulation took place over the course of approximately 15 minutes and an additional 15-20 minutes was used for debriefing of the simulation. Students were then asked to provide feedback (via an anonymous survey) regarding most and least liked aspects of the simulation, as well as potential suggestions for improvement. Several students commented on the opportunity that SL may offer for enhanced learning ("*It is a real good practice system and since it resembled real life scenarios, it gives almost the same experience one would have received in ordinary simulations.*"). Least liked aspects of the simulation related to technical difficulties (i.e., limited computer access and delayed response time) and the inability to recognize nonverbal responses by others ("*... chat format lacks a lot of the non-verbal communication that are conveyed in real life ...*").

Suggestions for improvement identified by the students included: (1) increased participation by students ("*... more of an ability to have group par-*

ticipate ... have student set up their own avatars ..."); (2) incorporation of voice responses, rather than text ("... *computer was too slow ... talking it through would have been better* ..."); and (3) enhancement of avatar capability ("*if these characters can also do the work ordered by the text, if possible ... i.e., give medication*").

Based on the feedback received in the beta testing, several modifications to the scenario (and subsequent implementation) were made: (1) clearer character identification to ensure immediate recognition of various characters (i.e., doctor, family member, charge RN, etc.); (2) incorporation of voice chat, rather than text; (3) separate room for observers and those participating in the scenarios; (4) addition of bed numbers above the bed to ensure quick identification of patient rooms; and (5) character specific outfits (scrubs for the nurse, lab coat for nurse manager, lab coat for physician, name tags).

STUDENT PREPARATION

Using SL to educate students requires some planning to avoid some of the pitfalls inherent in using any new technology. Students require training to use the SL platform and some additional directions in how to successfully participate in simulation scenarios. Simulation can be intimidating to students because it requires them to integrate the knowledge they should have received in their coursework with a novel patient situation and make decisions quickly to provide "care" to their "patients" or to interact with others in the situation before them. Pfeil, Ang, and Zaphiris (2009) studied usage patterns of 3D virtual worlds and identified a number of sociopsychological, pedagogical, and technological challenges faced by educators and students. These challenges included getting students into SL, orientation and navigation, identity, communication patterns, teaching activities and accessibility.

Some recommendations to prepare students to address these challenges will be outlined here. Nursing students typically do not have experience in the SL environment and many have little or no experience using computers games other that basic games, although many do use social networking sites. It is helpful to get students into the SL environment prior to the simulation to learn how to navigate their avatars and become familiar with the environment in which the simulations will take place. This allows the students to focus on their performance in the simulation and not be distracted with how to move their avatars or interact with other avatars, objects and/or the environment. Orientation issues at UMSON were addressed in two ways. The nursing students using SL for virtual scenarios received a digital presentation with detailed screen shots on how to download SL, set up their avatar, and change their appearance. A podcast was also developed to show basic information such as how to move their avatar, use the chat function, voice chat, interact with objects and receive note cards. Students are directed to spend some time in SL moving around the environment and to navigate to Wolverine Island.

Addressing the identity issues was a bit more challenging. The benefit of using avatars allows students to preserve some anonymity but also allows students to play with their identity (Pfeil et al., 2009); however, since avatar representation was part of their professional training certain expectations needed to be addressed. In particular student avatar attire needs to align with the same expectations as required by the mannequin based simulations that the students also participated in as part of their clinical training. As a result students are required to wear nursing appropriate attire during the simulation. To facilitate this, a set of nursing scrubs were designed and given to each of the avatars for use during the simulations. Creating some ground rules or "classroom" etiquette around appropriate attire during the simulation aligns with the professional nature of nursing and prevents the students from becoming distracted during the simulation. This formalism may be different for other types of learning experiences in virtual worlds where creative expression is part of the atmosphere.

Communication in SL consisted initially of text chatting so the chat logs could be saved for debriefing and evaluation, however, because the students felt that text chatting was cumbersome and voice chat was a viable option, they choose to switch to voice. Avatar gestures should also be considered, especially as virtual simulations become more complex so students can behave in a more realistic manner during the simulations.

There are many different teaching opportunities that can be used in SL in the area of nursing education. Students can visit many of the in-world support groups to learn more about how patients and families cope with illness, they can practice technical skills such as listening to and identifying heart or lung sounds, and they can practice using virtual simulations. The possibilities are many and varied.

ISSUES AND CHALLENGES

SL virtual scenarios can be an effective learning experience for nursing students at both the undergraduate (prelicensure) and graduate level. Some of the challenges faced and solutions are presented here. As noted in the beta testing phase text chat was initially used during the scenarios to enable the faculty to save the chat logs to use during the debriefing and after to evaluate student performance. As an alternative, screen capture programs could also be used to capture the dialogue. Students complained about the time it took to use the chat function, as voice provides a more natural and consistent flow in communication. However, using voice has a few limitations.

Students used their own computers in various environments to access SL, so finding an appropriate quiet place to use voice is a consideration. Students must find somewhere quiet enough for them to hear the conversation and speak without disrupting others. Another limitation is spatial voice projection. In using voice chat, students need to be aware of how far away they are from other avatars because local chat has a limit of 20 meters for hearing the speaker. If students move out of that range, they can easily miss out on the conversation. Other issues with using headsets need to be factored in. Specifications of the type of headset to use should be shared with students to minimize compatibility problems, such as the right software driver must be loaded for the headsets to work properly. This inconvenience becomes problematic when students use university owned computers in a computer lab and cannot access the programs to make these software changes. Students using university owned computers may not be able to use headsets to communicate effective, and this problem can frustrate students trying to use the voice features. This issue creates challenges when designing synchronous SL simulations. Once such problems are overcome, the ability to use voice chatting enhances the learning experience.

Ensuring the current version of the SL viewer is downloaded on institutional computers requires faculty to monitor software versions, prior to conducting scheduled activity in SL based simulations. Further, ensuring training materials are consistent with the latest version of the SL viewer is also critical. Such incompatibility may require updating materials periodically. Providing technical support to students when they are having difficulties using SL can become part of the faculty responsibility if the IT department is not familiar with SL. As virtual worlds such as SL become more widely used this issues could move to the IT department to assist students. Until then, faculty must accommodate for technically supporting students entering into virtual worlds for the first couple of times.

CONCLUSIONS

SL provides exciting new educational opportunities for nursing students as well as other students and practicing nurses. Virtual world simulation environments offer a unique and cost-effective method of teaching nursing competencies related to leadership and management skills. Although computer driven mannequin based simulations have become an essential component to nursing education, virtual simulations can add value to the overall simulation experience as they leverage social and communication skills over a distance. Nursing students are challenged with learning both technical and nontechnical skills to achieve competency in their future roles as practicing nurses and the availability of virtual worlds holds promising for engaging nursing students in effective ways. Both the CoI (Garrison, Anderson, & Archer, 2000) and Ericsson's (2004) work provides guidance in addition to the five step scenario development process. Nursing skill development will need to be reinforced through participation in many types of simulations. SL provides a less expensive training environment and can be used to practice many nontechnical skills such as clinical judgment, teamwork, leadership and communication. It can also provide an opportunity for nurses to participate in simulations from home or anywhere they can access the Internet, providing educators with a new, flexible option for training. In addition, this method of simulation may maximize opportunities for multidisciplinary teams to interact and practice necessary skills for high quality of care. Adding virtual simulations to well established curriculum that includes mannequin based simulations can be an added benefit and allow for expanding learning opportunities for students.

APPENDIX

Lesson Plan

Nurse-Physician Communication and Conflict Management Virtual Simulation

Concept

Managing difficult situations involving nursing physician interactions

Goal

To provide an opportunity for students to practice effective communication and feedback strategies in a challenging situation.

Objectives:

1. The student will recognize that the communication event is confrontational in nature.
2. The student will move the conversation away from the patient's bedside.
3. The student will engage necessary resources to effectively resolve the situation.
4. The student will use effective communication strategies in responding to the situation.
5. The student will use effective feedback skills in following up on the situation.

Prelearning Activities

The student will be exposed to didactic materials about effective communication strategies, giving feedback and managing conflict prior to the simulation. This can be in lecture or self learning module/reading assignment.

Required Materials

Virtual hospital unit in Second Life complete with a hospital bed that an avatar can lay down in, a chair at the bedside for a family member and a conference room close by; hospital scrubs that can be given to student's avatars; and notecards to pass to avatars.

Time

45-55" total time

Prep-5" to assign roles, give instructions, pass out scrubs and notecards (see notecard details below)

Scenario- 15-20"" to run the scenario

Debrief-15-20" to facilitate debriefing

Wrap up-5" to get feedback from students on the scenario/debrief

Roles

Physician-played by facilitator or trained designee
Nurse-played by a student
Charge Nurse- played by a student
Patient - played by a student
Patient Family member- played by a student
Patient Care Assistant- played by a student

Scenario Directions

Scenario Progression	Expected Actions by students	Cues
1. Nurses, charge nurse and patient care assistant are sitting at the conference room table waiting to get report and scenario directions.	Nurse will go to see Mr. Jones and his wife.	Facilitator: If necessary remind nurse to go to Mr. Jones to begin teaching.
2. Patient and patient family member are in the room. Patient is in bed and the family member is sitting in a chair at the bedside.	Nurse may begin teaching or state she/he will be back as soon they check on their other patients to begin teaching.	
3. Physician is on the unit but out of sight from the nurses and patient room.		

4. Facilitator states: "We are just beginning the day shift. There are currently 8 patients on the unit. You have been assigned your roles. Please listen to the following notes that should assist you with determining next steps. Staff nurse # 1, you have just received report on your four patients (bed 1-4). At the conclusion of report, you decided the first priority was to see Mr. Jones (Bed 2). He is being discharged and his wife still needs tracheostomy teaching. His wife is currently in and will be leaving very shortly. How will you proceed? Charge Nurse, you also have just received report on your patient assignment (beds 5-8) and should begin your assessments. How will you proceed? Please begin."	Charge Nurse visits other patients Patient Care Assistant waits in conference room or can wander in hallway but does not interact unless someone starts a conversation with him/her.	
5. MD walks in room and begins a loud, hostile conversation regarding an incomplete stat order (script to be written) "How come Ms. Jones ate? I wrote an order yesterday stating she was NPO for this test? I told the night RN, and yet when I went in the room Ms. Jones said she just a wonderful egg and cheese omelet. I can't believe how incompetent the staff is here."	The nurse should ask the physician to step out of the patient's room and go to a more private area to discuss/ resolve this. Nurse should continue to insist upon this even leaving the room if necessary. The nurse should say to the patient/family I will be back after I take care of this issue.	Family member: if conversation doesn't move away in 5 minutes, then say. "You're making my father very anxious." Facilitator: if nurse is not doing well handling the situation you could suggest they get others to help out such as charge nurse.

(Scenario directions continues on next page)

Scenario Directions Continued

Scenario Progression	Expected Actions by students	Cues
Physician will continue to interact with the nurse at the bedside in a loud and rude manner and will not leave the bedside until the nurse or the charge nurse is insistent they take the conversation outside of the patient's room to a more private area. The physician will eventually calm down and become reasonable if the nurse and /or charge nurse uses effective conflict management/communication strategies and will end in a successful resolution. If the nurse and/or charge nurse does not use effective conflict management strategies the conversation will continue to deteriorate and the physician will continue to be confrontational and angry.	Charge RN (who overheard) would be available and would facilitate further discussion if the nurse asks the charge nurse for help or if the nurse is having difficulty managing the situation. Nurse/charge nurse should use conflict management and effective communication strategies (learning in class or pre-learning activities) to try and resolve the situation.	Facilitator: if charge nurse does not seem aware of the interaction cue them to go to the room.
6. Facilitator will cue the physician to leave the area and tell the nurses we are moving into the feedback phase of the simulation. The facilitator states: "You were trying to figure out where the miscommunication occurred for Ms Smith (i.e. receiving tray). It has come to your attention that the nurse aide gave her the tray despite knowing she was NPO. How will you proceed?"	Nurse or charge nurse will go to the patient care assistant and ask him/her about Ms. Smith. She will then use effective feedback skills to ensure the patient care assistant understands to ask for clarification in the future when he/she has conflicting information.	Facilitator: may need to encourage the nurse or charge nurse to follow up with the patient care assistant.
Facilitator: At the end of 15-20" stop the scenario and begin debrief process.		

Note Cards

These are developed ahead of time by the facilitator and kept in the facilitator's inventory folder in Second Life. A copy is given to the appropriate avatar at the beginning of the simulation.

Patient note card

You are Mr. Jones. You need to get into the bed and lay there. You are not able to speak because you have a tracheostomy in. Your wife is at the bedside with you. You are supposed to go home today.

Patient family member note card

You are the wife of Mr. Jones who is being discharged within an hour. The nurse comes in to the room to start to give you discharge instructions and tracheostomy education, you just go along with what she says as if you understand everything. A physician will come in to the room and start "yelling" at the nurse. Wait about 2-3 minutes and if the nurse doesn't get the physician to stop yelling or leave the room either with her or without her, please say "You are making my husband very uncomfortable." If they don't stop and/or leave, state it again in 1-2 minutes."

Patient Care Assistant notecard

Sit at conference room table or wander in the hallway. Don't interfere with anything going on in Mr. Jones room. If the nurse asks you to do something, go ahead and do it. The nurse will interact with you the second half of the scenario about feeding Ms. Smith. During this interaction, respond, "I got report from the night aide. She said she was NPO but I didn't see any documentation of that so I fed her when she asked this morning."

Charge Nurse note card

After report go and visit your patients. When you overhear the conversation in Mr. Jones room you can wander by the room and see if the nurse asks you for help. You can decide at what point you want to intervene or not.

Debriefing Phase

Start the debriefing phase by asking the students who played the role of charge nurse and nurse the first two questions then open it up to every one to respond after that.

Questions to consider

1. What went well?
2. What would you do differently if you could do this all over again?
3. What conflict/communication strategies were used during the conversation and where they effective?
4. Was the feedback to the patient care assistant handled appropriately?
5. How did you feel when the physician was behaving verbally aggressive?

REFERENCES

American Association of Colleges of Nursing. (2009) *The essentials of baccalaureate education for professional nursing practice faculty toolkit.* Retrieved from http://www.aacn.nche.edu/Education/pdf/BaccEssentials08.pdf

Benner, P., Sutphen, M., Leonard, V., & Day, L. (2009). *Educating nurses: A call for radical transformation*. Princeton, NJ: Carnegie Foundation.

Burgess, M. L., Slate, J. R., Rojas-LeBouef, A., & LaPraire, K. (2010). Teaching and learning in second life: Using the community of inquiry (CoI) model to support online instruction with graduate students in instructional technology. *Internet and Higher Education, 13,* 84-88.

Conradi, E., Kavia, S., Burden, D., Rice, A., Woodham, L., ... Beaumont, C. (2009). Virtual patients in a virtual world: training paramedic students for practice. *Medical Teacher, 31,* 713-720.

Cronenwett, L., Sherwood, G., Barnsteiner, J., Disch, J., Johnson, J., ... Mitchell, P. (2007). Quality and safety education for nurses. *Nursing Outlook, 55(3),* 122-131. doi:10.1016/j.outlook.2007.02.006

Day, T., Iles, N., & Griffiths, P. (2009). Effect of performance feedback on tracheal suctioning knowledge and skills: Randomized controlled trial. *Journal of Advanced Nursing, 65*(7), 1423.

Ericsson, K. A. (2004). Deliberate practice and the acquisition and maintenance of expert performance in medicine and related domains. *Academic Medicine: Journal of the Association of American Medical Colleges, 79*(10), S70-81.

Fisher, R., & Ury, W. (1991). *Getting to yes: Negotiating an agreement without giving in* (2nd ed.). New York, NY: Random House Business Books.

Fanning, R. M., & Gaba, D. M. (2007). The role of debriefing in simulation based learning. *Simulation in Healthcare, 2*(2), 115-125.

Garrison, D.R., Anderson, T., & Archer, W. (2000). Critical inquiry in a text-based environment: Computer conferencing in higher education. *The Internet and Higher Education, 2*(2-3), 87-105.

Garrison, D. R., Cleveland-Innes, M., & Fung, T. S. (2010). Exploring causal relationships among teaching, cognitive and social presence: Student perceptions of the community of inquiry framework. *The Internet and Higher Education, 13,* 31.

Jeffries, P. R. (Ed.). (2007). *Simulation in nursing education*. New York, NY: National League for Nursing.

Kirriemuir, J. (2010). UK university and college technical support for second life developers and users. *Educational Research, 52*(2), 215.

Maithel, S., Sierra, R., Korndorffer, J., Neumann, P., Dawson, S., ... Callery, M. (2006). Construct and face validity of MIST-VR, endotower, and CELTS - are we ready for skills assessment using simulators? *Surgical Endoscopy and Other Interventional Techniques, 20*(1), 104-112.

Minocha, S., & Reeves, J. (2010). Design of learning spaces in 3D virtual worlds: An empirical investigation of Second Life. *Learning, Media and Technology, 35*(2), 111-137.

Moss, S. E., & Sanchez, J. I. (2004). Are your employees avoiding you? Managerial strategies for closing the feedback gap. *Academy of Management Executive, 18*(1), 32-44.

Oblinger, D. (2006). *Learning spaces*. London: EDUCASE.

Pfeil, U., Ang, C. S., & Zaphiris, P. (2009). Issues and challenges of teaching and learning in 3D virtual worlds: Real life case studies. *Educational Media International, 46*(3), 223-238.

Rosenthal, M. E., Adachi, M., Ribaudo, V., Mueck, J. T., Schneider, R. F., & Mayo, P. H. (2006). Achieving housestaff competence in emergency airway management using scenario based simulation training - comparison of attending vs housestaff trainers. *Chest, 129*(6), 1453-1458.

Rudolf, J. W., Simon, R., Dufresne, M. S., & Raemer, D. B. (2006). There is no such thing as non-judgmental debriefing: A theory and method for debriefing with good judgment. *Simulation in Healthcare, 1*(1), 49-55.

Schon, D. A. (1983). *The reflective practitioner: How professionals think in action*. San Francisco, CA: Jossey Bass.

Tschannen, D., & Aebersold, M. (2010, October). *Second Life: Innovative simulation development-making it REAL!* Paper presented at Meaningful Play, East Lansing, MI.

Van Sickle, K. R., McClusky, D. A., Gallagher, A. G., & Smith, C. D. (2005). Construct validation of the ProMIS simulator using a novel laparoscopic suturing task. *Surgical Endoscopy and Other Interventional Techniques, 19*(9), 1227-1231.

Wayne, D. B., Didwania, A., Feinglass, J., Fudala, M. J., Barsuk, J. H., & McGaghie, W. C. (2008). Simulation-based education improves quality of care during cardiac arrest team responses at an academic teaching hospital: A case-control study. *Chest, 133*(1), 56-61. doi:10.1378/chest.07-0131

Warburton, S. (2009). Second Life in higher education: Assessing the potential for and the barriers to deploying virtual worlds in learning and teaching. *British Journal of Educational Technology, 40*(3), 414-426.

Wong, S. H., Ng, K. F., & Chen, P. P. (2002). The application of clinical simulation in crisis management training. *HKMJ, 8*(2), 131.

Youngblood, P., Harter, P. M., Srivastava, S., Moffett, S. Heinrichs, W. L., & Dev, P. (2008). Design, development, and evaluation of an online virtual emergency department for training trauma teams. *Simulation in Healthcare, 3*, 146-153.

Zull, J. (2002). *The art of changing the brain: Enriching the practice of teaching by exploring the biology of learning*. Sterling, VA: Stylus.

CHAPTER 12

USING SECOND LIFE TO TEACH AND RESEARCH VIRTUAL ECONOMY

Teemu Surakka and Sami Ahma-aho

The greatest benefits of 3D virtual environments are currently in teaching and training applications. In order to tap into these benefits, an innovative approach to combine teaching and research was developed, with the goal of enhancing learning objectives in the collaborating business and marketing courses and improving the quality of research data on the electronic economy. This approach used the virtual world Second Life as a platform for its versatile business activity. This chapter presents the approach that was developed and the technologies that were utilized, as well as examples of the exercise outcomes and the problems that were encountered. Overall, students found the exercise challenging in certain aspects and the research outcomes were somewhat disappointing. Consequently, the paper also discusses future plans for applying emerging technologies to teaching and research.

Engaging the Avatar: New Frontiers in Immersive Education, pp. 339–363

USING SECOND LIFE TO TEACH AND RESEARCH VIRTUAL ECONOMY

Most contemporary business education is comprised of basic courses in marketing, accounting, management, perhaps some international business courses and, in some cases, entrepreneurship. These subjects provide a certain amount of background with which students can engage in business and understand the discussions in these topics. However, the business education program in general, as well as the various level business degree schools, has come under scrutiny recently regarding their relevance (Poul & Rask, 2008). The fundamental question is to identify the basic goal of business education and the best way to achieve it. A full discussion of this task, obviously, requires more than one book chapter. In this chapter we try to outline whether the promises of virtual worlds, in terms of teaching new business development, accomplish the challenges noted above.

Imagine you are entering the market in Second Life as a new virtual entrepreneur. Having explored a new market, you eventually reach the conclusion that there are surprisingly few services related to your undertaking. A variety of religious communities are present, but there is only one real undertaker to service the whole market. There are many cemeteries, but they are mainly used as graves for pets or as ambiance for night-time parties. Still unfamiliar with the culture of this market, as well as the various subcultures, you start to dig deeper into possible customer segments, business partners, business locations and the skills needed in the daily operations of your start-up. You follow up with a feasibility plan and consider setting up your business. Venture capitalists offer you the initial capital, with a condition that you report your findings about the market, possible financial projections and the use of their money. Having convinced them of the possibilities of your business idea, you are given more money to implement your plans. With some more hard work and a little luck, you make your first customer contacts in no time and, eventually, you have managed to turn your initial idea into profits within 12 weeks and only used a part of the venture capital you were given.

The scenario described above is the test case for this exercise, a virtual company called Final Plans Funeral Home, which we set up to test in our research setting before offering it to business plan development or marketing educators. Within this setting, we offered a small amount of venture capital for students to start their business. While the initial capital was risk-free for the students (invested capital is not collected back and no returns are collected), the time that the students invested raises the perceived risks of the business development and makes the setting more realistic.

During the first year, we collaborated with three educators from various Finnish universities and a total of 10 student groups signed up for the exercise. The number of students in the groups was limited to four because larger groups would have influenced the research setting on knowledge discovery (a topic that is not discussed in this chapter). Most of the participating groups had four students in them.

Background—VinCo Research Project

This chapter is based on experiences gained from the teaching and research approach described above, which was developed in a research project called VinCo—Innovative Business and Collaboration in Virtual Environments. The VinCo project explored the interfaces between virtual and traditional environments and investigated virtual worlds as centers for collaboration and business.

The focus areas of this research project—knowledge intensive value creation and virtual worlds—were chosen because they incorporate the two most recent evolutionary economies: the knowledge economy and the digital economy. The dynamic and evolutionary aspect of these economies led to the use of the exploratory research approach, which is described in greater detail later in this chapter. From an ethical perspective, mainly to ensure that the research setting did not cause any financial harm to the research subjects, a decision was made to collaborate with established courses using proper learning management systems. In order for the collaboration to work properly, it was necessary to create a win-win-win situation; this entailed a research setting that enhanced existing business education, motivated the students to participate in research and provided the researchers with rich data on the subject.

This study looked into one particular virtual world—Second Life. Compared to other, more "game-like" virtual worlds, it is the lack of stated and unified goals in Second Life that makes the economy in it versatile and gives it a strong resemblance to other, equally real economies in the world. For example, users can own virtual land, organize events, use or perform services and produce virtual goods and sell them in markets.

In addition, the economy in Second Life has become noteworthy. In Q1 2010, resident-to-resident transactions totaled approximately U.S.$160 million and there were an average of 65,000 unique users with "positive monthly Linden dollar flow" during the same period. While these resident to resident transactions are a good reference point, some proxy data on economic activity in Second Life can also be obtained from the virtual land sales data. During August of 2010, roughly 1.5 million m^2 of virtual land was auctioned. The corresponding average monthly figure during 2009 was 1.7 million m^2 (Linden Lab, 2010). Although these figures only provide a crude reference point, they also highlight the fact that the market is not static.

The number of possibilities is growing in terms of researching virtual economies and collaboration in different virtual environments. However, at the start of this research project the choices were limited and only one platform was used to research these different aspects of virtual environments.

Learning and Researching Entrepreneurship in Virtual Worlds

The goal from an educational viewpoint was to enhance business education. Theory-based conditions for effective learning are one marker to which new learning environments can be compared. The listing of theory-based conditions by Alavi, Wheeler, and Valacich (1995) is particularly useful for this purpose because it has been previously used to evaluate IT-based learning in business education. This listing comprises (1) active learning and construction of knowledge, (2) cooperation and teamwork in learning and (3) learning through problem solving.

These conditions can, arguably, be identified as effective factors for learning in business due to their resemblance to "first life" requirements needed in business. The research setting was created with the above mentioned conditions for effective learning in mind:

Construction of knowledge: Virtual worlds are a subgroup of contemporary social media platforms. Despite the differences in the mode of interaction, the participants behind different virtual identities are always human beings. The social, economic and psychological patterns of human interactions seem to translate to these new platforms of interaction, as suggested by Castronova (2005), just as they have translated to other online services previously. From an economic perspective, therefore, the virtual worlds do not differ fundamentally from the "first life" (e.g., the valuations that humans place on virtual property resemble those on "first life" objects (Lehtiniemi, 2008)). While the students receive normal teaching from collaborating courses and a tutorial for business in Second Life from researchers, the emphasis in this approach is on the process of inquiry and constructing new knowledge that can be applied in a new context.

Teamwork: One of the main reasons people use virtual environments and collaborative spaces is to collaborate on a topic from distributed locations. All the courses involved had distributed qualities, either in terms of multiple campuses or in the mode of participation. Additionally, two of the collaborating courses involved students with quite different backgrounds; this leads to perhaps the most interesting feature from a teaching point of view: the same setting can be used to teach business plan development, marketing and the use of social media in business. This enhances the use of multidisciplinary teams in teaching.

Problem solving: One of the earliest features requested by the participating educators was the students' ability to fail in the exercise and learn from the failure. From the students' perspective, the virtual economy serves as a rapid testing ground for their entrepreneurial skills. Failing does not necessarily mean automatic bankruptcy and setting up new ventures or expanding old ones is fairly straightforward. Furthermore, the concept of being able to anticipate and identify possible failures resonates with the current theory base and history on entrepreneurship (Mullins, 2004).

Our ideology, with regard to virtual environments as teaching grounds for entrepreneurial courses, is not to imply that operating in virtual worlds is the only training needed for entrepreneurship. Instead, we argue that most virtual worlds add new aspects to teaching entrepreneurial skills. One of these aspects is the continuous checking and rechecking of student's understanding and underlying mental models—the use of metacognitive strategies to ensure a deeper level of understanding (Bransford, Brown, Cocking, & National Research Council, 2000). This can be accomplished when students have to solve problems that arise when their previous understanding is not compatible with real and dynamic markets in the virtual world.

Research on the topic of virtual entrepreneurship is still scarce and concentrated on certain topics. There are still a noticeable number of similarities in the fundamentals on virtual world business with "first-life" business. Many people, who participate in business activities, either in goal-driven virtual environments such as massively multiplayer online games or in environments without a unified goal, can actually draw from those experiences in real-world business cases. Over the past few decades, firms have started operating in a more multinational setting because of the need to pull resources from multiple markets and also to enter attractive markets in multiple countries. Virtual entrepreneurs face similar problems as a result of having widely distributed team members, members from different cultures and a market that comprises agents from several countries (Teigland, 2010).

There are three reasons why we think virtual worlds are relevant it the study of entrepreneurship: (1) virtual platforms provide quite accurate and up-to-date data on the markets, which keeps the virtual world economics quite dynamic by increasing the information available to the economic agents. (2) Consumers in virtual worlds spend a certain amount of time, sometimes up to several hours a week, in virtual worlds, thereby valuating virtual goods in respect to time spent in virtual worlds. (3) Cyclical changes in markets can be found in many of the virtual worlds; as noted by Hamari and Lehdonvirta (2010), some of them coincide with cyclical components in "first-life" but some are also virtual-world-specific.

Collaborating Business Courses

To date, the setting has been conducted in cooperation with three different Finnish universities: Aalto University, Lappeenranta University of Technology and the University of Eastern Finland. Both sides in the setting—the researchers and the collaborating course teachers—have depended extensively on each other to deliver the expected outcome for each participating party. All of the participating courses dealt with the main topics of business or entrepreneurship and they all did so using a practical approach. Therefore, although the participating courses varied somewhat in terms of the course topic and participant background, the three collaborating courses were basically quite similar (Table 12.1). Participants from the Lappeenranta School of Technology were adult students; otherwise, the backgrounds of the students were generally alike. The course at Aalto University School of Science & Technology was the first one conducted, which meant that there was slightly less collaboration time than with the other two. For reasons of practicality and convenience, the collaboration with the University of Eastern Finland was conducted online. This was the best approach given that the course participants themselves participated from a wide range of locations.

Cooperation Characteristics

Students of courses that concentrate on electronic commerce could be expected to have had previous information about the topic or to be somewhat familiar with the user interface of Second Life. In our experience,

Table 12.1. Cooperation Characteristics

	Lappeenranta University of Technology	*Aalto University*	*University of Eastern Finland*
Guidance	Hands-on	Hands-on	Online
Collaboration with course patron	Extended	Brief	Extended
Participants	Adult students	Full-time students	Full-time students
Course topic	Development of Technological Business Idea	New Venture Development	Virtural Business World

this did not play a notable role in their behavior, although it may have increased their experimental attitude with regard to the virtual environments. Overall, the student business ideas were based on needs or experiences from the "real world"; these included ideas such as market research services for small and medium-sized enterprises (SMEs) companies, visualization services for the construction sector and additional exercises for driver's education. None of these ideas proceeded to real customer interaction because the customers of these student groups were mainly existing companies that had little or no knowledge about the possibilities of virtual worlds.

In order to compensate for possible differences in the background information of the students, an optional tutorial on Second Life Business was organized in the start of each business course. This tutorial was designed to provide rudimentary background knowledge about what is popular in Second Life, the business concepts that different entrepreneurs have tried and what is not suitable in Second Life due to the policies of Linden Lab. The material used in the tutorial was added to the Business Exercise home page (http://www.elisanet.fi/vincoproject/) for students who were unable to participate in the face-to-face event. Many of the business concepts in the tutorial are not suitable for the exercise by themselves, because our setting has limited resources and time. Also, the tutorial was not meant to give students building skills in Second Life or ready information sources to use during the exercise because the study emphasized problem solving in the setting and focus on knowledge acquisition and the use of service providers. The tutorial was carried out in the form of a field trip and students were required to create their avatars beforehand so that as many as possible could participate in the trip with their own avatar. This way, during the tutorial students could be instructed about basic movement and camera controls if necessary. Depending on the focus of the collaborating course, the tutorial could be tailored to the students' interests. Nonetheless, the following elements were always present in the tutorial.

1. **Basic business concepts.** Examples of business concepts aimed at utilizing an asset that the student group may already have. For example, the value proposition of property development in Second Life is to provide residents or companies with well designed (usually themed) and maintained environments populated with similar-minded residents or companies.
2. **Marketplaces.** Examples of business concepts aimed at making connections between buyer and the seller. Again, these business concepts may not all be viable in the time given to the students, but these are also places where students can acquire some of the

resources they need or sell and market their own products or services. Examples include web-based services and shopping malls inside Second Life.

3. **Networking services.** Examples of business concepts aimed at making connections between different parties. These usually overlap with the marketplace concepts, but also include many of the most innovative marketing concepts inside Second Life, such as b-places.com, various teleportation systems, and themed self-paced trips around Second Life shops with free gifts in each shops (a concept called "hunt").
4. **Marketing concepts.** Examples of business concepts that enhance marketing and help entrepreneurs receive attention in a virtual environment where the real battle is for people's attention. These can be part of a business idea, related to popular activities in the virtual world, B2B solutions, or just ideas about the importance of the location of your business in a virtual world.
5. **Border-spanning concepts.** Examples of business concepts that are aimed at helping "first life" companies enter the virtual world or monetize their presence in the virtual world. These business ideas usually come into the minds of students first and these examples are included to show that their initial idea may not be so unique after all.
6. **Not possible in "first life."** Examples of business concepts that include some unique aspect of the virtual world. Following the train of thought from the previous category, these examples are included to encourage students to think twice about their market and familiarize themselves with the virtual world before they decide what to do in the exercise.
7. **Business ideas not allowed in the exercise.** A list of business ideas that we have ruled out of this exercise based on Linden Lab's Terms of Service and Linden Lab's policies.

VIRTUAL BUSINESS EXERCISE

In the actual virtual business exercise, groups of students are given the task of forming a business plan that is suitable for Second Life and then implementing the plan within a given budget and time. The budget consists of venture capital that is provided in stages by researchers and of stipends received from Linden Lab. The total budget amounts to L$20,000; at an exchange rate of L$257 = U.S.$1, this amounts to U.S.$77.50 or about €58 (as of November 18, 2010). Based on our experience, this budget is enough to set up a small business in Second Life. The actual timing of the exercise is negotiated with the course organizers, but the aim is to give the student groups 12 weeks to familiarize themselves with the platform and the market, plan their approach, test their ideas, conduct marketing and reach the customer interface. This budget and timeline does not allow groups to implement the most capital-intensive business models, but there is plenty of room to generate original business ideas and even change plans during the exercise. For example, while testing the research setting, the researchers set up a full-service funeral home called Final Plans Funeral Home. Figure 12.1 (see next page) presents this test case and the exercise.

Figure 12.1. Timeline, phases, cumulative venture capital and deliverables used in the Virtual Business Exercise. The different phases, the cumulative amount of venture capital given in these phases and the deliverables requested from the groups are described with the help of the test case.

The exercise is divided into four distinctive phases, as shown in the left-hand column of Figure 12.2.

Orientation: This stage is for learning the ropes in the virtual world and searching for possible market opportunities. A small amount of money is given to the group account for orientation purposes.

Search: Time dedicated to idea generation and the formulation of an initial business idea. The participating groups are given a small amount of money for information-seeking purposes (identifying possible partners, objects, scripts, marketplaces) and time to form their initial business idea. Before this money is provided, however, groups are asked to fill in the market potential description form based on their findings in the previous phase. During this phase, student groups should also fill in the financial projections spreadsheet for the first time. These deliverables are described later.

Prototyping: The groups are given more money to prototype their product or service and time to plan the implementation of their business idea. Before the groups receive any venture capital, they should formulate a short description of their initial business idea. At this stage, the researchers will provide brief feedback on the business idea. During this phase, student groups should fill in the financial projections spreadsheet for the second time.

At this stage, the group leader is required to upgrade his or her Second Life account to a premium account and handle any possible income from this point forward. The premium account fee is refunded to the group leader, who will be asked about the means for refunding this money at this stage.

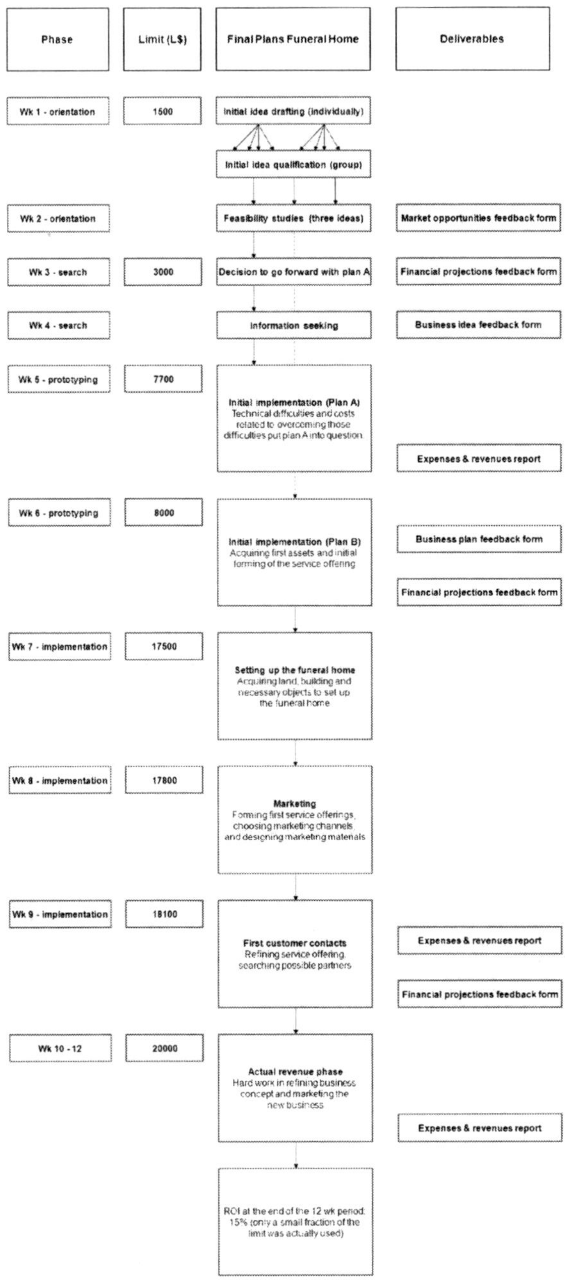

Figure 12.1. Timeline, phases, cumulative venture capital and deliverables used in the Virtual Business Exercise.

	week															
	-2	-1	1	2	3	4	5	6	7	8	9	10	11	12		
Stage	Enrollment		Orientation		Search		Prototyping		Implementation						Total L$	Total $
Limit		500	1000		1500		4400		9200						16600	
Account							10 $				10 $					20 $
Stipend							300	300	300	300	300	300	300	300	2400	
Signup													1000		1000	
															20 000 L$	20 $

Figure 12.2. Phases of cumulative amount of venture capital.

Implementation: Participating groups are given more money for land rental, outsourcing, marketing and other implementation costs. At the beginning of this stage, groups are required to briefly describe the implementation plan for their business idea. This period is the actual revenue phase, where the quality of the business idea is truly tested in a virtual economy. During this phase, groups fill in the financial projections spreadsheet for the final time, although they continue reporting expenditures and revenues to the researchers.

The deliverables mentioned above are used to monitor the progress of the group work, and also as background information for the research. During the first course, these report forms were implemented as Word documents that were made available from our website according to the exercise schedule. However, we soon changed many of these deliverables to web-based surveys that could be completed from Second Life with the current version of the client. The first report, the market potential description, is meant to encourage students to familiarize themselves with Second Life and its residents. This report asked, for example: What market segment has consumers with unmet needs or problems that your business might resolve? Does the market provide entry to other markets?

The second report, the initial business idea, builds on the first report, as long as the plans of the student group have not changed during this time. This exercise was not intended to be a writing exercise and students were instructed to focus on the essentials when writing the reports. The basic intention of the initial business idea report was to check whether the student group's idea is viable for Second Life and for this exercise, with limited resources and time. For this reason, they were asked to write a short description about the business environment of the product/service and its value proposition, the characteristics of the intended customers, the intended distribution channels, competitive solutions in the market, and technological know-how and risks related to the idea. Based on this information, the authors provided brief feedback to the group based on their knowledge of Second Life.

The finalized business plan was reported by elaborating the answers in previous reports or by re-writing the business plan, if it had changed from the initial business plan. Because this report was due after the prototyping phase, students were essentially given the chance to fail during the prototyping phase and learn from their failure.

The groups also delivered periodical financial plans. These reports included the most prominent goals of their business and the amount of capital that was used to pursue them. The financial projection also tracked participants' subjective confidence levels regarding their expected returns on the use of capital.

The research setting described above is designed to answer many research questions (not all of which are covered in this chapter). For example, from the point of view of knowledge-intensive value creation, we looked into the services and knowledge sources that these groups utilize during the exercise. For this reason, we mainly marketed the exercise to groups that were fairly new to this particular virtual world and needed additional information to complete the exercise. The main data gathering method in this part of the research is a modified version of the in-world tracking script originally created by Nick Yee and Jeremy Bailenson (2008), which makes it possible to form the social network of any given student group and analyze it dynamically across the entire timeline of the exercise. Combining this dynamic social network with information from the groups' expenditures enabled the authors to analyze the interaction with different value-creating entities in Second Life, as well as the value creation of the groups' businesses.

In addition, the participants were followed with regard to their financial actions in order to gain insights into their entrepreneurial decision making and learning. The participants' could take no action and not spend any capital setting up their entrepreneurial activity. This was a common decision among the participants in our exercise. According to the theory presented by McMullen and Shepherd (2006), a measure for a person's indecisiveness in terms of entrepreneurial action is formed by the person comparing the possibilities of being successful in a market to a state in which no successful opportunities are present. Because the markets within virtual worlds are relatively new to the participants, they might not perceive how successful business could be achieved in a virtual setting. In theory, this would lead the person to believe that there are no possibilities for anybody in the market, which could also increase their indecisiveness.

However, if the participant decides to take action on a continuing basis, it will provide the researchers with information about their expenditure and the motivations behind it. The participants reported the details of their expenditure of capital and filled out periodical questionnaires regarding their use of capital and the motives that drove their business decisions. Later on in the exercise, participants reported whether they felt the payoffs of their actions corresponded to their expectations. The theory behind the entrepreneurial part of the exercise was influenced by Arthur (1991) and Minniti and Bygrave (2001), who argued that entrepreneurs act on the basis of their information and evaluate and update that information as time goes on. However, empirical findings show that entrepreneurs may act contrarily to that theory, which would mean that the entrepreneurs would actually not discard the choices that brought them poor payoffs. For example, the work done by Cassar (2010) provides empirical evidence that nascent entrepreneurs can be overly optimistic in terms of their expectations about their business.

LESSONS LEARNED

The authors started to apply this setting in the fall of 2009 and the setting has evolved and matured in every collaborating course after that, reflecting the evolution of technology. In many respects, this has also provided some learning experience for the authors of this paper, and these are outlined below in somewhat chronological order.

From the start, we felt that we needed to test our research setting ourselves before we invited any students to participate in it. The test case described above had two goals: (1) To figure out the minimum amount of venture capital and time needed to implement a business plan and to reach real customers in Second Life, and (2) to test our data gathering methods so that the workload of the student teams in our exercise would be comparable to the teams participating in normal exercises in the same course. As researchers, we had some prior knowledge of Second Life but we did not know about aspects such as the detailed costs related to different services or marketing a business. In this sense, the test case proved invaluable. In addition, the test case helped us balance the workload of the students, although it did not help us anticipate all the different problems involved in starting the actual exercise, as we had already used Second Life ourselves.

The business exercise usually followed some weeks after the actual business course had started. Prior to starting the business exercise, we held tutorial sessions on virtual business, as these were felt to be important for teaching the basics skills and knowledge needed in Second Life and for minimizing the problems encountered in starting the exercise. Familiarity with virtual platforms and technology might have also impacted the number of groups that signed on to the exercise in the first place. The aim during these sessions was to give a 90-minute tutorial on technical issues related to Second Life, and another tutorial of similar length for the virtual business tutorial presented earlier in this chapter. In reality, there were always difficulties fitting two of these sessions to course schedules and the two sessions were eventually combined into one. Students were asked to create their avatars beforehand and we made sure that we had enough resources to guide those students that were totally new to the environment. However, it was not possible to ensure that students had done what was asked from them and we ended up helping some of them create Second Life accounts. This kind of on-site account creation has several associated problems that may hinder valuable learning experience:

(1) When several accounts are made at the same time, Second Life may block any additional attempts from that IP range, (2) Simple things, such as installing the client or confirming the validity of an e-mail address to Linden Lab, may take an unusually long time, (3) Second Life requires the use of a recently manufactured computer with a relatively powerful graphic card, and (4) Firewalls are a potential issue when using Second Life, especially when utilizing voice chat. Second Life version 2.x uses three software components, all of which must have full access to servers in Linden Lab to enable a full virtual experience.

We used web pages to provide support, detailed information about the schedule of the exercise, and the business tutorial, especially for students who were unable to attend the tutorial session. This was by far the most efficient solution, as we did not have access to the various educational support systems used in the collaborating universities, and we could not be sure that every member of the student teams used Second Life regularly. Unfortunately, this also created the need for multichannel communication during the exercise, as we were not able to restrict the communication to Second Life or require the use of Second Life from all the students in the groups. Although we managed to anticipate some of the implications of our research setting and increased the use of triangulation in data gathering, multichannel communication and the desktop study methods used by some of the students may have had secondary implications, which are discussed later.

During the exercise, we arranged weekly Q&A sessions in Second Life. The aim of these sessions was to provide mainly technical support, as information retrieval and acquisitions was one of our research interests. Nonetheless, most of the participants' questions during these sessions were related to the topic of how Second Life acts as a marketplace, and it became clear that the participants were taken aback by how the community inside Second Life operates. This reaction was expected, as it resembles the shock the participants felt when they entered into a new market place and into something of a new culture.

In order for the research setting to function properly, it was necessary to motivate the participants to take active part as virtual entrepreneurs. Students in business and entrepreneurship courses are expected to be willing to try entrepreneurship—especially in a risk-free setting in which the invested capital is not collected back and returns are not expected on the investment. In fact, the main risk to the participants is losing some of their time to run the virtual business endeavor. Participants would also be further motivated by receiving credits for participating in the exercise. For the reasons mentioned above, the participants' sense of risk aversion was not seen as an immediate problem in our setting. Still, problems were encountered related to the participants' motivation to proceed with the implementation of their plans in Second Life. Although the sample of participants is quite small, and is definitely not a representative cross-section of the student population, the low participation level is most likely related to the fact that we did not have the proper means of motivating the students.

It is also worth touching briefly on the topic of monetary compensation as incentives in our setting. The groups received an initial capital of U.S.$77.50 in Linden dollars to run their business. This means that a group participating would be rewarded monetarily with profits from their business and would also gain the invested capital, which was not collected back.

Most of the groups formed plans on virtual business but still hesitated to move forward by using capital in order to set up the virtual enterprise. The business plans ranged greatly in terms of their profit potential but, in general, they were valid enough to be implemented. The theoretical foundations for this type of behavior were addressed briefly earlier in the chapter. The feasibility of the business plans, in turn, leaves us to speculate that the threshold for participants may not have been the lack of incentives but virtual worlds as new market environments. It is also possible that some of the students based the business plans on second-hand knowledge and did not spend enough time in the actual market.

One final learning point for us was the fast pace of change in the technology platform and in the business landscape. Between the courses, we basically had to update all the information related to both technical support and virtual business tutorial. These updates are necessary partly because of the frequent upgrades in the client technology of Second Life, which may influence the user interface or add new features to the technology platform, upgrades that must be taken into account in the support material. Also, these upgrades are periodically made mandatory for users, which means that older client versions installed in computers used by students must also be upgraded. The other aspect of these changes is the possible new features that enable new business concepts. Along with fast-paced business landscapes, in which companies born and die in multitude, these frequent changes also influenced our business tutorial.

Trust and Ethics in Virtual Worlds

There are many parallels between the early days of the internet and the 3D virtual worlds of today. The sense of trust and security that a user of an interactive system experiences must be taken into account when designing business courses in Second Life. Mennecke et al. (2007) considered two different characteristics of trustworthiness when they discussed the barriers for large-scale adoption of virtual world commerce. First, the establishment of trust is largely experiential and due to past performance. The trustworthiness of an entity is established through consistent positive experience with that entity, which means that the entity develops a reputation over time. There may be referral systems in place that enhance the building of trust. However, in Second Life in particular, which involves a high degree of anonymity and residents who may have multiple accounts this is something that must be taken into account in courses that have a limited amount of time to establish trust. Second, trustworthy entities operate within a larger support system that sets a context for trustworthiness. Basically, the trustworthiness of an entity is enabled and sustained by the laws and legal framework within which that entity operates.

The policies concerning in-world banking provide a good example of this issue. Trust was clearly breached with a number of Second Life residents when a virtual "bank" changed its policy to restrict withdrawals and eventually vanished from the virtual world. The result was a loss of some U.S.$700,000 to those affected. This resulted in a policy by Linden Lab that virtual banks should offer "proof of an applicable government registration statement or financial institution charter" (Talbot, 2010). This requirement basically shut down all the Second Life's in-world banks and illustrates the second aspect of trustworthiness in a good and bad sense, since the viability of an entire industry may be subject to new policies by operators in these environments. This resembles government policy changes implemented in "first life" economies. Thus the risk to agents in the virtual economy is similar to "first-life" economies. It must be noted that political systems are rarely implemented in virtual worlds, but policy changes are immediate and depend only on the consideration of the provider.

The issue of trust also affected the design of our exercise. From an ethical point of view, our research was planned in a way that it would not cause any harm to human subjects. The fact that our research design was exploratory in nature and was not heavily influenced by the "first life" identities of the research subjects made it relatively easy to respect the privacy and anonymity of research subjects, intellectual property developed during the exercise, and the autonomy of the research subject. However, we could not limit the potential financial harm to research subjects while, at the same time, respecting the autonomy of the research subjects. In effect, we had to ban the use of students' own money during the exercise. In addition, to prevent misuse of the resources we collected data on the spending of the invested capital from several sources.

Similarly, entrepreneurs in virtual worlds must note that they do not operate in an ethical vacuum. There are few examples where economic activity in a virtual world has led to ethically questionable results in "first-life" or in the virtual world. Among the most noted examples are places called "gold farms," in which people are paid often low wages in return for their labor input in virtual worlds. The actual work of a "gold farmer" refers to a person who collects assets (originally virtual gold coins) in a virtual world and then resells them for currency in the "first-life." In essence, the demand for such a service originates from the scarcity of time that virtual world residents have at their disposal. The collection of these virtual assets is sometimes very labor-intensive, which means that some players are willing to trade their "first-life" currency for virtual currency. "Gold farming" is usually present in virtual worlds that are goal-driven and have virtual items that include game mechanics.

We are not arguing here that unethical actions are commonplace in virtual worlds, but rather that the ethical choices present themselves in virtual markets as they do in "first-life" markets, which means that educators must be aware of them.

FUTURE PLANS

The development of this research setting has been a long journey and we have gathered experiences from different courses in a relatively short period of time. The journey is not over yet and, based on our experiences, we are planning to launch our own course in 2011. The detailed lesson plan of this course is provided as an appendix to this chapter.

There are several reasons for developing our own course, many of which have been covered in the lessons learned section; the main reason is the motivation of the students. Our own course would make it easier to balance the incentives and expectations of the students with the actual workload in the exercise. Our own course would also allow us to provide access to Second Life and information through our own educational support systems in more robust way; or to concentrate communication and activities purely on Second Life, if this enhances both education and research done with our setting.

We will continue collaborating with other universities, and we have developed a new program—the venture capital program (http://www.elisanet.fi/vincoproject/entrepreneurship/), for entrepreneurs and groups that are more experienced with this environment. This program is mainly marketed for in-world educational institutes in Second Life, but also for "first-life" business and marketing courses that have an established presence in Second Life. The timeline for this program is shorter than in the virtual business exercise described above, as the groups can skip the Orientation and Search phases when developing their business plans. However, in this program the students are allowed to use their own capital in addition to the L$20,000 of risk-free venture capital. With this arrangement, we are hoping to obtain more research data on the actual implementation phase and entrepreneurial decision making. In effect, we are now dividing the focus in these settings to different research questions and are hoping to have motivated teams for both parts of the process. We still consider virtual worlds and Second Life to be the best place for our research. As the technology and user experience improves, and our knowledge about different possibilities in education increases, we are seeing more and more added value in education as well.

ACKNOWLEDGMENTS

This research was supported by Multidisciplinary Institute of Digitalization and Energy (MIDE) at Aalto University. The authors wish to acknowledge the support of the MIDE program and their coworkers in the VinCo project.

APPENDIX: LESSON PLAN FOR THE VIRTUAL BUSINESS EXERCISE

Grade Level

Higher education/Graduate level

Subject Areas

Entrepreneurship, marketing, social media

Duration

Four hours of contact teaching (creating avatars, short business tutorial). Minimum 100 hours of online presence plus reporting and offline planning (58 hours). Total of six credits (162 hrs). The length of the course is two periods or 14 weeks, of which the exercise is 12 weeks.

Lesson Description

Students make small groups (3-4 students), with participants preferably representing the different schools of Aalto University (School of Economics,

School of Art and Design, School of Engineering, School of Chemical Technology, School of Science, and School of Electrical Engineering). The students are given a short introduction to business in Second Life and some time to familiarize themselves with the market. Students then create an initial business plan and test their ideas inside Second Life within the given budget and timeline of the course.

Required Skills

As this is a graduate-level course, students are expected to have basic knowledge of their respective disciplines (business administration, marketing, visual design, computer science, etc.). Students are not required to have any prior knowledge about Second Life; instead, they are encouraged to have an entrepreneurial mindset in the course and familiarize themselves with the Second Life market.

Rationale for Exercise

The exercise gives participants first-hand experience of entrepreneurship and encountering a real market that is driven by real consumer needs, but without the paralyzing risk of large investments. The preferred method of learning is by encountering the issues in practice. Although the setting we use is in a virtual world, many aspects of entrepreneurship in virtual worlds are similar to those in other economies.

Materials and Resources for Teaching

- Reading material (one copy in the library): (1) The Entrepreneur's Guide to Second Life: Making Money in the Metaverse by Daniel Terdiman. (2) The Unofficial Guide to Building Your Business in the Second Life Virtual World: Marketing and Selling Your Product, Services, and Brand In-World by Sue Martin Mahar & Jay Mahar.
- Technology: During contact teaching, a classroom equipped with Second Life-compatible computers, data/video projector and overhead projector. For the rest of the course, one or two computers with Second Life access in each campus. In addition, accounts to a virtual teller machine in Second Life.
- Teaching resources: For contact teaching, an additional instructor (course assistant) to help groups to get started and to solve prob-

lems. Presentations entitled "How to create an account on Second Life" and "Business in Second Life".

- Communication capabilities: Web pages, SLoodle, official learning management system (Oodi in Aalto University), web forms for making progress reports.

Materials and resources for student group

Each group requires L$20,000, of which they will receive L$3400 in stipends from Linden Lab when the group leader upgrades his/her account to a premium account. The upgrading cost is U.S.$20 for the last two months of the exercise. The total cost is U.S.$97.50 per group (at the 18.11.2010 exchange rate of L$258/U.S.$1) (see Figure 12.2).

Limit is the amount of Linden Dollars (L$) that the group is given at different stages of the exercise. Account is the amount of US Dollars ($) (not including VAT) that the group leaders have to pay to Linden Lab during the exercise. This money is refunded to the group leader upon receiving certification/receipt of the payment. Stipend/Signup is the amount of Linden Dollars (L$) that the group leader receives from Linden Lab after upgrading to a premium account. This money is counted for the group budget.

Content Outline

The exercise is divided into different stages.

Enrollment. Time dedicated for the contact teaching (creating avatar and short business tutorial), forming groups and practicalities. Once the groups are formed, a small amount of Linden dollars is given to the group leader for learning purposes.

Orientation. This stage is intended for learning the ropes in the virtual world and for searching for possible market opportunities. A small amount of Linden dollars is given to the group account for orientation purposes. By the end of this stage, the student groups are required to return their first progress report, entitled "Market Opportunities," by answering short questions in a web form.

Search. Time dedicated for idea generation and the formulation of an initial business idea. The participating groups are given a small amount of Linden dollars for information-seeking purposes (identifying possible partners, objects, scripts, marketplaces). During this stage (by the end of week 3), groups are asked to fill in the Financial Projections spreadsheet for the first time. By the end of this stage, the student groups are required to return their second progress report, "Business Idea".

Prototyping. Student teams receive brief feedback on the viability of their business idea at this stage. After the feedback, groups are given more Linden dollars with which to prototype their product or service and time to plan the implementation of their business idea. The group leader is required to upgrade to a premium Second Life account and to handle possible income from this point forward. Since premium account fees cannot be paid on behalf of students, this fee is refunded to the group leader upon receiving certification/receipt for the payment. Also during this phase (by the end of week 5) groups are asked to fill in the Financial Projections spreadsheet for the second time.

Implementation. Participating groups are given more Linden dollars for land rental, outsourcing, marketing and other implementation costs. At the beginning of this stage, groups are required to briefly describe the implementation plan of their business idea (the "Business Plan" report). This period is the actual revenue phase, in which the quality of the business idea is tested in the virtual economy. During this phase, groups are asked to fill in the Financial Projections spreadsheet for the last time (by the end of week 9), but also to continue reporting expenditures and revenues to the course assistant on a biweekly basis.

Final report/ gala: A free-form multimedia presentation about the birth of the company and its unique capabilities in this particular market. An in-world public gala will also be organized, which will give student groups the opportunity to receive feedback on their ideas.

Student Products

Six short progress reports (Market Opportunities, Business Idea, Business Plan and three Financial Reports), end report in multimedia format, course evaluation form, and a real operational company in Second Life.

The "Market Opportunities" report is designed to encourage groups to familiarise themselves with the market and identify the most interesting market segment for them. The "Business Idea" report is meant to provide enough information about the risks involved and the possibilities of the business idea in order to enable the venture capitalists to make an informed investment decision. During the prototyping phase, groups still have the chance to change plans; if they do not change their business idea, the "Business Plan" report builds on the previous report.

The objective of the financial reports is to clarify the participants' thoughts on how they expect their company to generate revenue and costs during the exercise. The periodical structure of the reports gives participants the opportunity to follow their own predictions and generate a detailed view of their thought process.

The end report and the operational company are expected to display the group's multidisciplinary working style.

Instructional Learning Objectives and Assessment

By the end of this exercise, participants will be able to accomplish the affective understandings (AUs), cognitive understandings (CUs), and cognitive skills (CSs) listed below.

- Express increased awareness about different aspects of entrepreneurship (AU).
- Evidence: The course evaluation will ask participants if they feel more confident, less confident or about the same as they did before this exercise about becoming entrepreneurs and how this exercise has changed their confidence.
- Appreciate the importance of learning from failures and through problem-solving (AU).
- Evidence: This will also be addressed in the course evaluation. The main focus will be on whether groups changed their plans during the exercise or experienced problems that they could not anticipate based on their previous knowledge of their respective disciplines.
- Appreciate the different skills and viewpoints of the multidisciplinary group (AU).
- Evidence: In the course evaluation, participants will be asked to explain, in a short paragraph, the experiences they had during the multidisciplinary group work.
- Describe one's contribution to the unique skill set of the company (CU).
- Evidence: In the "Business Idea" and "Business Plan" reports, the groups are required to form strategies based on their unique skill sets and to describe their own role in the implementation of their strategies. The groups are also required to give voice to all the different viewpoints in their multimedia end report – to use their different skills and strengths in building the end report about their business venture.
- Demonstrate basic business administration knowledge by contributing to the progress reports and financial projections (CU).
- Evidence: In the three Financial Reports, the groups are required to convert their unique capabilities into financial projections, thereby utilizing basic business administration capabilities.

- Demonstrate effective group work and entrepreneurial skills in getting their first actual customer before the end of the exercise (CS).
- Evidence: Reaching customers and generating revenue within 12 weeks.
- Demonstrate basic marketing skills by creating a collaborative multimedia presentation about the birth of their company in Second Life (CS).
- Evidence: Participants will create a 15-minute multimedia presentation about the birth of the company and its unique capabilities in their particular market.

Safety Precautions

- Access to Second Life should be checked every week from student computers.
- Students are required to familiarise themselves with Second Life's Terms of Service and Community Standards to ensure the viability of their business plan.
- Students cannot use their own money to build up the business.

REFERENCES

Alavi, M., Wheeler, B. C., & Valacich, J. S. (1995). Using IT to reengineer business education: An exploratory investigation of collaborative telelearning. *MIS Quarterly, 19*(3), 293–312.

Arthur, W. B. (1991). Designing economic agents that act like human agents: A behavioral approach to bounded rationality. *The American Economic Review, 81*(2), 353–359.

Bransford, J., Brown, A. L., Cocking, R. R., & National Research Council. (2000). *How people learn: Brain, mind, experience, and school*. Washington, DC: National Academy Press.

Cassar, G. (2010). Are individuals entering self-employment overly optimistic? An empirical test of plans and projections on nascent entrepreneur expectations. *Strategic Management Journal, 31*(8), 822–840.

Castronova, E. (2005). *Synthetic worlds: The business and culture of online games*. Chicago, IL: University of Chicago Press.

Hamari, J., & Lehdonvirta, V. (2010). Game design as marketing: How game mechanics create demand for virtual goods. *International Journal of Business Science and Applied Management, 5*(1), 14–29.

Lehtiniemi, T. (2008). Macroeconomic indicators in a virtual economy. *Virtual Economy Research Network. March*. Retrieved from http://www.mendeley.com/research/macroeconomic-indicators-in-a-virtual-economy/

Linden Lab. (2010). *Economic statistics (raw data files) second life.* Retrieved from http//secondlife.com/statistics/economy-data.php?d=2010-09-24

McMullen, J. S., & Shepherd, D. A. (2006). Entrepreneurial action and the role of uncertainty in the theory of the entrepreneur. *Academy of Management Review, 31*(1), 132-152.

Mennecke, B., Roche, E. M., Bray, D. A., Konsynski, B., Lester, J., ... Rowe, M. (2007). *Second Life and other virtual worlds: A roadmap for research.* Twenty-eighth International Conference on Information Systems (ICIS), Montreal, Canada.

Minniti, M., & Bygrave, W. (2001). *A dynamic model of entrepreneurial learning.* Hoboken, NJ: Wiley-Blackwell.

Mullins, J. (2004). *The new business road test: What entrepreneurs and executives should do before writing a business plan.* Upper Saddle River, NJ: FT Press.

Poul, H. A., & Rask, M. (2008). Taking action: New forms of student and manager involvement in business education. *Marketing Intelligence & Planning, 26*(2), 145-165.

Talbot, D. (2010, January 10). Second Life closes banks. *Technology Review.* Retrieved from http//www.technologyreview.com/biztech/20037/

Teigland, R. (2010). Born virtuals and avapreneurship: A case study of achieving successful outcomes in peace train—a Second Life organization. *Journal of Virtual Worlds Research, 2*(4). Retrieved from http://journals.tdl.org/jvwr/article/viewArticle/866

Yee, N., & Bailenson, J. N. (2008). A method for longitudinal behavioral data collection in second life. *Presence Teleoper. Virtual Environ., 17*(6), 594-596.

CHAPTER 13

THE SIMESCUELA

An Innovative Virtual Environment for Teacher Training

Antonio Santos

The present chapter describes the dynamic relationship between educational theory, instructional design process, and the actual construction of the virtual environment called the SimEscuela in Second Life. This virtual learning environment was defined from the beginning as an online community of practice formed by Mexican teachers coming from different geographical areas in Mexico, institutions and levels, with the common domain of interest of sharing each others' professional experiences to acquire teaching skills and develop higher order thinking skills by engaging collaboratively in problem-solving activities. The set of instructional strategies that was actually used to develop this learning environment along with its theoretical underpinnings will be discussed. Also, the chapter explains in detail how these instructional strategies were applied to build the first prototype version of the SimEscuela.

Engaging the Avatar: New Frontiers in Immersive Education, pp. 365–390

INTRODUCTION

The objective of this chapter is to describe a practical application of an educational virtual world called the SimEscuela (*escuela* is the Spanish word for school) built in Second Life. The SimEscuela is a virtual environment which simulates a traditional Mexican K-12 school. This virtual environment was conceived as a place where Mexican teachers could share each other's professional experiences, acquire teaching skills, and develop higher order thinking skills by engaging collaboratively in problem-solving activities. The aim of this chapter is to explain the dynamic relationship between educational theory, instructional design process, and the actual construction of virtual objects, that take place during the development of this virtual environment and its associated courses.

The first section of the chapter will analyze the theoretical background that laid the foundations for the SimEscuela project. A detailed explanation follows of the teaching and learning strategies that resulted from the analysis and were used as the instructional base. The paper describes and illustrates how these teaching and learning strategies were used to design and build the educational virtual world and its corresponding learning experiences. Finally, we include the complete lesson plan of one of the first courses that was designed to be taught in the SimEscuela. We state the objectives, instructional methods, the exercises and the virtual objects that were created.

THEORETICAL BACKGROUND

Traditionally, in Mexico, teachers develop professionally through attending a variety of courses and workshops offered by the ministry of public education or through private institutions. These courses and workshops are predesigned by experts to acquire new knowledge to increase teacher achievement as educational professionals. Course instruction is lecture based, and content focuses on educational concepts and methods for practice. Often, the courses are run to simply provide access to educational reforms. The instruction tends to be one-sided and the faculty does not engage in activities during these exchanges. Instead they are taught theory. Two problematic situations arise with this method of instruction:

1. Experience is not integrated. The everyday experience, activities, relationships in the classroom is seldom systematically integrated

into those courses. Consequently, learning content is decontextualized, and there is a negative impact on the quality of learning because, as social beings, our knowledge building processes are intricately related to our engagement in everyday activities (Wenger, 1998). As one of his principles of instruction, Merrill (2002), remarks that learning is hindered when the world of the student is not integrated into the process.

2. Learning is not transferred. It is assumed that when learning occurs during regular courses and workshops, the skills will almost automatically transfer to their everyday classroom practices (Helleve, 2010). This is in contrast to the situated paradigm of learning, which claims that knowledge built in a decontextualized way remains rather inert in the students' heads and, thus, transfer is hindered. This happens because, although students do learn declarative and procedural knowledge, learners do not develop a mental set for transfer (Sternberg & Frensch, 1993) since they have not developed links between the new content and their own every day contexts.

It the field of teacher education we are in need of an epistemology of practice based more on what the teachers actually do (Schön, 1983). This epistemology of practice could be applied in the design and development of most of the courses and workshops that teachers take so that the quality of their learning and the probability of transfer to their everyday practice are enhanced. Once teachers understand this methodology as the practice for their learning, likely they will deliver the same kind of instruction to benefit their own pupils. One of the ways to unite the learning experience in class with the need to provide content for teachers was to look at virtual worlds as a way to engage the community.

A virtual world, SimEscuela, was built in Second Life to serve as a place where teachers could meet to build community around their learning experience and have access to educational topics and instructional methodologies. Similarly, the virtual world could be used as a way to integrate rich previous teaching experiences lived everyday in their classrooms into live places of practice. To design and build both the virtual world and the courses to be conducted in this learning environment, a set of strategies was created derived from the Three Contexts Methodology (TCM) developed by Santos (2007, 2009). The situated paradigm and the TCM will be succinctly explained in this paper to guide readers to understand the design decisions that informed the SimEscuela build and its corresponding courses.

THE SITUATED PARADIGM OF COGNITION

There is a large gap between how students acquire and use their knowledge in everyday life and the way they learn in schools. Outside school, students engage in authentic social activities to solve meaningful problems; but inside the classrooms, most of the time they passively learn abstract information (Resnick, 1987), which tends to remain inert inside their heads and is seldom generalized into the world (Hendricks, 2001). This is a problem because students tend to keep both worlds completely apart not relating them at all together, and thus, do not transfer what they learn in school into their lives. A situated learning paradigm conceptualizes the nature of human learning differently and proposes a different way of teaching.

The situated learning paradigm assumes that human cognition is influenced by the context in which it is being used (Brown, Collins, & Duguid 1989); or, as Clansey (1997, as cited in Driscoll, 2000) expresses, human thoughts are adapted to the context. During the process of engaging in a learning situation we are learning many aspects of the context where the event is taking place. A context is a rather complex concept, because it is not only defined by the walls that form it (e.g., a classroom) or by the physical objects it includes, but also by the dynamic interactions that take place inside of it between humans, ideas, signs and symbols, information, values, ways of doing things, like solving problems or using tools, and so on. Because all of these context's components are in turn coming from different cultures, when we learn in a given context we are also learning more general cultural aspects. In other words, we are engaged in a process of enculturation. For example, when a group of students is participating in a geography class they are not only learning about regions on a map, but also are learning those cultural aspects related to the professional community of practice known as geographers.

For some other authors coming from a more anthropological perspective, learning is conceptualized not as a mere individual experience, but more in terms of social, cultural and historical aspects. That is, the process of learning is extended beyond the intrapersonal cognitive abilities and merged into the complex process of contributing to the knowledge of a community of practice (Lave, 1991; Lave & Wenger, 1991; Wenger, 1998; Wenger, 2004). This view basically considers that knowing, learning, and cognition are social constructions, expressed in the actions of people interacting within communities" (Wilson & Meyers, 2000, p. 59), and "seen as a dialectical process of interaction with other people, tools, and the physical world" (p. 71). Communities of practice are defined by Wenger and Snyder (2000) as "groups of people informally bound together by shared expertise and passion for a joint enterprise" (p. 139).

For Lave and Wenger (1991), the process of learning in a community of practice is characterized by what they call legitimate peripheral participation, which explains the social relations that naturally emerge within the community among novices and experts. All members of a certain community have different levels of expertise in relation to the joint enterprise that binds them together. They are all interested in learning; however, novices participate marginally at the beginning, and, as they interact with more expert members, gradually increase their participation slowly becoming experts themselves. Thus, different members in a community of practice naturally assume different social roles according to their level of expertise in the abilities needed to perform a certain task.

Because learning in a community of practice translates into being engaged in all sorts of activities, we can say that "knowing is *in* our action" (using Schön's concept, 1983, p. 49) and that, consequently, knowledge built during an activity is frequently tacit. While members in the community are aware of what it is that they know, they need to stop and reflect about what they are doing. They need to engage in an ongoing process of reflection.

The situated learning paradigm, framed by the enculturation process, was used to build the SimEscuela community and the delivery of its courses in context of the shared experience the users would be having. To provision the sim build with a model of learning, we employed the three contexts methodology described below to include a more grounded pedagogical dimension to the instructional strategies.

THE THREE CONTEXTS METHODOLOGY (TCM)

Parting from the situated cognition notions discussed in the previous section, the three contexts methodology (Santos, 2007, 2009) basically assumes that during any learning situation three different contexts are in fact present (see Figure 13.1): the original context, the present context, and the future context.

Context 1: the original context. The original context refers to *where the content to be learned was originally created*. This could include content created solely by the instructor or by any community of professionals, such as those formed by engineers, physicists, or culinary chefs. These original communities produce information to basically describe how they build their distinctive professional body of knowledge. They speak often in terms of the type of problems they address, the methods they use, the tools they employ, their professional structures and hierarchies, etc. The content is usually written from their level of expertise, and therefore

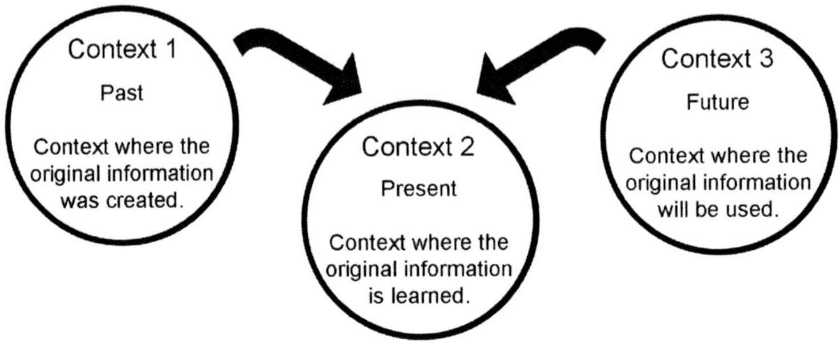

Figure 13.1. The three contexts methodology assumes that in a given learning situation three different contexts are present.

appeals to learners at the same level as themselves. However, when a student is learning that information, he or she is in fact operating "Within the context of the social construction of knowledge of discourse communities" (Streivel, 1994, p. 15). For example, when a student is learning calculus she is not only learning content, but also learning how to think and act as a mathematician. Therefore, in a learning situation it is important to include components of this original context; the TCM refers to these components as *original contextual indices*.

Context 2: the present context. The present context of the learning situation is *where students learn the information* (renamed as content) created by the original communities of professionals in Context 1. Considering that the situated paradigm of learning asserts that learning is affected by the context where it happens (Brown et al., 1989), this context should be carefully planned so that the original contextual indices of the professional community that created the content are also present during the learning situation. Therefore, content should not be stripped of those contextual indices if we want to facilitate a situated knowledge construction typified as "enculturation or adoption of a group's way of seeing and acting" (Wilson, 1996, p. 4).Thus, from a situated pedagogical perspective, the students should learn inside a learning environment which somehow (there are different ways of accomplishing this) simulates the original professional community.

Context 3: the future context. The future context refers to *where students will use the information that they are learning* in Context 2 (the learning situation). Context 3 represents where students live and also the future context where they will perform their own professional activities. Therefore, it is expected that the learning that occurs in the classroom eventually transfers into this context where they will naturally engage in their everyday problem-solving activities. To foster the possibilities of this transfer of knowledge actually happening, the TCM recommends also including contextual indices of this Context 3 in the learning situation (see Figure 13.1).

To apply these ideas in a learning situation, the TCM proposes three instructional strategies:

Strategy 1. A learning situation should contain contextual indices from the original community of practice that created that information. In this way, students will also learn to think and behave like that original community of practice. In the classroom, this strategy is achieved by allowing students to do authentic tasks; that is, the type of tasks that those original practitioners usually perform.

Strategy 2. To better apply Strategy 1, a learning situation should be designed as a situated learning environment. There are several design models to develop a learning environment provided by the constructivist literature (see, e.g., Jonassen, 1999; Perkins, 1991; Wilson, 1996). The strategy suggest that a learning environment should: (1) be designed starting from an ill-structured problem that is initially presented to the learners, and (2) that the environment should include components to support learning, like tools, information and technical resources, related cases and ways for the students to prove their solutions.

Strategy 3. A learning situation should contain contextual indices from the contexts where students are expected to apply what they are learning. In the classroom, this strategy can be achieved by asking students to do tasks that require they bring information from their everyday contexts related to the problems they are solving.

INSTRUCTIONAL STRATEGIES TO DEVELOP THE SIMESCUELA

The virtual environment known as SimEscuela is a 3D site built in the immersive world of Second Life. It was designed to serve as a learning space for Mexican teachers working at the K-12 level (See Figure 13.2 for a snapshot of this environment). Its main pedagogical goal is to serve as a virtual learning environment where teachers from any state in Mexico can gather virtually to engage in training courses to develop their teaching abilities. From the beginning, SimEscuela was established to foster learning as congruent with the situated learning paradigm.

Our approach was to create applications in the virtual world. Each application would represent a different instruction strategy and use a theoretical base identified in the situation learning paradigm (SLP). Also each application is related to the three context methodology (TCM) discussed above. Each instructional strategy, identified in the applications below, are built using theoretical links to the situated learning paradigm (SLP) and the three contexts methodology (TCM). Some strategies are grouped because they emerged from the same SLP idea. After stating the instructional strategies, an explanation will be given to clarify how these instructional strategies were applied in SimEscuela.

APPLICATION ONE

1. Learning experiences in the SimEscuela should be developed to simulate the culture and context where the community of Mexican teachers performs its everyday activities. So, the environment one encounters in Mexico is simulated in the virtual world experience. Students are engaged with each other in activities and participate in groups.

    - The SLP states that, in everyday life, human knowledge is constructed during the active participation of individuals and groups in activities performed within authentic situations.
    - The TCM proposes that a learning activity (Context 2) should include contextual elements from the original community of practice (Context 1) that created the learning content and also from the contexts where students are expected to apply what they learned (Context 3).

APPLICATION TWO

1. The main objective for a participant (a Mexican teacher) within a course in the SimEscuela must be to solve problems collaboratively and not to just learn declarative and methodological content. So, the objective of each activity in the SimEscuela is to solve a problem given to the students, by the teacher.

    - The SLP proposes that humans construct knowledge engaging in activities within communities of practice. Frequently, these activities are related with solving problems that are relevant for that community of practice.

- The TCM proposes that a learning situation (Context 2) should be conceived as a constructivist learning environment designed around an ill-structured problem.

APPLICATION THREE

1. Participants perform authentic problem-solving activities that characterize the system of activities of the community of Mexican teachers. Teachers create activities in which students must work together to identify the problem, identify the information required to break apart the task into activities that help to structure the problem, and assign tasks to each member of the team to begin using strategies and tools to accomplish the task's goal.

    - The SLP states that each community of practice has its own way of solving relevant problems; they do it according to a system of activities characterized by a set of interests, rules, norms, division of labor, tools and outcomes.
    - The TCM proposes that a learning activity should include contextual elements from the original community of practice that created the learning content.

APPLICATION FOUR

1. Each of the courses taught in the SimEscuela should declare explicitly the common domain of interest around which they are started and that give coherence to the practices and actions performed by their participants. Students are given information directly related to their problems through discussion, demonstration, reference to information on the Internet, or other sources to contextualize their learning. Content from the original community of practice is introduced to users to explore as elements for considering variables that might be explored.

    - According to the SLP, what gives coherence to a community of practice is that its members share a common interest.
    - The TCM proposes that a learning activity should include contextual elements from the original community of practice that created the learning content.

APPLICATION FIVE

1. The SimEscuela should include the necessary resources so that participants can build and maintain their new community of practice around a given course.

    - For the SLP, "All doing is knowing and all knowing is doing." (Maturana & Varela, 1987, p. 27); that is, the unity of analysis to understand cognition is the activity, which includes people, tools and context. Thus, a community of practice is constantly engaged in activities that require their own types of tools (physical and psychological) and their own ways of using them.
    - The TCM proposes that it is expected that the learning that occurs in the classroom eventually transfers into the future context of where the student will naturally engage in their everyday problem-solving activities. So, utilizing the environment of the island in Second Life, and allowing the students to create their own environments helps to build a community in which the students identify with the future outcomes.

APPLICATION SIX

1. The SimEscuela respects and value the different types of social roles that participants naturally assume within a learning experience. So, in the SimEscuela participants are encouraged to set their own goals and identify a possible course of action to achieve them. The participants are active members of the community and interact openly with other members of the community. The teachers strive to keep the participants active and make them feel they are needed in the learning environment. As part of the instructional practices within a certain course in the SimEscuela, teacher and assistants model the professional practices that participants are expected to learn. Teacher and assistants constantly supervise the participants' performances giving the necessary feedback; but, as participants become more experts, the supervisory actions are gradually retired.

    - According to the SLP, learning in a community of practice is characterized by legitimate peripheral participation; that is, different members in a community of practice naturally assume different social roles according to their level of expertise. For example, novices participate marginally at the beginning, but, as they interact with more expert members, they model their behavior and beliefs and, gradually, become experts themselves. The SLP states that frequently knowledge built performing activities in a community of practice is tacit knowledge. That is why members, to be aware of what it is that they know and how they know it, need to engage in an ongoing process of reflection.

- The TCM suggests that the pedagogy should promote participants' reflection processes during and after their learning activities so that they can "know how they know". When the students begin to participate as instructors of their solutions, and help to guide and consult others on achieving their goal, they have developed their own future contexts.

How Were the Instructional Strategies Applied in the SimEscuela?

In this section each instructional strategy will be discussed in terms of how it was applied to the creation of the SimEscuela and its courses. When appropriate, snapshots will be included to show how these features were finally instrumented in Second Life.

It is important to remark here that, to the present, only a smaller prototype of the SimEscuela has been built to test the basic aspects of the proposed instructional methodology; thus, some of the following descriptions of the applications of the stated instructional strategies are emerging and will be in full operation in the development of SimEscuela over time. This principle is at the core of achieving expertise and ranks the SimEscuela project as an actual exemplar of the very techniques it is prescribing. Currently, due to the small size of the prototype, the SimEscuela only allows for a limited number of teachers and courses; however, it is expected that it will grow to become a whole island in Second Life and that will be capable of offering all the educative services described here.

The SimEscuela curricular dynamics centers around participants (the term participant refers to a Mexican teacher studying in the SimEscuela) taking predesigned courses and encouraging proposal of new courses to extend the list according to their own needs. Thus, the in-world virtual school offers two types of courses: (1) predesigned courses conducted by the school's educational staff; and (2) proposed courses initiated by the participants themselves with guidance from the school's staff. Both types of courses declare a domain of interest from the start and are treated as communities of practice; however, predesigned courses also declare more specific learning outcomes and are taught following particular activities to achieve them. In the courses proposed by the participants, a less controlled social environment is fostered and supported. The premise behind this course organization is that most of the Mexican teachers/participants do not have much experience in learning in online unconstrained communities of practice due to their rather inexperienced technological literacy. Thus, they will first enter a more traditional controlled learning environment (i.e., predesigned course) and, once they are more proficient in learning in a virtual world like the SimEscuela, they will be motivated by the school's staff to create their own courses more in accordance with their own professional needs (i.e., proposed course).

Learning experiences in the SimEscuela should be developed to simulate the culture and context where the community of Mexican teachers performs their everyday activities.

The SimEscuela was carefully designed to look as similar as possible to a traditional school used in the Mexican public system (see Figure 13.2). Of course, there are many different types of school buildings depending on size and geographical location; however, a more rural type of construction was selected because it was thought that it better typifies most of the existing schools around the country. The idea for doing this is not to copy the real world into Second Life or to perform training activities just as we would in the real world. This would be a misuse of the possibilities that the technology of virtual worlds affords. The purpose of building a virtual school similar to the ones actually used in the K-12 level was, according to the TCM discussed before, to contextualize the learning experience by simulating the context where this type of professional community performs its everyday activities.

The main objective for a participant within a course in the SimEscuela must be to solve problems collaboratively and not to just learn declarative and methodological content. Most of the design decisions taken to build the SimEscuela mainly looked to promote collaborative activities between novice participants and experts. The main objective was to facilitate virtual spaces where participants could learn from each others' experience as professionals of the teaching community of practice. Along the process of interacting with fellow teachers, participants develop abilities to solve problems using many different types of tools and resources and, of course, also learn different types of content. In fact, the knowledge that participants are expected to build is either originated in the oral and physical interactions between them during their social activities, like informal chatting or collaborative learning activities, or in different sources like in-world note cards, textbooks, articles or the web.

Figure 13.2. The SimEscuela.

When participants enter the virtual environment, they are welcomed by staff members and explained that their main mission within the SimEscuela is to socially interact with other participants to solve teaching and learning problems, and that, to support those activities, there is a staff for guidance and consultation, different physical spaces and information resources.

In predesigned courses, the problems to be solved by a certain group of participants are stated by the course's teacher, but in the proposed type of courses the group of participants determines the problem, under the supervision of the course's teacher.

Participants perform authentic problem-solving activities that characterize the system of activities of the community of Mexican teachers. The objective here is to increase the probability of transfer by contextualizing learning and making sure that the knowledge built during the experience does not remain inert in the participants' heads. Predesigned courses are created around a common domain of interest of the community of Mexican teachers. Although these domains of interest are pre-selected by the SimEscuela educational experts, they are not organized around certain content that teachers should know, but around the cognitive and methodological skills that are needed to solve certain problems. These problems are considered authentic because they have been previously gathered by the SimEscuela designers' own experience observing and working with Mexican teachers at their own schools.

In proposed courses, participants are encouraged to create new courses parting from their expressed professional needs, maybe stated during a predesigned course. To this objective, participants, both novices and experts, state problems that they face each day when teaching in their different classrooms. When some of these real problems gather enough consensuses among them, they become the focal point of a new course organized to solve that real given problem. These problems are considered authentic because they are full of contextual indices from the teachers' daily professional activities.

Each of the courses taught in the SimEscuela should declare explicitly the common domain of interest around which they are started and that give coherence to the practices and actions performed by their participants. As explained before, the SimEscuela was conceptualized as a large online community of practice formed by teachers, coming from different geographical areas in Mexico, institutions and levels, with the common domain of interest of sharing each others' professional experiences to acquire teaching skills and develop higher order thinking skills by engaging collaboratively in problem-solving activities. Once teachers understand how the SimEscuela works in general and its pedagogical purpose and methodologies, they are asked to participate in smaller communities of practice around smaller domains of interest, corresponding to the different predesigned and proposed courses participating in.

Both types of courses offered in this learning environment are considered subcommunities of practice with their own common domain of interest. Some domains of interest could be rather simple and others more complex; for instance: "How can we create learning outcomes for our courses?" would probably be shorter and more straightforward than "How can we create constructivist learning experiences." The first domain of interest selected to start a predesigned course is "How can we integrate the use of technology in our teaching practices?" which was designed to test some of the SimEscuela's basic pedagogical propositions.

In the next section of this paper the lesson plan that was created for this course will be described. The SimEscuela should include the necessary resources so that participants can build and maintain their new community of practice around a given course. This is accomplished by offering the following four types of resources: information resources, virtual physical resources, communication resources to sustain collaborative work, information storage resources

Type 1: Information Resources

The SimEscuela has a library containing: (1) general reading materials relevant to the SimEscuela's stated general objective of teachers education; (2) more particular reading materials related to the learning goals of the courses; (3) specific learning resources and activities for the predesigned courses, which participants are free to use to start their learning process; and (4) technical manuals to assist participants in how to live and interact in Second Life, like how to build new objects, different ways of moving or how to find and teleporting to a certain place in this virtual world.

Type 2: Virtual Physical Resources

The SimEscuela is divided into the following areas (see Figure 13.3): (1) a room including the library and a meeting area where participants can get together to plan their activities; (2) a simulated classroom, which participants can use as a practicing area; (3) a building area (sand box) to fabricate new 3D objects; (4) a presentation area where participants can display information for the rest of the teachers; and 5) a garden area where participants can relax and chat.

Type 3: Communication Resources to Sustain Collaborative Work

The SimEscuela provides different in site resources so that participants can engage in collaborative practices, share information and ideas, and get to know and trust all members' identities. Using the technical tools available in Second Life, like actual avatar interactions or voice communication and chatting, participants can perform different social and learning activities and keep communication active during their synchronous periods of learning. Asynchronously, participants are encouraged to contact each other via e-mail, discussion forums, and so on, so that they can get to know each other better. In this way, the avatars they encounter in the virtual world have a real identity, which participants know and trust.

Type 4: Information Storage Resources

The SimEscuela includes technological resources that participants can use to document their practices and be able to build a cogenerated information base that

Figure 13.3. Snapshots of the physical resources.

can be stored and consulted by others. For this purpose, each course offers a wide variety of technologies where participants can document and store information in different formats: wikis, blogs, web pages, in-world note cards, transparencies, machinima production (films constructed inside the virtual world), and so on. Also, for this objective, early in the process, participants are shown how to use their personal inventory, a very valuable resource when interacting in Second life.

The SimEscuela should respect and value the different types of social roles that participants naturally assume within a learning experience. The SimEscuela promotes that novice participants (that is, teachers considered beginners in a given topic) interact with more expert participants. To this end, there are no special requirements to enter any of the courses; in fact, it is encouraged that teachers with different previous levels of knowledge register for a certain course so that novices and experts have the opportunity to come together, interact and learn from each other.

The SimEscuela encourages participants to set their own goals and identify a possible course of action to achieve them. In both types of courses participants are encouraged to follow their own ways of building knowledge, like employing their personal thinking strategies, interacting with other participants according to their own personalities and belief systems, and so on. In the predesigned courses, where learning goals, problems to be solved, and activities are preselected, participants are more constrained; however, they are still encouraged to use and share their own ways of thinking and solving problems.

In proposed courses, this strategy is easier to apply because participants, acting as novices and sometimes as experts are asked to collaboratively contribute to the knowledge base of the community, set their own domains of interest, learning goals, the problems they must solve, meeting agendas, and working plans.

The SimEscuela encourages participants to remain as active participants of the community of practice for as long as they feel necessary to achieve their personal learning goals. It is expected that the duration of the courses created by the teachers vary, depending in the type of the stated learning goals and in the interest or intrinsic motivation of the participant. This presents a logistic problem to the schools' administrative staff, similar to the one a real school would face, in terms of resources and time periods.

Depending on the stated domain of interest, predesigned courses do have a preplanned time structure. However, the hypothesis here is that participants, during their learning process in these predesigned courses, will share or discover other common professional needs or domains of interest that can be further explored starting a new proposed course. Thus, along any predesigned course, the teacher is always promoting this possibility.

Once a proposed course is on its way, the SimEscuela supplies the required resources so that the community of practice, created around that course, can remain actively working for as long as it feels is necessary. When these communities of practice start to function in the SimEscuela in the future, a great opportunity will arise to investigate issues related to how this type of communities of practice work, like what motivates members to stay active in a certain community.

As part of the instructional practices within a certain course in the SimEscuela, teacher and assistants model the professional practices that participants are expected to learn. In the SimEscuela modeling is considered to be a very important instructional strategy. Two types of modeling exist during the learning activities: behavioral modeling, where proper performance is demonstrated to the students, and cognitive modeling, where, through reflection, appropriate reasoning for a certain task is articulated to the students (Jonassen, 1999). Teachers, assistants, and even participants (they are also experts in certain topics), when asked, can perform certain teaching abilities to demonstrate the right way to achieve them. Also, if the participants articulate the reasoning process they are going through, covert cognitive processes are also modeled. Thus, whenever considered necessary, a teacher demonstrates a certain performance describing, at the same time, her thinking process. Also, she can ask the SimEscuela technical staff to produce a real-time recording of her demonstrations using the machinima technique, which is a filmmaking procedure used in 3D worlds like Second Life. The idea here is to build a reservoir of films in the school's library that participants can download any time they need during their problem-solving activities.

The SimEscuela pedagogy promotes participants' reflection processes during and after their learning activities so that they can "know how they know." It is expected that participants get involved in all sorts of in-world collaborative learning activities related with the profession of teachers. During these tasks, participants engage in tacit knowing-in-action (Schön, 1983); that is, they acquire new skills, like being able to discriminate between good and bad teaching practices, but remain ignorant of how they acquired that knowledge. Thus, as a relevant aspect of the SimEscuela pedagogy, participants should also engage in different types of reflection activities during and after their learning practices. The aim is that they are more aware concerning the actions they are performing and also about the outcomes of those actions. It is important to recover this type of knowledge to increase their quality of learning and foster even more the possibilities of transfer.

To accomplish this objective, the teacher intersperses specific activities or tasks, like including everyone in general meetings, or asking for a written paper to contribute to the library. Also, casual conversation is encouraged to be recorded in certain dialogues with the teachers. This could occur when asking the teachers questions like: How are you learning right now? What thinking paths did you follow when doing the task? How did you solve the problem? Additionally, the teachers are asked to perform a certain task like evaluating a written essay, describing out loud during his thinking process, participating in a survey, or contributing to an activity with others involved.

Teacher and assistants constantly supervise the participants' performances and provide the necessary feedback; but, as participants become more experts, the supervisory actions are gradually retired.

In both kinds of courses offered in the SimEscuela, predesigned and proposed, supervision is an important issue; in fact, they share some supervisory methods and differ in others. In all of the courses, the supervisory actions are gradually retired as participants become more capable. However, in the predesigned courses, because they are more closely controlled by a course's teacher and participants perform predesigned tasks, supervision is more traditional in the sense that it is used to give close coaching and feedback when needed regarding the stated learning objectives. However, in the proposed courses, because participants are more free to choose their own learning goals and the paths to achieve them, supervision is given as part of the ongoing novice-expert relationships that is expected to develop in this form of learning. For instance, a teacher considered skilled on the community's domain of interest will be assigned to act as an expert member in that community and, as such, will be available for guidance and consultation for all members.

Also, as part of their supervisory activities, the SimEscuela's teachers make an effort to document their findings regarding the participants' performance in the different tasks and activities they are engaged in. Thus, this information can be used as a kind of qualitative evaluation reporting resource that participants can read at any time; and, in this way, know how they are performing within the SimEscuela.

Conclusion

Although many educational researchers are nowadays working in the topic of learning in virtual worlds like Second Life, we still do not fully comprehend how we can apply what we know about the teaching and learning process in the design of an in-world learning experience. That is why we must continue to systematically investigate this issue using various research and development methodologies. In the present chapter, one of these efforts has been described. First, the theoretical underpinnings of the SimEscuela were explained so that all the instructional decisions that followed had a foundation for discussion. The actual learning environment prototype was described and depicted in a series of snapshots to clarify the links between the theory and how it was applied. Finally, a step by step lesson plan was presented to exemplify how the entire instructional proposal is presently being grounded in the virtual world SimEscuela.

It is desired that, in the future, the findings obtained experimenting with the SimEscuela could be of great help to other educators interested in teaching in immersive worlds. Educators reading this chapter are definitely encouraged to adapt and apply the proposed instructional strategies in their own teaching projects so that they can be further refined. In this way, we will be able to develop a set of practical strategies to foster good teaching in virtual worlds and, most importantly, we will be able to improve the background theory in which the SimEscuela was based.

APPENDIX: THE SIMESCUELA: AN INNOVATIVE VIRTUAL ENVIRONMENT FOR TEACHER TRAINING

A Complete Lesson Plan

In this section, a lesson plan for a predesigned course called "Unity 1" is presented. This course is planned for a group of six participants. By the time they take this course, it is assumed that they are already reasonably familiar with the basic knowledge needed to be in Second Life, like knowing how the move, how to sit and stand, how to move objects, etc.

Domain of Interest of Unity 1

How can we integrate the use of technology in our teaching practices?

Learning Outcome

As a result of the learning activities, participants will be able to design a learning activity where they integrate the use of technology as a cognitive tool.

The course's activities are organized in two tasks:

Task 1: The differences between learning *from*, *about* and *with* technology.

Learning outcome: participants will be able to explain the differences between learning *from*, *about* and *with* technology.

Task 2: Applying the concept of cognitive tool.

Learning outcome: Participants will be able to teach a class in the SimEscuela, where they integrate the use of technology as a cognitive tool.

Procedure for Task 1

Step 1

Participants are asked by the course's teacher to go to the meeting area and open the notebook *Unidad 1* located in the big bookcase, where they must open and read a Notecard (see Figure 13.4).

Step 2

After reading the previous note card, participants click on Task 1 and receive a second note card, where they read what they have to do to accomplish the task on the note card. The instructions they read in this second note card are:

Task 1

To do this Task 1, please follow the next steps:

1. First of all, form three working teams with two persons each.
2. Once you know in which team you are, click the drawer that corresponds to your team in the computer desk next to this bookcase.
3. When you click your corresponding drawer a folder will be added to your personal inventory, which has clear instructions for your

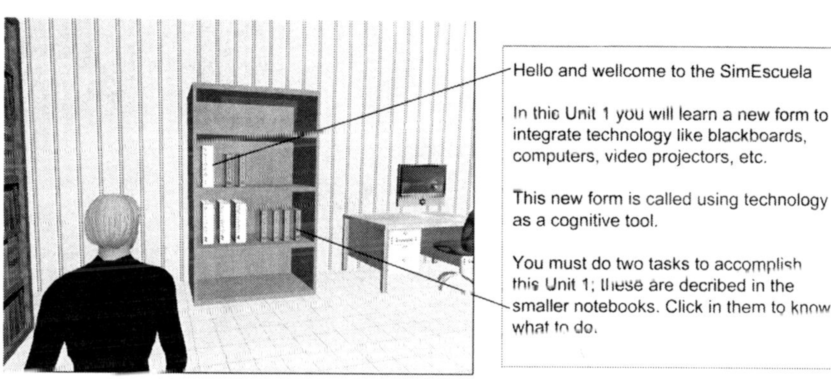

Figure 13.4. Participants click Unidad 1 to start.

team and all the materials you will need to accomplish Task 1. Please do not tell other teams what you are being asked to do.

Team 1 receives the following instructions:

1. Go to the classroom next to this meeting area.
2. There, you should teach a small class about "Democracy."
3. You should act as teachers and the rest of the group as your students.
4. To teach your class on Democracy you should use the three images that were added to your inventory in the folder named "Team 1."
5. Prepare your class showing these three images in the classroom's blackboard and when ready teach the class.

Once the three teams teach their classes, the six of you please go back to the meeting area and, in group, discuss your experience and try to identify what are the differences between the three classes in the way they each chose to deliver instruction.

After stating the differences, each one of you must send an e-mail to the course's teacher explaining what you discover while doing this Task 2.

The three images that Team 1 should display are shown in Figure 13.5. Team 2 receives the following instructions:

1. Go to the classroom next to this meeting area.

Figure 13.5. Images used by Team 1 depicting information in Spanish about the topic of democracy.

2. There, you should teach a small class about the correct graphic characteristics that images that are shown on the classroom's blackboard must have.
3. You should act as teachers and the rest of the group as your students.
4. To teach your class about the blackboard, you should use the three images that were added to your inventory in the folder named "Team 2."

Once the three teams teach their classes, the six of you please go back to the meeting area and, in group, discuss you experience and try to identify what are the differences between the three classes. After stating the differences, each one of you must send an e-mail to the course's teacher explaining what you discover while doing this Task 2.

The three images that Team 2 should display are shown in Figure 13.6. Team 3 receives the following instructions:

1. Go to the classroom next to this meeting area.
2. There, you should teach a small class. You should act as teachers and the rest of the group as your students.
3. When teaching your class you should ask your students to complete a jigsaw puzzle. Tell them its parts are next to the blackboard and that they should carefully move them to make the puzzle over the blackboard.

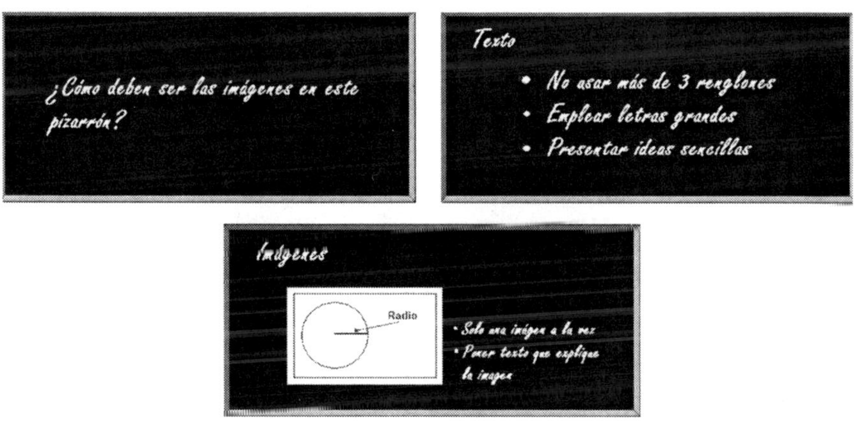

Figure 13.6. Images used by Team 2 depicting information in Spanish about the proper graphic characteristics that images that are shown on the classroom's blackboard must have.

Once the three teams teach their classes, the six of you please go back to the meeting area and, in group, discuss you experience and try to identify what are the differences between the three classes. After stating the differences, each one of you must send an e-mail to the course's teacher explaining what you discover doing this Task 2.

Figure 13.7 shows the jigsaw puzzle that should be made over the blackboard.

Procedure for Task 2

To avoid being too repetitive, Task 2 is not explained in great detail. It is believed that going through Task 1 was clear enough to understand how both tasks were actually implemented in the SimEscuela.

Step 1

Working in teams, participants are asked to teach a class in the SimEscuela classroom using its blackboard.

Step 2

The different teams teach their classes.

Figure 13.7. The jigsaw puzzle being made over the classroom's blackboard.

Step 3

The course's teacher explains to all that they are using the blackboard just as a vehicle of information and not as a cognitive tool.

Step 4

The teacher shows a class where the blackboard is used as a cognitive tool and explains why she is using it as a cognitive tool.

Step 5

The teacher asks the different teams to produce the same class again, but that the challenge now is that they should use the technology as a cognitive tool.

Step 6

The teams present their classes again and receive feedback from the group and from the teacher.

Step 7

The teacher tells to all of the participants that they should write an essay where they explain the concept of cognitive tool and propose the design of a class where they integrate video as a cognitive tool.

REFERENCES

Brown, J., Collins, A., & Duguid, P. (1989). Situated cognition and the culture of learning. *Educational Researcher, 18*(1), 32-42.

Driscoll, M. P. (2000). *Psychology of learning for instruction.* Needham Heights, MA: Allyn & Bacon.

Helleve, I. (2010). Theoretical foundations of teachers' professional development. In J. O. Lindberg & A.D. Olofsson (Eds.), *Online learning communities and teacher professional development* (pp. 1-19). Hersey, PA: Information Science Reference.

Hendricks, C. (2001). Teaching causal reasoning through cognitive apprenticeship: What are results from situated learning? *The Journal of Educational Research, 94*(5), 302-311.

Jonassen, D. H. (1999). Designing constructivist learning environments. In C.M. Reigeluth (Ed.), Instructional-design theories and models (2nd ed., pp. 215-239). Mahwah, NJ: Lawrence Erlbaum Associates.

Lave, J. (1991). Situating learning in communities of practice. In L. B. Resnick, J. M. Levine, & S. D. Teasley (Eds.), *Socially shared cognition* (pp. 63-82). Washington, DC: American Psychological Association.

Lave, J., & Wenger, E. (1991). *Situated learning: Legitimate peripheral participation*. Cambridge, England: Cambridge University Press.

Maturana, H., & Varela, F. (1987). *The tree of knowledge.* Boston, MA: Shambhala.

Merrill, M. D. (2002). First principles of instruction. *Educational Technology Research and Technology, 50*(3), 43-59.

Perkins, D. N. (1991). Technology meets constructivism: Do they make a marriage? *Educational Technology, 31*(5), 19-23.

Resnick, L. B. (1987). Learning in school and out. *Educational Researcher, 16*, 13-20.

Schön, D. (1983). *The reflective practitioner: How professionals think in action.* New York, NY: Basic Books.

Santos, A. (2007). *Using technology to bring reality into the classroom*. In C. Vrasidas, A. Stilianou-Georgiou, & E. Panastosiou (Eds.), Proceedings of Educational Media and Innovations Practices: Challenges and Visions, ICEM 2007 (pp. 42-49). Nicosia, Chipre: International Council for Educational Media.

Santos, A. (2009). Three contexts methodology: Strategies to bring reality to the classroom. In H. Song and T. Kidd (Eds.), *Handbook of Research on Human Performance and Instructional Technology*. Hershey, PA: Idea Group Reference.

Sternberg, R. J., & Frensch, P. A. (1993). Mechanisms of transfer. In D. Detterman & R. Sternberg (Eds.), *Transfer on trial: Intelligence, cognition, and instruction* (pp. 25-38). Norwood, NJ: Ablex.

Streivel, J. M. (1994). Misattributions about situated learning. *Educational Technology, 34*(8), 14-16.

Wenger, E. (1998). *Communities of practice: Learning, meaning and Identity*. Cambridge, England: Cambridge University Press.

Wenger, E., & Snyder, W. (2000, January-February). Communities of practice: The organizational frontier. *Harvard Business Review*, 139-145.

Wenger, E. (2004). Knowledge management is a donut: Shaping your knowledge strategy with communities of practice. *Ivey Business Journal, 68*(3), 1-8.

Wilson, B. (1996). *Constructivist learning environments*. Englewood Cliffs, NJ: Educational Technology.

Wilson, B., & Meyers, K. M. (2000). Situated cognition in theoretical and practical context. In D. Jonassen & S. Land (Eds.), *Theoretical foundations of learning environments* (pp. 57-88). Mahwah, NJ: Lawrence Erlbaum Associates.

ABOUT THE AUTHORS

Michelle Aebersold PhD, RNI has a background in simulation using high-fidelity simulators and virtual reality simulation using Second Life. She has extensive experience in simulation including using it to measure student and staff nurse performance. She is lead faculty and have developed a comprehensive simulation program and embedded it in the undergraduate curriculum. Michelle is coinvestigator on an ongoing study to determine the feasibility of using Second Life virtual reality simulations with undergraduate nursing students. She also has 20 years of nursing management experience in both acute and intensive care settings.

Sami Ahma-aho, MSSc, is a payments expert at Finland's Slot Machine Association (RAY) and previously a researcher at Aalto University, School of Science and Technology, Finland. His academic interests lie in the field of economics. His professional background involves expertise on different payment methods as well as experience in trade promotion in multicultural environments operating with different scopes of firms and corporations.

Caroline Avey is a respected learning strategist with two decades of experience in traditional instructional strategy, design, and course development for Fortune 500 clients and government agencies. As an early pioneer in 3D and gaming, Caroline is best known for her innovative work developing employee learning and engagement strategies as well as the design of practical applications for proficiency, collaboration and reinforcement in 3D-virtual worlds. She has designed hundreds of high yield learning and development experiences in a variety of modalities, currently working with a various government and corporate organizations on integrating virtual and interactive learning into their broader learning and development strategies. Caroline holds the title of Sr. Director of Innovative Learning Strategy at ACS, a Xerox Company and is a published author and regular speaker at events including Chief Learning Officer Magazine, DevLearn, SALT, ASTD and other "virtual" events.

Thomas A. Bryer is an assistant professor in the School of Public Administration at the University of Central Florida. His teaching and research focus on public participation with government, citizen engagement, cross-sector collaboration, and ethics. His work has appeared in *Public Administration Review, Journal of Public Administration Research and Theory, Administrative Theory & Praxis, Public Performance and Management Review, Journal of Public Affairs Education, American Review of Public Administration, International Journal of Public Participation, Journal of Homeland Security and Management, Public Administration and Management,* and *International Journal of Organization Theory and Behavior.* He also has published chapters in peer reviewed books and award winning teaching simulations.

Mary Anne Clark holds a PhD in biology from Bryn Mawr College. Since 1979, she has been teaching at Texas Wesleyan University, where she is professor of biology. She teaches genetics for both biology majors and no majors, in addition to participating in cross-disciplinary and honors courses. Dr. Clark acquired her alter ego Max Chatnoir in 2004, when an artist friend suggested that Second Life might offer an interesting educational venue. Max plans someday to retire to Genome Island, which has a much nicer view than Dr. Clark's real life office.

Georgios Dafoulas is a Principal Lecturer-Business Information Systems Department, School of Engineering & Information Sciences, Middlesex University. George is program leader for the bachelor of science in Business Information Systems with Management and undergraduate BIS transitional programs, curriculum leader in pedagogy for Global Campus and member of the university's e-learning strategy group. He holds bachelor of science, master's in philosophy, and a PhD in Computation, PG certificate in academic practice and executive MBA. He was recently awarded a teaching fellowship and associate membership for the Institute of Work-Based Learning. Dr. Dafoulas authored several publications for refereed journals and peer-reviewed international conferences, in fields of software-engineering and computer-supported cooperative work. He is coauthor of six books and learning guides.

Sara de Freitas is director of research at the Serious Games Institute, professor of virtual environments. She leads the applied research team at the Serious Games Institute and the Serious Games and Virtual Worlds Applied Research Group at Coventry University. She holds a visiting senior research fellowship at the London Knowledge Lab, and is a Fellow of the Royal Society of Arts. In 2003 she founded (and continues to chair) the UK Lab Group, which creates stronger links between industrial and academic research. Voted the Most Influential Woman in Technology 2009 by US Fast Company, she chaired the IEEE Serious Games and Virtual Worlds conference in 2009, and is a regular speaker at international conferences. Her current lines of research are examining multimodal interfaces, experience design and perceptual modelling in games and virtual worlds. She publishes widely in the areas of: pedagogy and e-learning; change management and strategy development for implementing e-learning systems and serious games and virtual worlds for supporting training and learning.

Mike Eisenberg, PhD, is a dean emeritus and professor in the information school at the University of Washington. Mike is the founding dean of the information school, serving from 1998 to 2006. During his tenure, Mike transformed the school from a single graduate degree program into a broad-based information school with a wide range of research and academic programs, including an undergraduate degree in informatics, masters degrees in information management and library and information science (adding a distance learning program and doubling enrollment), and a doctorate degree in information science. Mike's current work focuses on information literacy, information problem solving in virtual environments (funded by the MacArthur Foundation), and information science education K-20. His "Big6 approach to information problem-solving" is the most widely used information literacy program in the world. Mike is a prolific author (nine books and dozens of articles and papers) and has worked with thousands of students—pre-K through higher education—as well as people in business, government, and communities to improve their information and technology skills.

Martin J. Eppler is a full professor of media and communication management at the University of St. Gallen (HSG), where he is also the managing director of the institute for media and communication management. He conducts research on knowledge management, knowledge visualization, and knowledge communication. He has been a guest professor at various universities in Asia and Europe. He has been an advisor to organizations such as the United Nations, Philips, UBS, the Swiss Military, Ernst & Young, KPMG, Swiss Re, Daimler and others. Martin

Eppler studied communications, business administration and social sciences at Boston University, the Paris Graduate School of Management, and the Universities of Geneva and St.Gallen. He has published more than 100 academic papers (in journals such as *Organization Studies, LRP, Harvard Business Manager, TIS, EMJ, IV,* and others) and 10 books, mostly on knowledge communication, management, and visualization.

Michelle K. Gardner is a coordinator of administrative services at the John Scott Dailey Florida Institute of Government at the University of Central Florida. She is currently enrolled in the Texts and Technology PhD program at the University of Central Florida. Primarily focusing on the tools of digital media and the WWW, Michelle's current research centers on the various ways in which people integrate media sharing websites such as YouTube, virtual environments or Flickr into their daily conversations.

Marian Glenn is on the faculty of the College of Arts and Sciences at Seton Hall University in South Orange, New Jersey. A member of the Department of Biological Sciences, she teaches Ecology and is a director of the Environmental Studies Program. Her current research activities are in integrative science and bridging the gap between the sciences and humanities.

Randy Hinrichs is CEO of 2b3d, a company dedicated to building Internet-based strategies and transformational solutions in education and business. 2b3d employs user experience research methods to drive metrics-based development for the 3D Internet, including virtual worlds and augmented reality. Prior to founding 2b3d, Randy formed the Learning Science and Technology Group in Microsoft Research, served in Microsoft's Advanced Strategies and Policy group in eScience and ePublishing, and marshaled in web-based learning at Sun Microsystems producing innovative books on web page design and multimedia. He holds several patents in Internet-based education, large-scale information collection and mining, and Internet-based medical record systems. Randy is also faculty at the University of Washington, where he developed and teaches the Certificate in Virtual Worlds in the iSchool. You will likely find Randy in a number of virtual worlds, usually as an avatar engaged in activity-based community.

Natascha Karlova is a PhD student in the information school at the University of Washington. Natascha studies how teams of online game players disambiguate between misinformation and disinformation. As a gamer herself, she loves researching player practices around games, and how they transform it. She is passionate about the roles that information plays in games and virtual worlds, and how information can shape the user experience. Natascha is advised by Prof. Karen Fisher. You can follow her on Twitter @nataschakarlova. She also works with Prof. Mike Eisenberg, Peyina Lin, and John Marino on the UW VIBE Project investigating trust and credibility in Second Life, a 3D, social virtual world. You can follow her avatar on Twitter @frejakrovac.

Peyina Lin is a PhD candidate in the information school at the University of Washington. She is interested in social structures; particularly how social relationships and affiliation in groups contribute to social identity, social behavior, and interactions via different communication technologies. This general interest spans various contexts and topics, including virtual worlds, social media, learning environments, choice of resources, and social influence. In her dissertation, she investigates the social ecology of high school students. She examines how social labels (like "jocks" and "nerds") cluster or are distant, how this distance relates to friendship choices and the mode and frequency of interaction with friends, and the implications for decisions on information diffusion and/or technology adoption. Her dissertation has won a 2009 Doctoral Dissertation Improvement grant from the National Science Foundation and a 2011 University of Washington Presidential Dissertation Fellowship. Her dissertation committee consists of Drs. Michael B. Eisenberg, David W. McDonald, and Katherine Stovel. Website: http://students.wahsington.edu/pl3 Twitter: @peyinalin.

Lyn Maize is a market strategist, workforce development, and social media consultant with an extensive background in the training industry. A former senior executive at ACS, ProQuest, PRIMEDIA and Harcourt Brace Publishers, Lyn has a long track record of bringing innovative solutions to workforce development and learning. She is an author and speaker specializing in innovation, social media strategies and engagement, and virtual worlds. She currently consults with organizations to help them develop effective, visionary strategies for utilizing innovative technologies for business development, and creating valuable interactive experiences in both 3D and collaborative workspaces.

John Marino is a PhD candidate in the information school at the University of Washington. His research interests include information behavior, information literacy, information problem solving, and K-12 library information and technology programs. He has been a teacher-librarian, educational technology coach, and a teacher of the deaf. John is advised by Dr. Michael B. Eisenberg; he is currently working with Dr. Eisenberg, Peyina Lin, and Natascha Karlova on the UW VIBE Project investigating information behavior and information problem solving in Second Life.

Grant Meredith, BIT, BComp is a multimedia/games lecturer at the School of Science, Information Technology and Engineering at the University of Ballarat, Australia and is a member of the Virtual Reality Simulation Laboratory (VRSL) located there. Grant is an active researcher within the use of simulations and group interaction within virtual worlds. Grant is the program leader for the Technologies for Empowering People for Participation in Society (TEPPS) program which is focused on designing, creating, evaluating and rolling out assertive technologies for people in need. Grant is a confirmed PhD student investigating the experiences of students who stutter engaged within the Australian Higher Education sector. Grant has published and presented nationally and internationally within his research fields. Grant has worked in education for a number of years and previous to tertiary employment has been very successful in the fields of web design and short film production. Grant is a proud person who stutters and is very active within the academic and social circles of people who stutter.

Charlynn Miller, BSc, Post-Bach (HRD), MEd VCU, PhD UVA—is the deputy head of school of the Graduate School of Information Technology & Mathematical Sciences at the University of Ballarat Australia. Charlynn's research primarily involves the enhancement of learning and teaching through the use of emerging technologies; specifically virtual worlds, social networking, and podcasting. Charlynn also conducts research in the area of cyber-safety. Charlynn has a number of publications and grants and regularly presents in the area of emerging technologies and the impact of these technologies on learning, society, and organizations. Charlynn worked in the private IT sector for a number of years as a consultant and manager before entering tertiary education. Charlynn is a member and computer professional with the Australian Computer Society and the ACS Women's Committee for Victoria.

Noha Saleeb is a researcher in the Business Information Systems Department, School of Engineering and Information Sciences, at Middlesex University and a lecturer in the computer science department, American University in Cairo, Egypt. Noha currently teaches undergraduate and postgraduate students delivering education and guidance in a variety of topics including programming, business information technology (BIT), architecture and graphics design. Her bachelor of science, master's of science and PhD degrees in engineering encompass all of the previously mentioned disciplines. She is also a practicing architect and graphic designer both in the physical and 3D virtual worlds. Noha has publications connected to education and design of 3D Virtual Environments in numerous peer reviewed international conferences, journals, workshops and book chapters.

Antonio Santos, EdD from Indiana University, is currently a professor at Universidad de las Americas Puebla, Mexico. Antonio Santos has been a researcher and consultant in the field of educational technology in several Mexican universities. He has also been a producer of instructional materials for teachers' training at different educational and industrial organizations. He has published several articles in the areas of constructivist learning environments, the use of technology in developing countries, and serious games and simulations.

Andreas Schmeil is a research associate and PhD candidate at the University of Lugano (USI), Faculty of Communication Sciences. His background is in Human-Computer Interaction, centered around the fields of Collaborative Virtual Environments and Augmented Reality. He has worked with Fraunhofer FIT and Fraunhofer IAO, Germany, and the Human Interface Technology Laboratory New Zealand, HIT Lab NZ. Visiting scholarships include positions at Simon Fraser University, Canada and the Palo Alto Research Center, California. His current research focuses on avatar-based collaboration and learning practices in 3D virtual worlds. He has published at conferences and in journals in the fields of HCI, knowledge management, design research, and computer science.

Greg Simmons BSc, MSc VU—is a senior lecturer in the Graduate School of Information Technology & Mathematical Sciences at the University of Ballarat Australia. Greg has worked at the University of Ballarat for the past 13 years in a number of different capacities including the development and delivery of courses around the convergence of technology. Greg's research covers a range of diverse areas such as statistical analysis, software engineering, open source software development, virtual teams and massive multiplayer online games.

Teemu Surakka, MSc, is a PhD candidate and a researcher at Aalto University, School of Science and Technology, Finland. His doctoral research topic is knowledge-intensive value creation in virtual worlds. Prior to his doctoral research, Surakka worked in several research projects that investigated how emerging technologies change existing business networks in different industries. He has published several papers at international conferences, related to his previous research and to his current doctoral research.

Dr. Joseph N. Trachtman received a doctor of optometry degree from the Pennsylvania College of Optometry, a master's of education degree from the Johns Hopkins University, and a master's in vision science from the State University of New York, State College of Optometry. He also holds a PhD in experimental psychology from Yeshiva University, New York, where the topic for his research was the use of biofeedback to reduce nearsightedness. Dr. Trachtman completed this segment of his education with a National Library of Medicine Post-Doctoral Fellowship in computers in medicine at Mt. Sinai Medical School, New York. Most recently, he was a member of the first class at the University of Washington's Certificate Course in Virtual Worlds. Dr. Trachtman has personally provided vision training for an array of patients ranging from world-class athletes to learning-disabled schoolchildren.

Heidi Trotta, MA, EdM, is an instructional designer at Seton Hall University in South Orange, New Jersey, assisting faculty with integrating instructional technologies into the classroom. She is also involved with a number of student technology initiatives as well as teaching undergraduate and graduate courses. Her particular academic interests change from time-to-time however presently she is interested in presentation design, real-world problem-based learning, digital storytelling and creating engaging e-learning modules. She completed her EdM in instructional technology and media from Columbia University.

Dana Tschannen PhD, RN research area has included work environment characteristics (as related to patient outcomes), informatics, interdisciplinary collaboration, patient safety, and educational use of Simulation. Currently, I am a coprincipal investigator on a CRLT grant involving the design of a virtual nursing unit using the platform of Second Life, to be used in undergraduate nursing education. Virtual simulations related to patient safety, conflict management, adverse drug events, and priority-settings are embedded into the senior-level curriculum in an effort to improve knowledge acquisition, knowledge transfer, and student/faculty satisfaction.

Charles Wankel, associate professor of management at St. John's University, New York, earned his doctorate from New York University. Charles is on the Rotterdam School of Management Dissertation Committee and is Honorary Vice Rector of the Poznañ University of Business. He has authored and edited about 30 books including the bestselling *Management*, 3rd ed. (Prentice-Hall, 1986), 10 volumes in the IAP series Research in Management Education and Development, the *Handbook of 21st Century Management* (SAGE, 2008), and the *Encyclopedia of Business in Today's World*

(SAGE, 2009), which received the American Library Association's Outstanding Business Reference Source Award. He is the leading founder and director of scholarly virtual communities for management professors, currently directing eight with thousands of participants in more than 70 nations. He has been a visiting professor in Lithuania at the Kaunas University of Technology (Fulbright Fellowship) and the University of Vilnius (United Nations Development Program and Soros Open Society Foundation funding).

CPSIA information can be obtained at www.ICGtesting.com
Printed in the USA
BVOW031702080412

287101BV00002B/2/P